Carnegie Learning®

INTEGRATED MATH II

Cognitive Tutor® Student Text

FOREWORD TO THE STUDENT

Congratulations! You are about to begin the exciting journey through the world of mathematics: the language of science and technology. As you sit in front of the television, computer screen or video game, ride in an automobile, fly in a plane, talk on a cellular phone or use practically any other tools of modern society, realize that mathematics was critical in its invention, design and production.

The modern workplace today demands employees who are technologically literate, work well in teams and are self-starters. So, we have designed a course that combines state-of-the-art computer software with collaborative classroom activities. While you use the Cognitive Tutor software, it actually learns about you as you learn the mathematics. You'll receive "just-in-time" instruction from the Tutor and it will always make sure you're ready for the next problem. In the classroom you will collaborate with your peers to explore and solve real-world problem scenarios. Working in groups, you'll be answering questions by using multiple representations to analyze and solve the problem. Then, as a group, you will present your solutions to the class. Throughout the whole process, your teacher will serve as a facilitator and guide in support of your learning. As a result, you will become a self-sufficient learner; moving through the software and text at your own rate, using the Tutor's extensive help system and discovering solutions to problems you never thought possible.

Have fun Learning by Doing!

William S. Hadley,
Author, Vice President of Educational Services and Chief Academic Officer
Carnegie Learning, Inc.

P.S. Have questions about the curriculum? Ask me directly by sending an email to billhelp@carnegielearning.com. I will be happy to send you a response within 5 to 7 days. I look forward to hearing from you!

CONTRIBUTING AUTHORS

William S. Hadley
David Dengler
Matthew R. Freedman
Corinne M. Murawski
Lora J. Shapiro

Jaclyn Snyder
Pittsburgh Public Schools

CONTRIBUTING EDITORS

Corinne Murawski
Kelly Clay-Slack

REVIEWERS

Theodore R. Hodgson
Dept. of Mathematical Sciences
Montana State University
Bozeman, MT

Signe E. Kastberg
IUPUI School of Education
Indianapolis, IN

ISBN 1-932409-14-9

Student Text

Printed in the United States of America.
1-2004-MW
2-2004-MW
3-2005-VH

Carnegie Learning, Inc. | Pittsburgh, PA
Phone 888.851.7094 | Fax 412.690.2444
www.carnegielearning.com

Table of Contents

Unit 1: Probability

Unit 2: Linear Functions

Unit 3: Parallel and Perpendicular Lines

Unit 4: Triangles and Quadrilaterals

Unit 5: Transformations and Symmetry

Unit 6: Similarity

Unit 7: Exponential and Logarithmic Functions

Unit 8: Quadratics

Unit 9: Higher Order Polynomial and Rational Functions

Unit 1: Probability

Contents

The Lady or the Tiger

input

Probability and uncertainty are present in many aspects of daily life from games of chance to predicting the weather. The story entitled *The Lady or the Tiger,* summarized below, represents a situation in which probability was involved in a life or death decision. If you have read the story, you know the premise. Reread the synopsis to refresh your memory. If you have not read the story, the summary will serve for the purposes of your discussion.

Long ago in a place far away lived a somewhat barbaric king. He had devised what he felt was a just method for trying those suspected of a crime. The "trial" was held in a public arena with two doors at one end of the arena. Behind one door waited a hungry and fierce tiger which would devour the suspect on the spot, if it were chosen. Behind the other door waited a beautiful lady who would be married to the suspect on the spot, if she were chosen. This system of justice seemed to work for the king because the verdict was decided and carried out in an instant.

Unfortunately, the king's daughter was in love with a handsome commoner. After the king discovered this affair, the young man was thrown into prison, only to have his fate decided in the public arena. The princes, while a very jealous person, wanted to save her love. Being very resourceful, she discovered which door held the tiger and which the lovely maiden. On the day of the "trial," after having spent many a sleepless night deliberating what to do, the princess managed to send a signal to her lover indicating which of the two doors to open. Which door did the princess indicate for her love to open?

Probability is defined as the likelihood or chance of an event occurring.

Adapted from: *The Lady or the Tiger* by Frank R. Stockton (www139.pair.com/re/Frank_R_Stockton/The_Lady_Or_The_Tiger)

The Lady or the Tiger

analysis

1. In your opinion, what is the likelihood the princess indicated the door hiding the tiger?

2. In your opinion, what is the likelihood the princess indicated the door hiding the young lady?

3. What is the theoretical probability the young man would pick the door hiding the tiger without any outside influences (the manner in which the king intends his system to work)?

The Lady or the Tiger

process: model

4. What is the lowest possible probability that the young man would pick the lady?

5. What is the highest possible probability that the young man would pick the lady?

6. What *should* be the sum of the probabilities you gave in questions 1 and 2? Explain why this is the only possible sum for those two probabilities.

7. Rate each of the following events according to the probability that the event will take place.

   a. It will rain within the next 72 hours.

0 - - - - - - - - - - - - - - - - - - - ½ - - - - - - - - - - - - - - - - - - - 1

   b. You will drive a car home from school today.

0 - - - - - - - - - - - - - - - - - - - ½ - - - - - - - - - - - - - - - - - - - 1

   c. You will sleep at least 4 hours tonight.

0 - - - - - - - - - - - - - - - - - - - ½ - - - - - - - - - - - - - - - - - - - 1

   d. The Chicago Cubs will win the World Series this baseball season.

0 - - - - - - - - - - - - - - - - - - - ½ - - - - - - - - - - - - - - - - - - - 1

analysis

The Lady or the Tiger

8. If the meteorologist on Channel 4 says there is a 0.75 (75%) chance of rain tomorrow, what does that mean?

9. If a sportscaster on the same newscast says there is a 20% (0.20) chance that the Cubs will win the World Series this fall, what does that mean?

10. If you will surely sleep from 11 PM until 6 AM, what is your response for 7c? Why?

11. If you do not have a driver's license, what should be your response for 7b? Why?

input

The Lady or the Tiger

Some situations that relate to probability have a small set of possible outcomes or possible end results. For example, in *The Lady or the Tiger*, the only possible results or outcomes are:

The young man chooses the lady. (Call it **L**.)
The young man chooses the tiger. (Call it **T**.)

As this story is written, there is no possibility for the cage behind the chosen door to be empty, and there is no possible way that both the lady and the tiger will be chosen.

The list of possible outcomes in any probability experiment is called the **sample space**. The sample space for *The Lady and the Tiger* is {**L, T**}. This sample space has two elements in it. An N is used to indicate the size of a sample space. In this case, $N = 2$.

Mathematically, the **probability** or likelihood of an event is the ratio of the number of desired outcomes to the total number of possible outcomes. Probability is often written as a fraction, but can also be written as a decimal or percent.

$$P(event) = \frac{number\ of\ ways\ the\ desired\ outcome\ can\ occur}{total\ number\ of\ possible\ outcomes}$$

process: evaluate

The Lady or the Tiger

You may have performed some probability experiments in the past with coins, number cubes, spinners, and decks of cards.

12. Based on your previous experiences and on the input from the previous page, what is the sample space for tossing one coin?

    a. What is the size of this sample space (*N*)?

    b. What is the probability that a coin will land with heads showing?

13. What is the sample space for rolling one number cube?

    a. What is *N* for this sample space?

    b. What is the probability that the toss of one number cube will show a 3?

    c. What is the probability that the toss of a number cube will show an odd number?

    d. What is the probability that the toss of a number cube will show a 1 OR a 6?

    e. What is the probability that the toss of a number cube will NOT show a 5?

The Lady or the Tiger

input

Often experiments in probability contain more than 2 to 6 elements in the sample space. When there are many elements, keeping track of the information in an organized manner is important. In general, a good technique is to list the events or outcomes.

Hal, a fairly organized guy, does not pair his socks before stuffing them into his drawer. One morning, Hal needed to pick a pair of socks from his drawer. The drawer had white socks, black socks, and a couple of red socks. If Hal was picking only two colored socks from his drawer, what was the sample space for the possible outcomes? To make an organized list of this sample space, pick a color of sock for Hal's first pick and match it with each of the three colors mentioned. Fill in the missing choice(s).

First Pick	Second Pick
W	**W**
W	**B**
W	_ __

Pick another color for the first sock and match it with the possibilities for the second.

First Pick	Second Pick
B	**W**
B	_ __
B	__ __

Complete the sample space by choosing the third possibility for the first pick and pairing it with all the possibilities for the second pick. You complete the list below.

First Pick Second Pick

Another way to list the sample space for Hal's picks is to use a **tree diagram**. A tree diagram makes use of branches to show all possible choices. The tree diagram for Hal's sock picks is as follows. Fill in the missing choices.

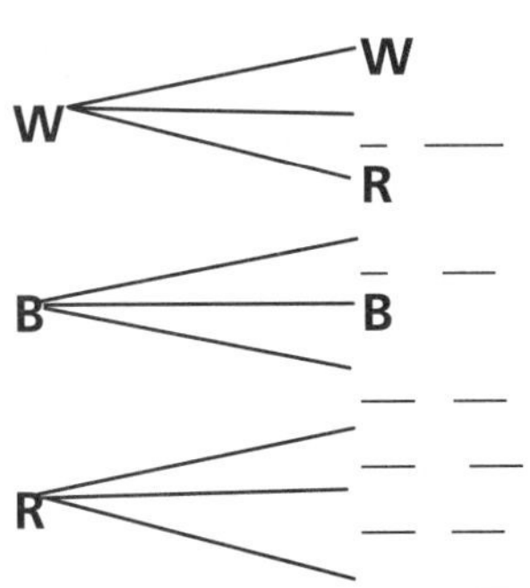

process: evaluate

The Lady or the Tiger

14. List the sample space for four successive basketball foul shots, using an organized list. What is the size of this sample space?

15. Provide the sample space for tossing four coins concurrently, using a tree diagram. What is the size of this sample space?

A third way to create a sample space when more than one choice is being made is to use a chart.

16. Use a chart to create the sample space for tossing two number cubes.

Outcome on first number cube (columns); Outcome on second number cube (rows)

	1	2	3	4	5	6
1	1, 1	2, 1				
2	1, 2					
3						
4						
5						
6						

process: evaluate

The Lady or the Tiger

17. Use the same chart to list the sums of the two number cubes when tossed.

Outcome on first number cube (columns); Outcome on Second Number Cube (rows)

	1	2	3	4	5	6
1	2	3				
2	3					
3						
4						
5						
6						

18. What is the sample size for the **sum** of two number cubes?

19. What is the sample size for rolling two number cubes?

analysis

The Lady or the Tiger

20. Compare and contrast the two charts you just completed for rolling two number cubes. How is the sample space for the sums similar to the sample space showing the numbers on each cube? How and why are these two sample spaces different?

Relative Frequency vs. Probability

preview

1. Quick! Choose a number between 1 and 10 and record it here.

2. Theoretically, what should be the probability of your number being chosen by another student in your class?

3. Tally the numbers chosen by your classmates below.

2	3	4	5	6	7	8	9

4. How do the results of this experiment compare with the theoretical results?

Relative Frequency vs. Probability

process: model

5. Create a bar graph to show the expected probability of choosing each of the numbers between 1 and 10.

process: model

Relative Frequency vs. Probability

6. Use your class results to create a relative frequency histogram, showing the actual rate at which the numbers between 1 and 10 were chosen.

analysis

Relative Frequency vs. Probability

7. Cite two similarities between the two graphs above.

8. Cite two differences between the two graphs above.

9. What is a possible reason for the differences between the two graphs above?

process: evaluate

Relative Frequency vs. Probability

Let's perform another experiment in order to compare probability to relative frequency. For this experiment, you will need two coins.

10. What are the possible outcomes when two coins are tossed, one after the other?

11. How many elements are in this sample space?

12. Are each of the outcomes equally likely when two coins are tossed? Why or why not?

13. What is the probability of getting HH when two coins are tossed? What is the likelihood of getting TT? What is the likelihood of getting HT? What is the likelihood of getting TH?

Now, let's change the experiment slightly. Instead of tracking the entire outcome, we will keep track of the number of heads that appear when the two coins are flipped.

14. What is the sample space for this experiment?

15. How many elements are in the newly defined sample space?

16. Are the outcomes for the number of heads that will appear when two coins are tossed equally likely? Explain.

process: evaluate

Relative Frequency vs. Probability

17. What is the probability of getting 2 heads when two coins are tossed? What is the probability of getting 0 heads when two coins are tossed? What is the probability of getting 1 head when two coins are tossed?

18. Toss two coins 10 times, keeping track of the outcomes in the tally chart below.

Number of Heads	2	1	0
Tally			

19. Create a relative frequency histogram based upon your group's first 10 trials. (Use a light pencil, because you will return to this histogram after running more trials.)

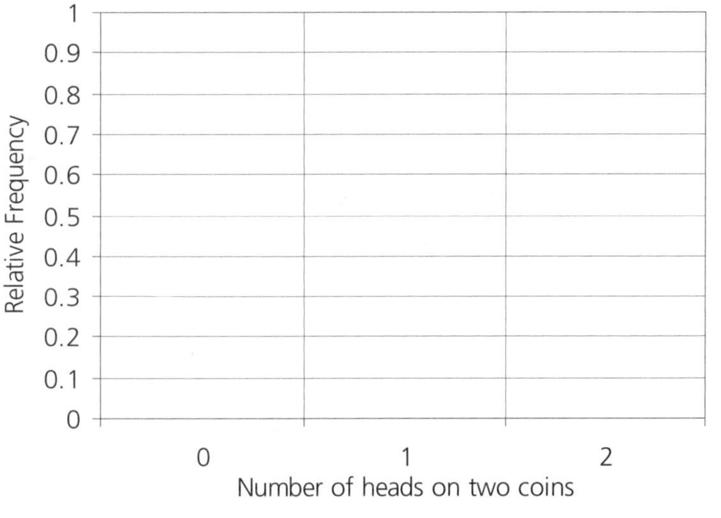

20. Continue the experiment by tossing the coins 10 more times. Add the outcomes to the tally chart above.
21. Return to your relative frequency histogram. Change it to reflect the new relative frequencies of the three possible outcomes after 20 trials.
22. Run the experiment 10 more times. Track the outcomes in the tally chart and change the relative frequency histogram after the experiment is run.
23. Gather the data from the rest of the groups in your class. Write the totals below your tally chart and change your relative frequency histogram to show class data.

analysis

Relative Frequency vs. Probability

24. How did the relative frequency of the three possible outcomes change as you included additional trials throughout the experiment?

25. Create a tree diagram to show the possible outcomes for tossing two coins.

26. Why is the branch showing H→T equivalent to the branch showing T→H in this experiment?

27. If there are four branches in the tree diagram, why are the relative frequencies in this experiment **not** all $\frac{1}{4}$?

Relative Frequency vs. Probability

analysis

28. If there are three possible outcomes in this experiment, why are the relative frequencies different from one-third?

29. In this experiment, how does the theoretical (expected) probability for each of the possible outcomes in the sample space compare to the relative frequency of that outcome in the experiment after 10 trials? after 20 trials? after 30 trails? after the class' data was gathered?

The Queen of Hearts

process: evaluate

1. What is the probability that you will draw a card with a picture of a KING when you pick a single card from a standard deck of 52 cards (no jokers)?

2. What is the probability that you will draw a card with a 3 when you pick a single card from a standard deck of 52 cards?

3. What is the probability that you will draw either a card with a KING *or* a 3 when you pick a single card from a deck of 52 cards? Show your work.

4. Describe how you calculated the probability in question 3.

5. Is there any possibility that you can draw a KING *and* a 3 with one pick?

input

The Queen of Hearts

Because there is no way that a card drawn from a regular deck of 52 cards can be both a KING and a 3 at one time, drawing a KING and drawing a 3 are called **mutually exclusive events**. **Mutually exclusive events** describe events that have no elements in common. You can represent these events with **Venn Diagrams**. (**Venn Diagrams** are visual representations for sets or events.)

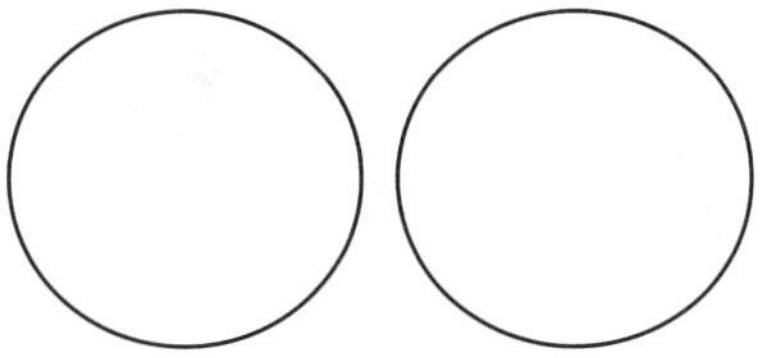

Drawing a KING Drawing a 3

Notice that there is no overlap between these two circles. This indicates that there are no elements in common between the two events.

If we change the two events to be:

Drawing a card showing a QUEEN from a deck of cards

Drawing a card with a HEART from a deck of cards

then there is a card that is both a QUEEN and a HEART. The fact that both events can occur simultaneously means the events are **not mutually exclusive**. When the Venn Diagram for these events is drawn, it will show some overlap between the two circles indicating that there is at least one element in the set of QUEENS that is also in the set of Hearts.

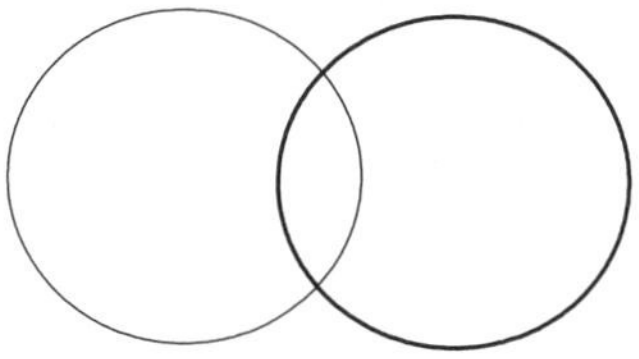

Drawing a QUEEN Drawing a Heart

The Queen of Hearts

process: evaluate

6. Shade the area of the Venn Diagram that shows the QUEEN of HEARTS.
7. How many elements are there in the overlap of the circles in the Venn Diagram above? Write this number in the overlap.
8. How many elements are there in the set of QUEENS?
9. How many elements are there in the set of HEARTS?
10. How many elements are there in the set of QUEENS that are not HEARTS? Record this number in its place on the Venn Diagram.
11. How many elements are there in the set of HEARTS that are not QUEENS? Record this number in its place on the Venn Diagram.

The Queen of Hearts

process: evaluate

12. Use the Venn Diagram to determine how many ways there are to draw a QUEEN *or* a HEART? Describe how you determined this answer.
13. Determine the probability of drawing a QUEEN *or* a HEART from the standard deck of 52 cards. Describe how you determined this probability.

input

The Queen of Hearts

The **union** of two sets contains all the elements that are in either one set or the other set. To symbolize a union of sets, use a symbol that looks like a capital U. For example, to show the union of the set of KINGS (call it set K) and the set of 3s (call it set T), you would show: $K \cup T$. The union of K and T has 8 elements in it, so $N(K \cup T) = 8$.

The **intersection** of two sets contains all the elements that are common to both of the sets. To symbolize an intersection of sets, use a symbol that looks like an upside-down U. For example, to show the intersection of the set of QUEENS (call it set Q) and the set of Hearts (call it set H), you would show $Q \cap H$. The intersection of Q and H has one element, so $N(Q \cap H) = 1$.

process: evaluate

The Queen of Hearts

14. How many elements are the in the set K? (What is $N(K)$?)

15. What is $N(T)$? (How many elements are in the set T?)

16. How many elements are in the intersection of K and T? (What is $N(K \cap T)$?)

17. How can you use the answers to questions 14 through 16 to find the number of elements in the union of the set of KINGS and the set of 3s?

process: evaluate

The Queen of Hearts

18. Previously in this activity, you found $N(\mathrm{Q})$ and $N(\mathrm{H})$,and you were given $N(\mathrm{Q} \cap \mathrm{H})$. What are those values?

19. How can you use those values to find $N(\mathrm{Q} \cup \mathrm{H})$?

process: practice

The Queen of Hearts

Use the spinner provided on this page to answer the next series of questions.

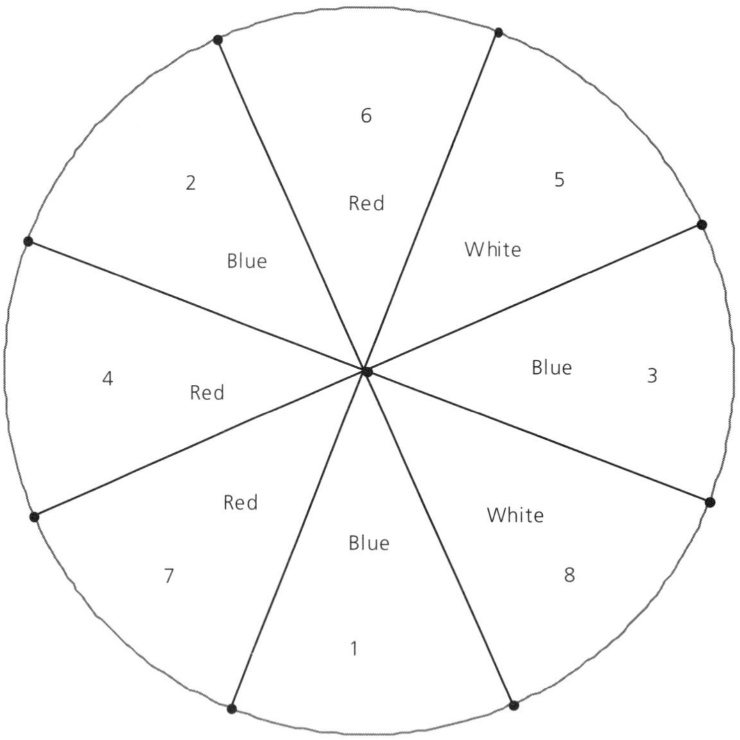

20. What is N(Red), N(Blue), N(White), N(odd #), and N(even #)?

process: practice

The Queen of Hearts

21. Tell whether the following events are mutually exclusive or not mutually exclusive.
    a. Spinning red and spinning blue
    b. spinning white and spinning a 1
    c. Spinning blue and spinning an odd number
    d. Spinning an even number and spinning red
    e. Spinning an even number and spinning an odd number

22. Model each of the following with Venn Diagrams.
    a. Spinning blue and spinning white
    b. Spinning red and spinning an odd number

23. Find the number of elements in the union of each of the sets mentioned in the previous question. Use the correct set notation to answer this question.

24. Find the number of elements in the intersection of each of the sets in question 22. Use correct set notation to answer this question.

process: practice

The Queen of Hearts

25. Using the results you have calculated so far, find the probability of spinning red, *P*(Red).

26. Find *P*(Odd #).

    a. How do you interpret *P*(Odd#) in words?

27. Find the probability of spinning red or blue. Show all of your work and use the correct set notation in your answer.

28. Find $P(\text{Red} \cap \text{Blue})$.

    a. How do you interpret $P(\text{Red} \cap \text{Blue})$ in words?

29. Find the probability of spinning Blue *or* an even number. Use the correct set notation in your answer and show your work.

30. Find the probability of spinning an even-numbered, blue sector. Use the correct notation in your answer and show your work.

The Queen of Hearts

analysis

31. In general, how can you find the number of elements in the union of two mutually exclusive sample spaces or two disjoint sets?

32. In general, how can you find the number of elements in the union of two sample spaces that are not mutually exclusive?

33. In general, how can you find the probability of picking an element if the sets involved are mutually exclusive?

34. How can you determine the probability of picking an element if the sets involved are not mutually exclusive?

analysis

The Queen of Hearts

35. Create a Venn Diagram showing three mutually exclusive events.

36. Create a Venn Diagram showing three events, none of which are mutually exclusive.

Use the following Venn Diagram to respond to the questions on this page and the next.

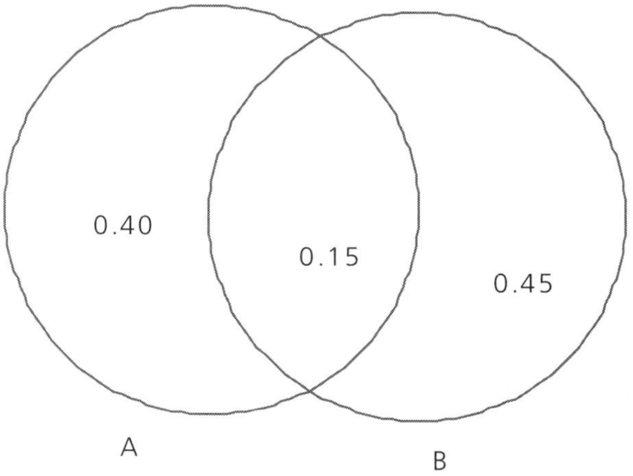

37. Are the two events, A and B, mutually exclusive events? Explain your response.

The Queen of Hearts

analysis

38. What is P(A)? Explain how you found your answer.

39. What is $P(A \cap B)$? Explain how you found your answer.

40. What is $P(A \cup B)$? Explain how you found your answer.

41. If the total number of elements in the sets is 540, what is $N(B)$?

To Be or Not To Be

input

In Shakespeare's *Hamlet,* the main character asks himself the question, "To be or not to be?" Were "to be" and "not to be" the only options open to Hamlet? In mathematics, the set of all possible options is called a set of **complementary events**.

Complementary events are mutually exclusive and fill the entire sample space. For example, being less than 18 years of age is the complementary event to being 18 or over. Drawing a red card from a deck is a complementary event to drawing a black card.

Any event called "A" has a complement called "not A." Together the event A and its complement fill the entire sample space. In order to signify this in a Venn Diagram, a rectangle is drawn around the circle(s). The rectangle encloses all the possible events. It symbolizes the **universe** for the events being considered.

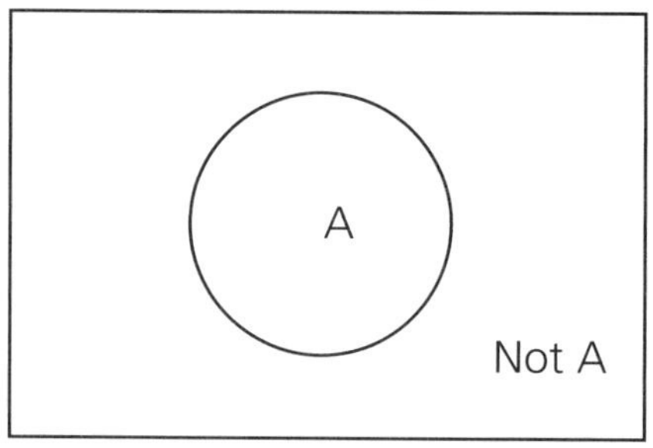

There can be more than 2 possibilities in a set of complementary events.

If you reach into your change-filled pocket to pick out a U.S. coin, what is the set of four complementary events that will fill the entire sample space?

To Be or Not To Be

analysis

1. In your own words, explain why A and not A are complementary events in the Venn Diagram pictured above?

process: model

To Be or Not To Be

2. What is the complement to each of the following events?
   a. Flipping a coin and getting "tails"

   b. Tossing a number cube and getting a multiple of 3

   c. Picking an even number

process: evaluate

To Be or Not To Be

3. What is the probability of tossing a number cube and getting a multiple of 3?

4. What is the probability of tossing a number cube and not getting a multiple of 3?

5. What is the probability for rolling a sum of 7 or less when two number cubes are tossed? (You may want to refer to page 1-9 for the sample space.)

6. What is the probability for rolling a sum over 7 when two number cubes are tossed?

To Be or Not To Be

process: evaluate

7. The probability for event B is $\frac{1}{8}$. What is the probability for event B not occurring? Explain how you determined this answer.

In your pocket, you have 19 U.S. coins. Thirteen of the coins are silver. Ten of them are quarters, and two are dimes.

8. What is the probability that you will choose a penny?

9. What is the probability that the coin will be a nickel?

10. What is the probability that the coin will be worth more than 5 cents?

11. Are these three events complementary? Explain.

analysis

To Be or Not To Be

12. If events are complementary, what is the sum of their probabilities? How do you know?

process: summary

To Be or Not To Be

13. What is $P(\text{A} \cup \text{not A})$? Justify your answer.

14. What is $P(\text{A} \cap \text{not A})$? Justify your answer.

15. What is $N(\text{A} \cap \text{not A})$? Explain.

process: summary

To Be or Not To Be

16. You are about to pick the winner in the class raffle. In front of you, there is a hat containing 100 ticket stubs, numbered 001 to 100. You will select one winner of the grand prize, one runner-up, and one consolation prize winner. Use correct notation when writing your answers to the following questions.

    a. What is the probability that the winning ticket will be a multiple of 5?

    b. What is the probability that the winning ticket will be a multiple of 6?

    c. What is the probability that the winning ticket will be both a multiple of 6 and 5?

    d. What is the probability that the winning ticket is neither a multiple of 5 or 6?

    e. Use a Venn Diagram to illustrate this situation.

Pizza and a Movie (Part 1: The Movie)

input

You and four friends decide to see the newest release in the *Lord of the Rings* series. The person you are "interested in" will be one of the four friends accompanying you, and you wonder how likely it is that the two of you will sit next to one another.

In order to determine this, you first need to determine how many different ways there are for the five of you to arrange yourselves in a row of five seats once you get to the movies. The names of the four people going with you to the cinema are: Alex, Chris, Daryl, and Kelly.

Pizza and a Movie (Part 1: The Movie)

process: model

1. Start an organized list to show the possible seating arrangements for the five of you in the movie theatre. Remember that the key word is "organized."

2. Perhaps it will be easier to use a tree diagram to track the possible seating arrangements for your group. Start a tree diagram in the space below.

Pizza and a Movie (Part 1: The Movie)

process: model

3. Were you able to complete the sample space using either an organized list or a tree diagram? If not, why were you unable to complete the sample space? If you did complete the sample space, how many possible seating arrangements are there for you and your friends?

   What made using an organized list or a tree diagram difficult in this problem?

   Use the model of five seats if you need some assistance in answering these questions.

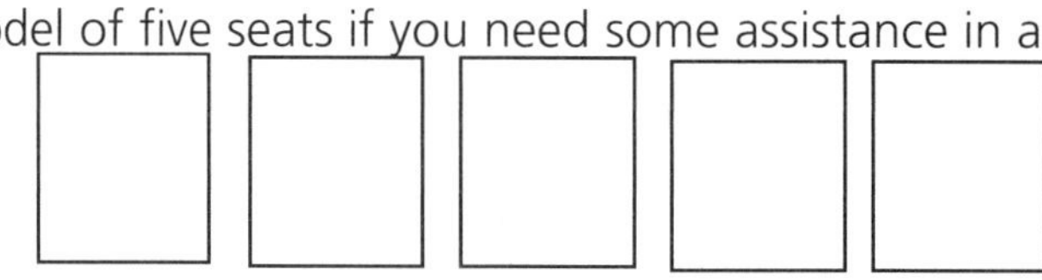

4. How many possible choices are there for the person who sits in the first seat? How do you know?

5. Assume that Alex is the person who sits in the first seat. How many choices are there for the person to sit next to Alex? Why?

6. Assume that you are the person who sits in the first seat. How many choices are there for the person who sits next to you?

7. If Chris is the person who sits down first, how many choices are there for the person who sits next to Chris?

8. If Daryl is the first person to sit down, how many choices are there for the person who sits next to Daryl?

process: evaluate

Pizza and a Movie (Part 1: The Movie)

9. Lastly, what if Kelly sits in the first seat? How many choices are there for the second seat?

So far, you have seen that for each of the five choices you have for the person who sits in the first seat, there are four choices for who sits in the second seat.

10. How many possible ways are there for your group of five to fill the first two seats?

11. How could you have calculated this number using multiplication? Base your answer on the number of choices you had for seat #1 and the remaining choices you had for seat #2?

12. Assume that the arrangement for the first two seats is: Alex, Chris. How many choices are there for the third person? What are those choices?

13. Assume that the arrangement for the first two seats is: Alex, Daryl. How many choices are there for the third person? What are those choices?

14. How many choices for the third seat are there for <u>each</u> of the possible arrangements in the first two seats? How do you know?

15. What mathematical operation should be done to calculate the number of ways to fill the first three seats? How many ways are there to fill the first, second, and third seats? Show any work.

process: evaluate

Pizza and a Movie (Part 1: The Movie)

Now, you have calculated that for each of the twenty (5 • 4) ways that exist to fill the first two seats in the theatre, there are 3 ways to fill the third seat. That means there are 20 • 3 ways to fill the first 3 seats.

16. For each of the 60 ways that exist to fill the first three seats, how many choices remain for filling the fourth seat? Explain how you found this number.

17. What mathematical operation should be done to calculate the number of ways to fill the first four seats? How many ways are there to fill the first, second, third, and fourth seats? Show any work.

18. Once the first four seats are filled, how many choices remain for the last seat in each of the arrangements?

19. Does filling the fifth seat with the last remaining person in the group change the number of possible arrangements that exist? Why or why not?

20. How many ways are there to fill all five seats in the movie with the five people in your group? Show your work.

input

Pizza and a Movie (Part 1: The Movie)

In determining the number of possible seating arrangements, you solved a problem based on **permutations**. A **permutation** is an arrangement of a set of objects. In your problem you had a choice of 5 people with a choice of sitting in 5 seats. In finding the result, you multiplied 5 • 4 • 3 • 2 • 1 to get a final answer of 120 arrangements.

The multiplication problem you performed is defined as a **factorial**. In general, the **factorial** of a number is that number multiplied by every positive integer less than itself. It is noted as the number followed by an exclamation point. For example, the problem you solved used 5! called "five factorial."

Using the definition above, $7! = 7 \bullet 6 \bullet 5 \bullet 4 \bullet 3 \bullet 2 \bullet 1 = 5040$, and
$n! = n \bullet (n-1) \bullet (n-2) \bullet \ldots \bullet 3 \bullet 2 \bullet 1$.

process: practice

Pizza and a Movie (Part 1: The Movie)

21. Calculate 3!. Show your work.

22. Calculate 6!. Show your work.

23. Calculate $\frac{6!}{3!}$. Show the full multiplication problem in the numerator and the full multiplication problem in the denominator and cancel any common factors before performing the multiplication problem.

24. Calculate $\frac{6!2!}{4!}$. Again, show the full multiplication problem in the numerator and denominator and cancel before performing the multiplication.

Pop Quiz

input

Your teacher is giving a surprise quiz in math class. The quiz is made up of 5 True/False questions. You have not done your assignment, so you are unprepared for the quiz. During the quiz, you find yourself wondering about how many possible ways there are to answer the five True/False questions and what the likelihood is that you can get a 100% based upon pure guessing.

You realize that the problem about seating arrangements in the movie theater has something in common with this problem because you have choices to make for five different answers on your paper. However, this problem is somewhat different than the problem from the movies. In this problem, an answer can be used more than once, but in the movie problem, once a person was seated, he or she could not take another seat. (Alex could not sit in the first seat and the third seat at the same time.)

Mathematically, the seating arrangement problem is a situation **without replacement**, while the test with true or false questions is a situation **with replacement,** since a choice can be used again.

Pop Quiz

process: evaluate

1. How many choices do you have for the first response on the pop quiz?

2. How many choices do you have for the second response on the pop quiz?

3. How many choices are there for the answers to the first and second answer? Show your work.

4. How many choices are there for the third question? How many choices are there for answering the first three questions? Show all your work.

process: evaluate

Pop Quiz

5. How many choices for the fourth question? … for the fifth question?

6. How many choices are there for answering all five questions? Represent your answer as an exponent.

analysis

Pop Quiz

After turning your paper over, you discover that in addition to the five True/False questions, there are three multiple choice questions with choices A, B, C, D, and E for each question.

7. Describe how you can find the total number of ways the three multiple choice questions can be answered.

8. Find the total number of ways to answer the three multiple choice questions. Give a numerical answer and represent your answer as an exponent.

9. Does this situation involve replacement or not? How can you tell?

analysis

Pop Quiz

10. Why do you need to use multiplication to find the total number of ways to answer the questions on the quiz or when finding the number of seating arrangements in the movie?

11. Can you use a tree diagram or an organized list to find the number of possible ways to answer the 5 True/False questions or the three multiple choice questions? If so, use a tree diagram to model the number of ways to answer either the set of True/False questions or the set of multiple choice questions. Use an organized list to model the number of ways to answer the other set of questions.

	Pop Quiz
analysis	12. Compare and contrast the situations involving replacement with those that do not involve replacement. Why does one involve factorials while the other involves exponents?

	Pop Quiz
process: evaluate	13. How many ways are there to answer all five of the True/False questions correctly? 14. What is the probability of getting a 100% on the True/False questions if you guess at all five of the answers? 15. How many ways are there to answer the three multiple choice questions? 16. What is the probability of getting a 100% on the multiple choice questions if you guess at all three answers? 17. How many ways are there to answer all 8 questions on this pop quiz? 18. What is the probability of getting a 100% on the quiz, if you guess at all the answers?

Pizza and a Movie (Part 2: The Pizza)

input

The Friday after seeing *The Lord of the Rings* movie, you and a group of five friends (a total of six people this time) go to see the latest horror movie and then out for pizza. Once again, the person you have your "eye on" is part of the group.

Fortunately or unfortunately, the movie you are seeing is quite popular. By the time you and your friends arrive, you cannot find six seats together. At most, you can find four seats together.

You start to panic because you think your chances of sitting next to that "special someone" have decreased. Try to determine if that is true.

Use the model of four seats pictured below, if you need some assistance in answering the questions.

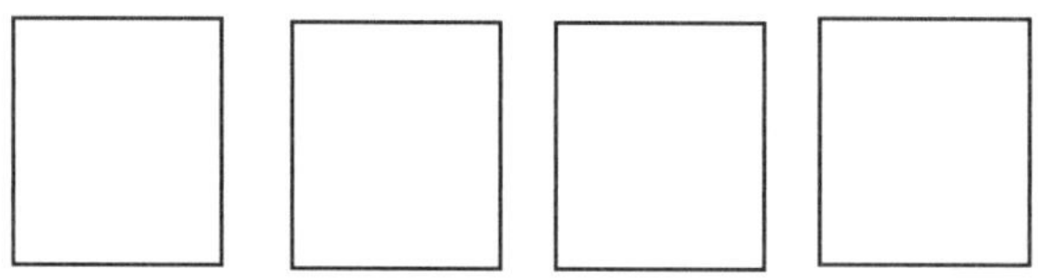

You might also use an organized list or tree diagram for help.

Pizza and a Movie (Part 2: The Pizza)

process: evaluate

1. How many ways are there to choose the person to sit in the first seat from the six people in your group?

2. Once you have determined who is sitting in the first seat, how many ways are there to choose the person who will sit in second seat? How do you know?

3. How many ways are there to choose the person who will sit in the third seat, once the first two positions have been determined?

process: evaluate

Pizza and a Movie (Part 2: The Pizza)

4. How many ways are there to choose the person who will sit in the fourth seat?

5. What operation must be performed with these numbers to determine the total number of ways to fill the four seats? Why?

analysis

Pizza and a Movie (Part 2: The Pizza)

6. Determine the total number of ways to seat four people from a group of 6 people.

7. Does the above problem involve a permutation? How can you tell?

8. Is the number of possible arrangements for your group in the four seats equal to 6!? Explain.

9. If you use 6! to help find the answer to this problem, you'll need to divide in order to get the correct result. What number must be in the denominator in order to determine the total number of ways of seating four people, when the size of the group is six? Show this fraction and the resulting answer.

input

Pizza and a Movie (Part 2: The Pizza)

This problem involves a permutation or arrangement of 6 objects, but you are not making 6 choices. Only 4 choices are being made. Therefore, only four numbers (6, 5, 4, and 3) were multiplied to determine the answer. In mathematical language, this is called a permutation of 6 objects taken 4 at a time. Because we did not make choices for the last two objects in the group, the last two numbers in the factorial of 6 were not used. So, if factorials are used to determine the answer to such a problem, you need to divide by 2 • 1 or 2!. Note that 6 – 4 = 2.

$$\textit{number of ways to arrange 6 objects 4 at a time} = \frac{6!}{2!} = \frac{6 \bullet 5 \bullet 4 \bullet 3 \bullet 2 \bullet 1}{2 \bullet 1} = 6 \bullet 5 \bullet 4 \bullet 3 = 360$$

In general, if a permutation involves *n* objects, but only *r* of them are being chosen, you will need to divide *n!* by *(n - r)!* because you do not make choices from the last *n – r* objects.

$$\textit{n objects r at a time} = \frac{n!}{(n-r)!} = \frac{n \bullet (n-1) \bullet (n-2) \bullet \ldots \bullet (n-r+1) \bullet (n-r) \bullet (n-r-1) \bullet \ldots \bullet 2 \bullet 1}{(n-r) \bullet (n-r-1) \bullet \ldots \bullet 2 \bullet 1}$$

There is also a special notation used for permutations. To indicate the answer to the seating problem, a permutation of 6 objects taken 4 at a time, you would write ${}_6P_4$ To read this notation, you would say, "A permutation of 6 objects taken 4 at a time." To denote a permutation of n objects taken r at a time, write ${}_nP_r$.

$${}_nP_r = \frac{n!}{(n-r)!} = n \bullet (n-1) \bullet (n-2) \bullet \ldots \bullet (n-r+1)$$

analysis

Pizza and a Movie (Part 2: The Pizza)

10. Write the permutation notation for 5 objects taken 5 at a time, and use the formula above to find the result. (Recall that you found this numeric result in the first movie problem because all 5 people in your group were seated.)

11. Based on the answer for a permutation of 5 objects taken 5 at a time and the formula in the input window, what must be the value of 0!? Explain your answer.

analysis

Pizza and a Movie (Part 2: The Pizza)

12. Compare and contrast ${}_nP_r$ and ${}_nP_n$.

input

Pizza and a Movie (Part 2: The Pizza)

The movie is over, and it's time to get the pizza. The group decides to go to a new pizza shop owned by the parents of another friend. The shop opened a little over a month ago, but has become known for having the best crust in town.

Since the owners did not have a lot of money when the shop first opened, and believed they had a great-tasting crust, they decided not to offer a lot of different toppings. In fact, the first week cheese pizza was the only pie on the menu. You could not get any toppings on your pizza that week.

The second week, the owners were able to offer pepperoni as a topping. The menu now included cheese pizza (no toppings) and pepperoni pizza (1 topping).

The owners added mushrooms to the list of available toppings for the third week. The menu now included a cheese pizza, either pepperoni or mushrooms on a one-topping pizza, and pepperoni *and* mushrooms on a two-topping pizza.

The fourth week, sausage was added to the list of available toppings, so now the selections included a no-topping pizza, a 1-topping pizza, a 2-topping pizza, or a 3-topping pizza.

This week (the fifth week of business), peppers were added to the topping choices.

process: evaluate

Pizza and a Movie (Part 2: The Pizza)

13. During the third week, how many different one-topping pizzas can you order?

14. During the third week, how many different two-topping pizzas can you order?

15. Is mushroom-pepperoni pizza any different than pepperoni-mushroom pizza? Does your answer to this question change your response to question 13?

16. During the fourth week, how many different 1-topping pizzas can you order? How many <u>different</u> 2-topping pizzas can you get? How many different 3-topping pizzas can you get? Recall that you can always get a 0-topping pizza (a cheese pizza).

17. This week, the fifth week of operation, how many different 0-topping pizzas are available? How many different 1-topping pizzas are available? How many different 2-topping pizzas are available? How many different 3-topping pizzas are available? How many different 4-topping pizzas are available? Show any and all work you do.

analysis

Pizza and a Movie (Part 2: The Pizza)

18. Describe any patterns in the number of pizzas available as the number of toppings changes.

19. As has become their custom, the owners of the pizza shop plan to add another topping to the list next week. In fact, they are conducting a poll to determine what it will be. No matter which topping is added to the list, how many 0-topping, 1-topping, 2-topping, 3-topping, 4-topping, and 5-topping pizzas will be available?

process: model

Pizza and a Movie (Part 2: The Pizza)

20. If you arrange the numbers you have found for the different available pizzas during weeks 1 through 6 in six separate rows, you can explore some interesting patterns. Arrange the numbers so they form an equilateral triangle. Here is the start.

Week 1 (0 toppings available)	1
Week 2 (1 topping available)	1 1
Week 3 (2 toppings)	
Week 4	
Week 5	
Week 6	

Pizza and a Movie (Part 2: The Pizza)

analysis

21. Describe any patterns that you see in the triangle above.

22. Complete the next two rows of this triangle using the patterns you noticed.

23. Find the sum of the entries in each row of the triangle, and write each row's sum along the right margin of the previous "process: model" window. What pattern do you notice in these numbers?

24. What does this sum symbolize in terms of the pizza problem?

25. How is the pizza problem different from the two movie problems?

Pizza and a Movie (Part 2: The Pizza)

input

The problems you encountered in both parts of the movie problem are permutations, whereas the problems you solved in the pizza problem are **combinations**. **Combinations** are like permutations because they are arrangements of objects, but the order of the objects *does not* matter with combinations. The order of the objects *does* matter with permutations.

For example, in the pizza problem, you found that if there are 5 toppings to choose from and you want a pizza with 2 toppings on it, you have 10 different choices. However, if 5 people go to the movies, but 2 seats are available, there are 5 • 4 or 20 different arrangements.

input

Pizza and a Movie (Part 2: The Pizza)

Like permutations, combinations also have a notation. The letter *C* is used for combinations instead of the letter *P*, which was used for permutations. For a combination of 5 toppings choosing 2 at a time, the notation reads $_5C_2$, and you say, "A combination of 5 objects taken 2 at a time." In general, $_nC_r$ means a combination of n objects r at a time.

The triangle you created on the previous page is called **Pascal's Triangle**. Blaise Pascal, a French mathematician in the 1600s, first created it. Pascal's triangle displays all the combinations of n objects taken r at a time. The first row shows the combinations of 0 objects. Recall that when only cheese pizza was available with no toppings, there was one choice for 0 toppings ($_0C_0$). The second row shows the combinations with 1 object, first taken 0 at a time ($_1C_0$) and then taken one at a time ($_1C_1$). This is akin to the situation when only pepperoni was available.

Which combinations are shown in the third row of Pascal's triangle? What is the notation for those combinations?

process: practice

Pizza and a Movie (Part 2: The Pizza)

26. Calculate $_7P_3$, showing your work.

27. Calculate $_7C_3$, showing your work or using Pascal's triangle.

28. Calculate or find $_9C_4$ and $_9P_4$.

process: evaluate

Pizza and a Movie (Part 2: The Pizza)

29. If you calculate ${}_7C_3$ from ${}_7P_3$, what do you need to divide by? Is this quantity the factorial of a number involved in the combination?

30. Recall from the "input" window on page 1-46 that the permutation of 5 objects taken 2 at a time was 20, while the combination of 5 objects taken 2 at a time was 10. If you calculate ${}_5C_2$ from ${}_5P_2$, what do you need to divide by? Can this number be viewed as the factorial of a number involved in the combination?

31. If you calculate ${}_9C_4$ from ${}_9P_4$, by what do you need to divide? Is this quantity the factorial of a number involved in the combination?

analysis

Pizza and a Movie (Part 2: The Pizza)

32. Based upon the three problems on the previous page, how can you find ${}_nC_r$ from ${}_nP_r$?

33. Knowing that the formula for ${}_nP_r$ is $\frac{n!}{(n-r)!}$, devise a formula for ${}_nC_r$.

34. How and why is the calculation of a combination different from the calculation of the permutation?

extension

Pizza and a Movie (Part 2: The Pizza)

35. Both times you went to the movies you hoped to sit next to the person you like, but each time it did not work out as you had hoped. What was the likelihood that you would sit next to a specific person when you attended the first movie? What was the likelihood that you would sit next to a specific person for the second movie?

Socks and Free Throws

input

When calculating the probability for two events, you need to know if the two events are mutually exclusive.

What does it mean for two events to be mutually exclusive?

However, mutual exclusivity is not the only condition that makes a difference in determining probability. Whether or not events are **independent** or **dependent** is also an important condition and makes a difference in determining probability. Sometimes the likelihood of an event changes if a particular occurrence precedes it. In such a case, the second event is **dependent** on the first event. If the result of one event has no bearing on the result of another event, these events are **independent events**.

For example, if a number cube is being rolled, the result of the first roll has absolutely no influence on the second roll. The likelihood of rolling any number on the cube is still $\frac{1}{6}$. Therefore, the first roll is independent of the second roll.

However, if a card is picked from a standard deck of 52 cards and kept, this will have an influence on the next card that is picked. The probability of picking each of the remaining cards has now changed from $\frac{1}{52}$ to $\frac{1}{51}$. The second draw is dependent upon the first draw.

Socks and Free Throws

process: model

Determine if the following events are independent or dependent and explain your choice.

1. Two spins on a spinner numbered from 1 to 8 and with three colors

2. Pulling two socks, one at a time, from a drawer to try to get a pair

process: model

Socks and Free Throws

3. The 10th and 11th tosses of a quarter

4. Rolling two number cubes and getting a 5 on one of them while getting a sum under 9

input

Socks and Free Throws

It's Monday morning, and you are running late for school. You go to your messy sock drawer to try to find a pair of socks. Since you did not do your laundry over the weekend, you have only six socks in the drawer. Two of the socks are black, and four of the socks are gray, but the socks are not put into pairs in the drawer. The old adage of "haste makes waste" proves true, and while you are hurrying you inadvertently put the first sock you take out of the drawer back before pulling the second sock out.

You wonder what the likelihood is that you can actually draw two gray socks in a row.

process: evaluate

Socks and Free Throws

5. What is the probability of pulling out a gray sock on the first attempt?

6. What is the probability of pulling out a gray sock on the second attempt?

process: model

Socks and Free Throws

7. Is drawing the first sock independent of drawing the second sock? Explain.

8. Use an <u>organized</u> list (or a tree diagram) to create the sample space for drawing socks from your drawer. Make sure you can tell which socks are gray and which are black in your list.

9. Create a Venn Diagram showing how these two events relate to one another. Let A represent drawing a gray sock on the first attempt. Let B represent drawing a gray sock on the second attempt.

10. Shade the area of the diagram that represents drawing a gray sock on both the first and the second attempt.

11. Use the correct set notation to describe the area you just shaded.

12. Using your organized list or tree diagram, what is the probability of drawing a gray sock on both of the attempts you made?

analysis

Socks and Free Throws

13. Is there an operation you could perform to calculate the probability of drawing gray socks on both attempts, $P(A \cap B)$, from the individual probabilities for drawing gray socks, $P(A)$ and $P(B)$?

process: evaluate

Socks and Free Throws

14. What is the probability of tossing two heads if you flip a coin twice? If you need to, create the sample space below using an organized list or a tree diagram.

15. What is the probability of tossing heads on the first flip?

16. What is the probability of tossing heads on the second flip?

analysis

Socks and Free Throws

17. Are these events independent or dependent? Explain.

18. Could you perform the same operation to find the probability of flipping two heads as you did when drawing two gray socks?

analysis

Socks and Free Throws

19. In general, given two independent events X and Y, is it possible to perform this same operation to find the probability of X and Y? Explain your response and provide an example.

input

Socks and Free Throws

For any two independent events, $P(A) \bullet P(B) = P(A \cap B)$. Now, you will determine if this property is true of events that are dependent.

The morning after the gray sock problem, you are again running late for school and quickly grab two socks from your drawer. You want a pair of black socks. In the drawer there are still two black socks, but only one pair of gray socks. (The pair you wore yesterday is in the wash.) If you make sure you keep the first sock you draw, what is the probability that you will draw a pair of black socks after pulling only two socks?

process: evaluate

Socks and Free Throws

20. What is the probability of drawing a black sock on the first attempt?

21. If a black sock is drawn on the first attempt, what is the probability of drawing a black sock on the second attempt?

22. If a black sock is NOT drawn on the first attempt, what is the probability of drawing a black sock on the second attempt?

analysis

Socks and Free Throws

23. Are these two events independent? Explain.

process: model

Socks and Free Throws

24. Use an organized list or a tree diagram to show the sample space for this experiment?

process: evaluate

Socks and Free Throws

25. Use your list or diagram to calculate the probability of drawing two black socks in two attempts?

26. For this experiment, is it true that $P(A) \bullet P(B) = P(A \cap B)$? Show your work.

input

Socks and Free Throws

Your school's basketball team is in the playoffs. You decide to attend the game. The lead goes back and forth between the two basketball teams throughout the game. The teams are evenly matched, and you know that determining the winner will come down to the end. At the buzzer, your team is losing by one point, but one of the players on your team is fouled! You quickly try to determine the likelihood that your team will win, knowing that the player going to the line has made 54 foul shots in 90 attempts throughout the regular season.

analysis

Socks and Free Throws

27. How likely is it that your team will win if the fouled player gets to take two foul shots?

28. How likely is it that your team will win if the fouled player gets to shoot a one-and-one free throw? (In a one-and-one, the player only gets to attempt a second shot, if the first shot is made.)

29. Are both of the situations described in questions 27 and 28 independent events? Explain your response.

Socks and Free Throws

input

Two useful techniques for finding conditional probabilities are:

- Using a "pipe" diagram similar to a tree diagram.
- Using a grid to help visualize the outcomes and their likelihood.

In this situation, the player who is fouled has a $\frac{54}{90}$ or $\frac{3}{5}$ (0.60) chance of making each shot. For the two-shot foul, here is how you might use a "pipe" diagram.

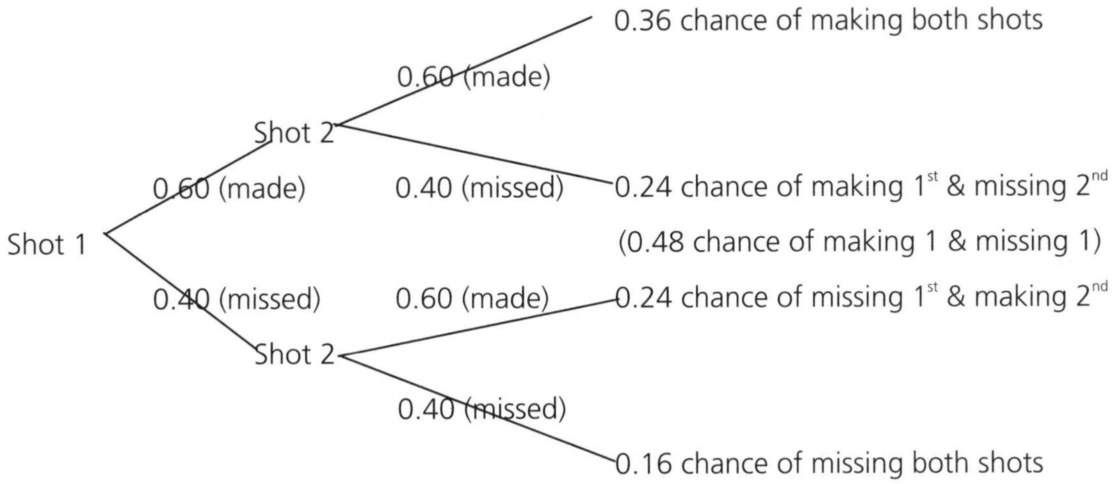

For the one-and-one free throw, here is how you might use a grid.

First Foul Shot	Second Foul Shot: Made (60%)	Second Foul Shot: Missed (40%)
Made (60%)	36%	24%
Missed (40%)	No option for a second shot, if the first one is missed	

process: summary

Socks and Free Throws

30. Your team won the game, after making both free throw shots. When you were replaying the final seconds of the game in your mind, you realized even though the player who made the foul shots was 60% from the line overall, the percentages of first shots made and second shots made were not 60% each. In fact, this player had made 30 out of 62 possible first shots, and 24 out of 28 second shots. Based on this, calculate the probability your team would win the game. Show your work or show any diagram you use to help with this calculation.

31. The following problem appeared in the weekly column called *Ask Marilyn* by Marilyn Vos Savant in *Parade Magazine*.

    A woman and a man, who are unrelated, each have two children. At least one of the woman's children is a boy, and the man's older child is a boy. Do the chances that the woman has two boys equal the chances that the man has two boys?

    Marilyn claimed that the man had a greater chance of having two boys than the woman. Her readers responded with letter after letter telling her she had made a mistake and that the man and the woman were indeed equally likely of having two boys.

    Does Marilyn owe her public a retraction? Show your work and explain your response. An organized list, pipe diagram, or grid might help.

32. Revisit the black sock problem at the beginning of this section, and make any changes you need to the answers you provided, based upon your new knowledge.

input

"No Fair!"

When Friday rolls around, you invite your friend to your house to play games. You plan to order pizza from the new shop that has all the different combinations of toppings.

You decide to play a game involving number cubes. In the first round, you will be the "roller" and your friend will be the "recorder." In each successive round, you will switch who is the "roller" and who is the "recorder." Only the "roller" gets to roll the number cubes in each round.

In this game, you each start with 10 points. If the "roller" rolls a sum of 7, the "recorder" must give three of his or her 10 points to the "roller." When the "roller" gets a sum that is not 7, he or she must give one point to the "recorder."

The winner of the game has the most points after ten rolls of the number cubes. (If someone runs out of points before 10 rolls, that person automatically loses, and the game is over.)

Before you begin playing the game, predict whether or not you think this game is a fair game and why. Record your prediction and your reasoning below.

process: model

"No Fair!"

1. Play at least two rounds of this game with a partner in your class. Use the table below to track the points in the game, when it is your turn to be the "recorder."

	Begin	Roll 1	Roll 2	Roll 3	Roll 4	Roll 5	Roll 6	Roll 7	Roll 8	Roll 9	Roll 10
Sum of 7?		Y/N									
Roller	10										
Recorder	10										

2. Who was the winner of your games, the "roller" or the "recorder?"

process: model

"No Fair!"

3. Gather the results of all the games played in your class in the table below.

	Team 1	Team 2	Team 3	Team 4	Team 5	Team 6	Team 7	Team 8	Team 9	Team 10	Team 11
Roller											
Recorder											

4. How many total games were played in the class? How many games were won by the "roller?" How many games were won by the "recorder?"

analysis

"No Fair!"

5. Is this game fair, meaning, does each of the players in the game have an equal chance of winning? Justify your answer based upon the data you gathered above.

6. Return to the table you created for the sums of two number cubes on page 1-9. What is the probability of rolling a 7? Use this information to justify whether or not the game is fair. (Recall that if a sum of 7 is rolled, the "recorder" must forfeit 3 points, and if a sum other than 7 is rolled, the "roller" must forfeit one point.)

"No Fair!"

analysis

7. Alter the game so that it will be fair.

"No Fair!"

extension

8. Because of your prowess in determining if games are fair, you have been hired for a part-time job at a company that manufactures games called Games-R-Us. The company takes pride in distributing fair games. One of the games that Games-R-Us makes has a pair of number cubes that must be rolled. You already know that all the sums are not equally likely when two number cubes are rolled, but your employer asks you to design a new pair of number cubes so that every sum from 1 to 12 is equally likely. Can this be done? If so, design the number cubes. If not, explain why not. (Sides of the number cubes can be blank.)

Unit 2: Linear Functions

Contents

input

Linear Equations and Inequalities

A fundraising event is going to be held by the high school to raise money for a variety of charitable organizations. The organizers decide they need a theme, decorations, and prizes. For the theme, they chose an old-fashioned carnival with games and rides. To set up the event, they will incur an initial cost of $950 for the rental of the rides and booths. The local merchants have offered to donate the food and prizes. Tickets will sell for $2.00 each. The organizers hope to sell enough tickets to make a profit.

process: model

Linear Equations and Inequalities

1. Write an algebraic equation to represent this situation. Define the independent and dependent variables and the constants.

process: evaluate

Linear Equations and Inequalities

2. Determine the profit, if the group sold:
   a. 500 tickets
   b. 1500 tickets
   c. 3000 tickets
   d. 100 tickets

process: evaluate

Linear Equations and Inequalities

3. If the group wants to make a profit of $1,000, how many tickets would have to be sold?

4. If there was a profit of $4,000,how many tickets were sold?

process: model

Linear Equations and Inequalities

5. Using the information collected, complete the following data table.

Labels		
Units		
Expression		

process: model

Linear Equations and Inequalities

6. Generate a graph using the data from the table. Label the axes and bounds.

Variable Quantity	Lower Bound	Upper Bound	Interval
x:			
y:			

Linear Equations and Inequalities

analysis

7. Where does the graph intersect the y-axis? How do you interpret this point?

8. Where does the graph intersect the x-axis? How do you interpret this point?

9. What information in this situation allows you to model it with a linear equation?

10. What represents the slope in this problem situation? How do you interpret this value?

analysis	**Linear Equations and Inequalities**

analysis

Linear Equations and Inequalities

11. What is the domain for this situation? Write an inequality to express which values make sense. Recall that the domain is the set of all possible values for the independent variable.

12. What is the range for this situation? Write an inequality to express which values make sense. Recall that the range is the set of all possible values for the dependent variable.

13. Do all the independent and dependent variable quantities listed for this problem so far make sense? Why or why not?

process: model

Linear Equations and Inequalities

14. Use a number line to graph the solution set for the number of tickets that have been sold.

15. Use a number line graph the solution set for the profit.

16. If the organizers expect the profit to be greater than $2,000, how would you represent this symbolically?

17. The organizers expect a minimum profit of $5,500 and a maximum of $5,500. Use inequalities to represent this statement algebraically.

Linear Transformations

input

Recall from your earlier studies of linear and quadratic functions in Integrated Math I that you were introduced to the concept of the parent function:

$$y = x$$

This parent function represents the most basic form for the function. Recall that any linear function can be constructed through a series of transformations to the parent function.

Linear Transformations

process: practice

1. Graph the following functions and indicate the transformations that were performed on the parent function to arrive at the transformed function.

a. $y = x$

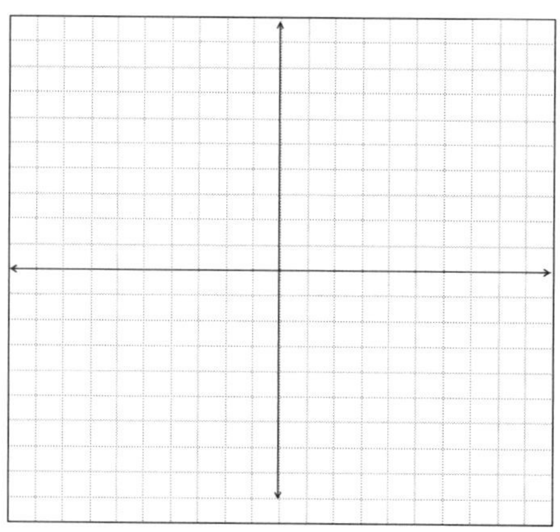

$y = x - 5$

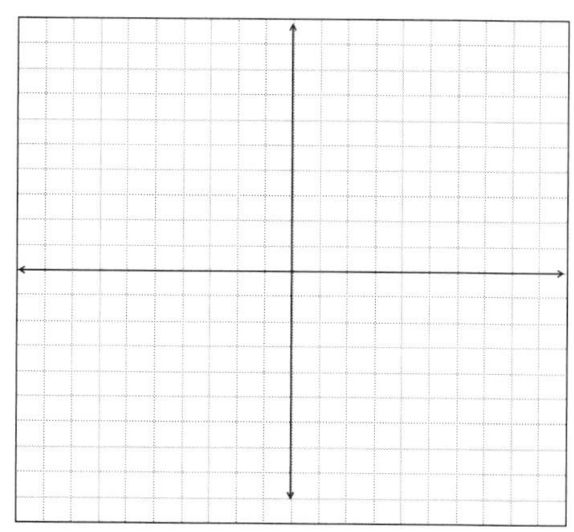

process: practice

Linear Transformations

b. $y = x$

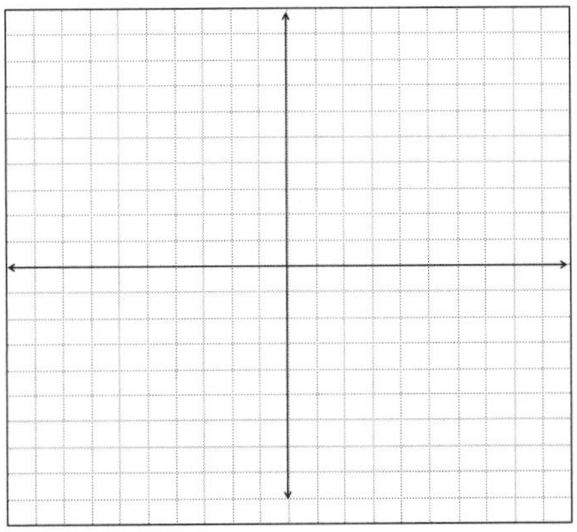

$y = 2x$

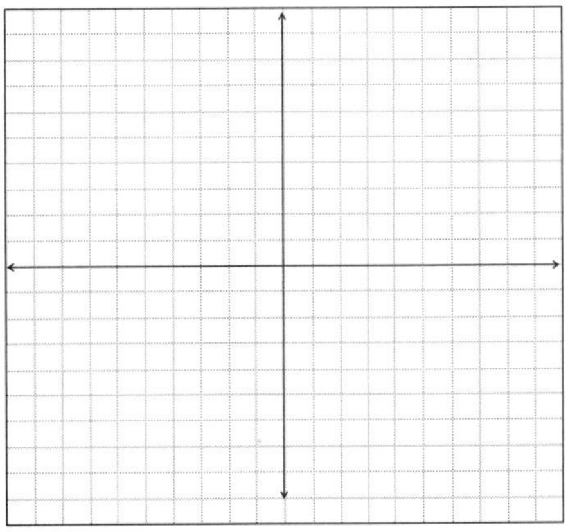

c. $y = x$

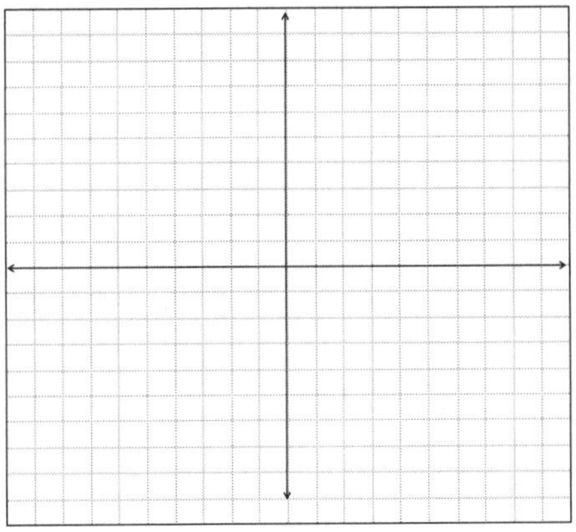

$y = 3x - 1$

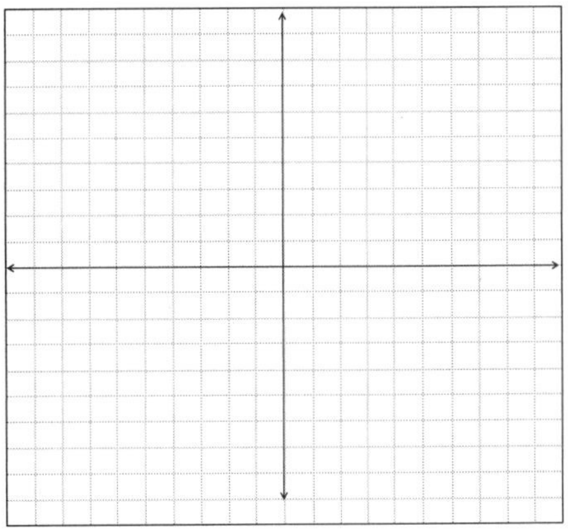

process: practice

Linear Transformations

d. y = x

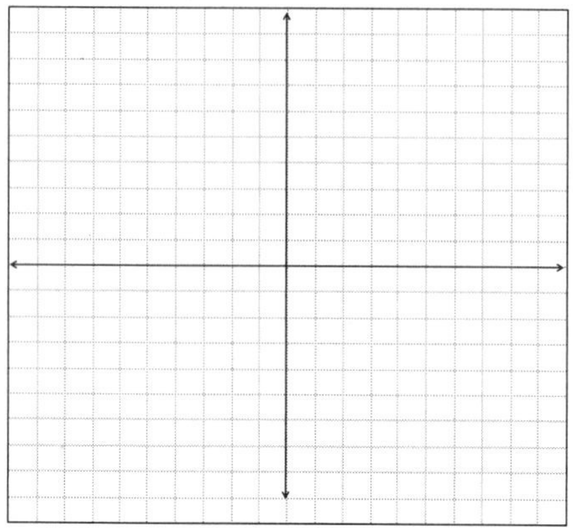

y = 4(x – 2)

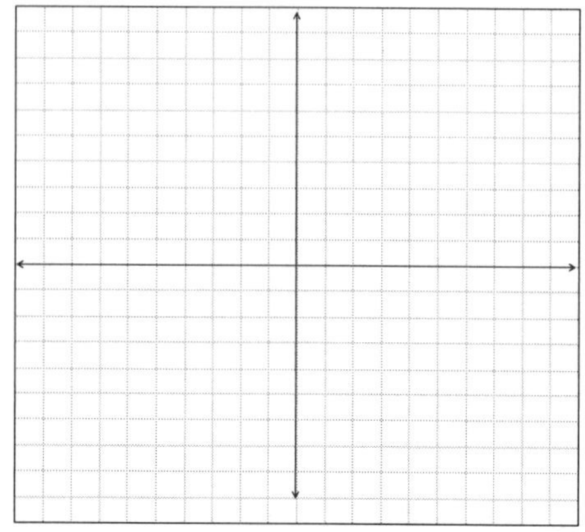

e. y = x

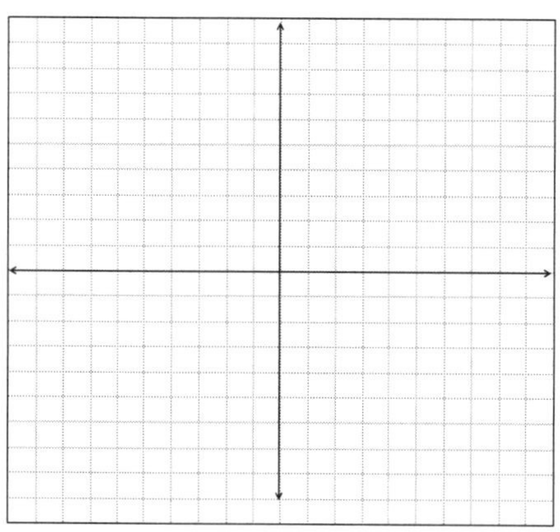

y = -3(x + 5)

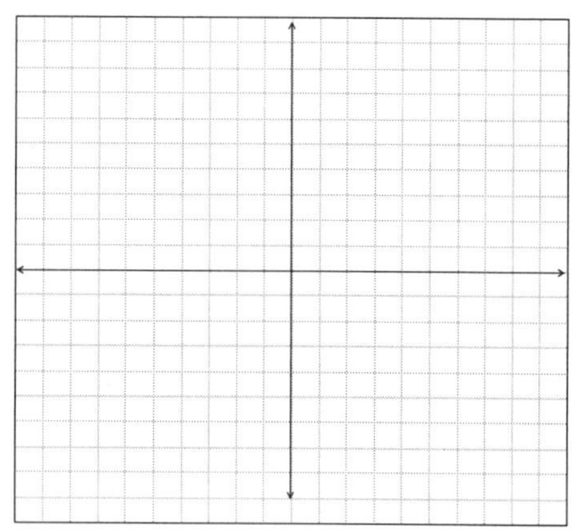

Linear Transformations

analysis

2. Look at the following interpretation of the two equations in terms of the transformation on the parent function.

$y = 3(x - 2)$	There was a horizontal shift two units to the right along the x-axis and a dilation by a factor of three.
$y = 3x - 6$	There was a dilation by a factor of three and a vertical shift downward by 6.

a. Compare the functions and the interpretations by drawing the graphs of the two equations on the same set of axes.

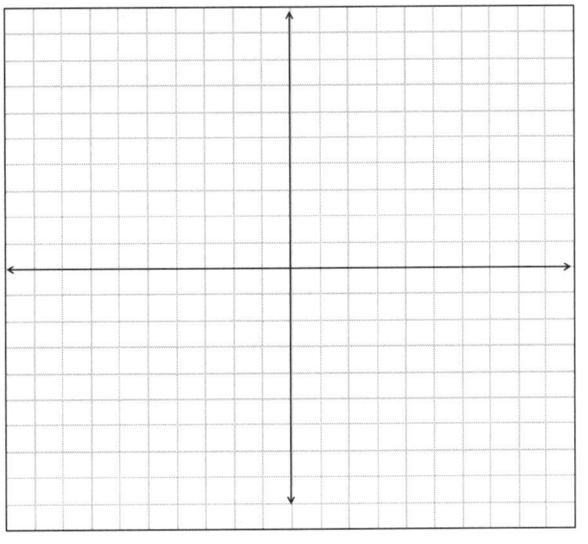

b. What does this tell you about the equations?

process: model

Linear Transformations

3. Write a scenario that matches each of the equations in question 2.

analysis

Linear Transformations

4. Indicate the transformations that were performed on the parent function for each of the linear equations below:

   a. $y = -3x - 4$

   b. $y = 2(x+1) + 2$

   c. $y = \left(\frac{x}{2}\right) + 5$

   d. $y = 0.5x - 10$

   e. $y = \frac{-2x}{3} + \frac{1}{4}$

extension

Linear Transformations

5. How would you defend the statement: The form in which a linear equation is written will determine how you interpret whether a shift is horizontal or vertical.

input

Linear Transformations

Read the two scenarios:

You have a babysitting job that pays $7.00 per hour. You save all your earnings. Determine how much money you will have at the end of the month if you work: 4 hours; 40 hours; 11 hours.

You have a babysitting job that pays $7.00 per hour. You have $50 in the bank. If you save all your money, at the end of the month how much will you have if you work: 4 hours; 40 hours; 11 hours?

analysis

Linear Transformations

6. Identify ways in which the two scenarios are the same and ways in which they are different.

Linear Transformations

process: model

7. Graph the parent function that applies to the babysitting scenarios.

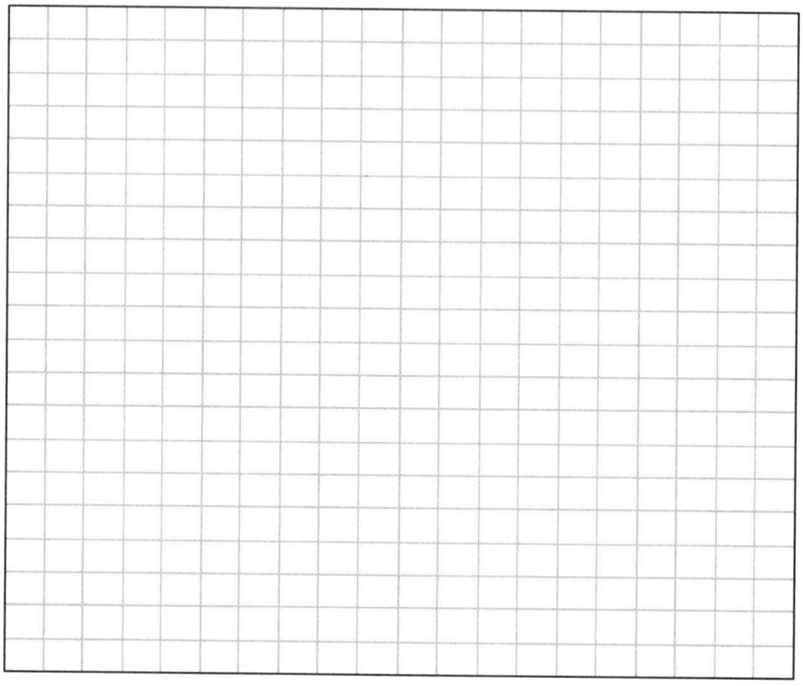

8. Write the equations representing each of the babysitting scenarios.

9. Graph the function that applies to the first babysitting scenario on the grid above.

10. Describe how the parent function has been transformed in the first function.

11. Graph the function that applies to the second babysitting scenario.

12. Described how the first function has been transformed.

input

Linear Transformations

G.P. has $165 in the bank. The money represents savings from a part-time job. G.P. uses the money in his savings to supplement his allowance. Each week he withdraws $5.00 from his savings.

process: model

Linear Transformations

13. Graph the parent function for this situation, y = x.

Graph each of the transformations below so that each one builds on the others.

a. Change the graph so that it represents withdrawing instead of depositing.

b. Change the graph so that is represents withdrawing $5 weekly instead of $1.

c. Change the graph to show that $165 is already in the bank.

14. Write the function that represents this situation.

analysis

Linear Transformations

15. Given all the possible situations that can be modeled by a linear equation, which statements are true and which are false? Explain why.

    a. The situations modeled by a linear function can all be represented graphically by a straight line.

    b. The rate of change of a linear function does not have to be constant.

    c. All linear functions can be described as a series of transformations from the parent function.

process: model

Linear Transformations

16. Given $y = x$, transform the function by performing a vertical shift upward by one unit. Write the transformed function.

    a. Graph the transformed function.

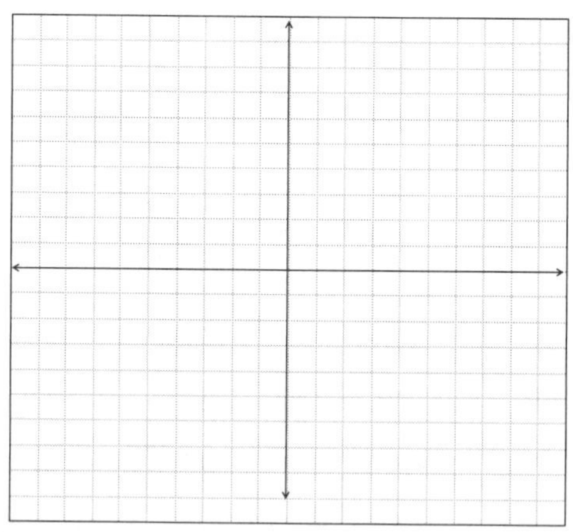

    b. Given $y = x$, transform the function by reflecting it through the y-axis and shifting upward by one unit. Write and graph the transformed function.

analysis

Linear Transformations

17. Describe the similarities and differences between the two transformed equations in question 16.

18. Using the graphs in question 16, how would you interpret the point at which the two lines meet?

Systems of Linear Equations: Producing and Selling Markers

input

Since you are interested in learning about business, you take a part-time internship at a company that produces color art markers. In particular, you will be studying production costs. You are told it costs $2 to manufacture each marker and, before any marker is produced, there is a $100 start-up cost for setting the color.

Systems of Linear Equations: Producing and Selling Markers

process: model

1. Write an algebraic equation that represents the cost to produce the markers. Indicate what each variable and constant represent in the equation.

Systems of Linear Equations: Producing and Selling Markers

process: evaluate

2. How much would it cost to produce 50 markers?

3. How many markers can be produced for $474?

Systems of Linear Equations: Producing and Selling Markers

process: model

4. Graph the equation representing the cost of production. Using inequalities, indicate the domain and the range for the function in this situation.

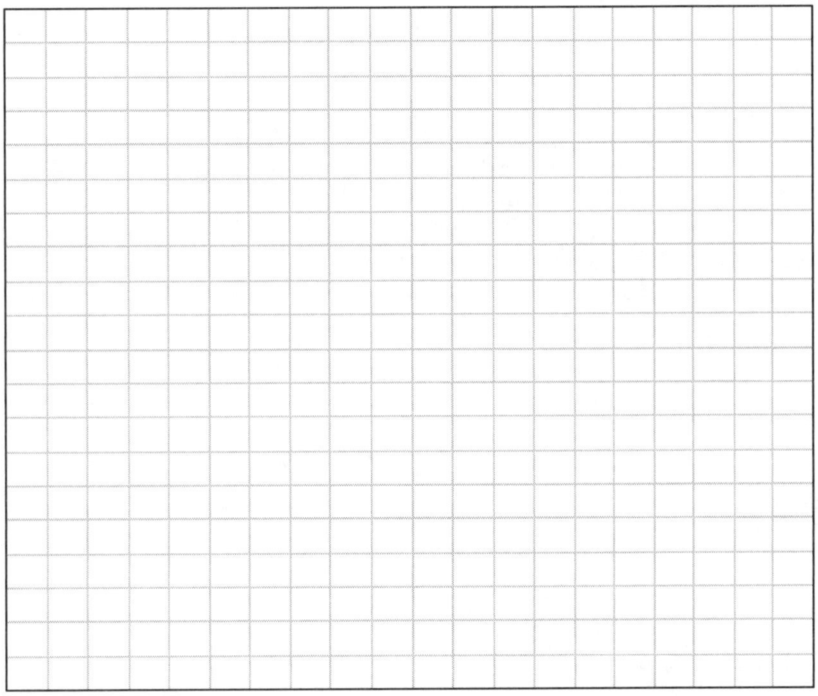

Domain: ____________ Range: ____________

analysis	**Systems of Linear Equations: Producing and Selling Markers**

5. To analyze the model for production cost more generally:

   a. Determine the value of the slope. Define what the slope means with respect to this problem situation.

   b. Determine the y-intercept. Define what the y-intercept means with respect to this problem situation.

   c. Determine the x-intercept. Determine if the x-intercept has meaning with respect to this problem situation. If it does, how do you interpret its meaning?

   d. Determine if the graph of the equation increases or decreases from left to right? What does this mean with respect to this situation?

input	**Systems of Linear Equations: Producing and Selling Markers**

In the next phase of your internship, you learn about income from sales. You are told that the company sells its markers to distributors for $3 per marker.

Systems of Linear Equations: Producing and Selling Markers

process: model

6. Write an algebraic equation that represents the company's income from the sale of its markers. Indicate what each variable and constrain represents in the equation.

7. Graph the equation representing the income from the sale of the markers. Indicate the domain and range using inequalities.

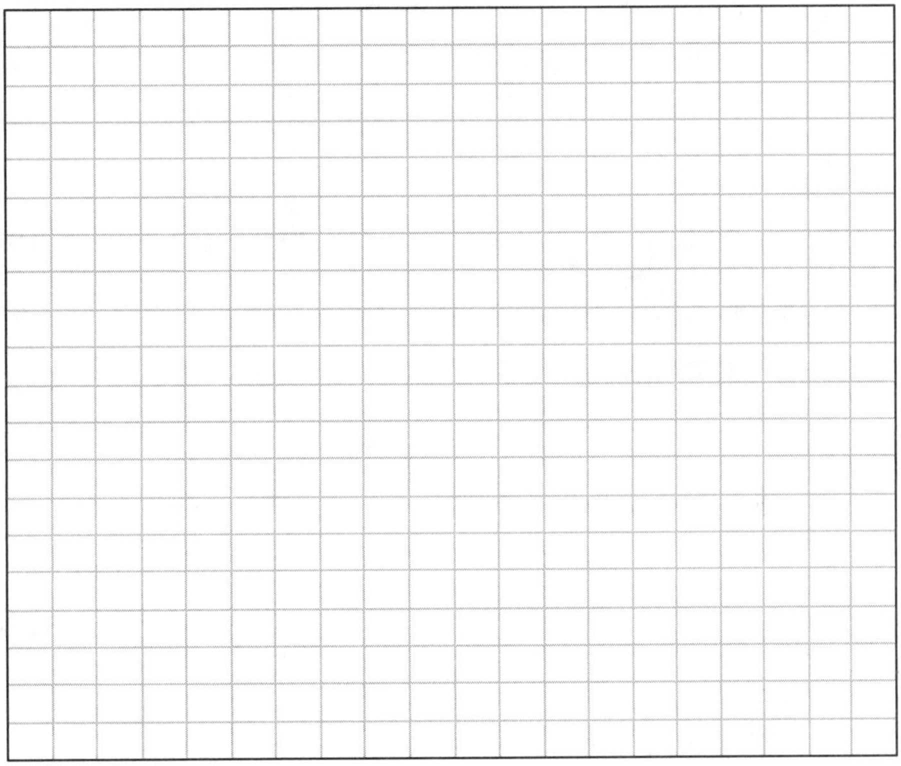

Domain: ________________ Range: ________________

Systems of Linear Equations: Producing and Selling Markers

process: evaluate

8. To get a better idea of the company's income based on its sales, determine:
   a. The income if 50 markers are sold

   b. The number of markers sold when the income is $474.00

Systems of Linear Equations: Producing and Selling Markers

analysis

9. To analyze the model for income more generally:
   a. Define the slope and tell what it means with respect to this problem situation.

   b. Define the y-intercept and tell what it means with respect to this problem situation.

   c. Determine the x-intercept. Does it have meaning with respect to this problem situation? If it does, how do you interpret its meaning?

   d. Determine if the graph of the equation increases or decreases from left to right? What does that mean with respect to this situation?

process: model

Systems of Linear Equations: Producing and Selling Markers

10. Using the information you know about production costs and the income, complete the following table.

Labels	**Number of Markers**	**Production Cost**	**Income**
Units	Markers	$	$
Expression			
	0		
	20		
	30		
	35		
	55		
	125		
	200		
	400		
	2500		

Systems of Linear Equations: Producing and Selling Markers

analysis

11. Examine the relationship between the production costs and the income by graphing both equations on the grid below. Make sure to label each axis and each line with its equation.

Based on the graph, what can you say about the relationship between the production costs and the income from sales?

input

Making and Selling Shirts

After working at U.S. Shirts and at the marker company, you decide to go into business for yourself. Your plan is to sell customized T-shirts. Since you know U.S. Shirts does a good job making the shirts, you will have them produce your shirts. U.S. Shirts will charge you $7.50 per shirt, plus a set-up fee of $22.50 for each new design. You will charge $8.25 for each shirt that you sell.

process: model

Making and Selling Shirts

1. Write an equation describing the cost of production. Indicate what the variables and constants mean in this equation.

2. Write an equation describing the income from sales. Indicate what the variables and constants mean in this equation.

process: evaluate

Making and Selling Shirts

3. How much will it cost to produce 15 shirts?

4. What will your income be if you sell all 15 shirts?

5. If your production costs are $165 for one type of customized shirt, how many shirts did you have produced?

6. If your production costs are $247.50, how many shirts did you have produced?

   a. If your income is $247.50, how many shirts did you sell?

process: model

Making and Selling Shirts

7. Graph the two equations. Label each axis and each line on the graph.

process: evaluate

Making and Selling Shirts

8. Based on the information you have about production costs, income from sales, and from the graph, complete the table below.

	Number of Shirts	Production Costs	Income Based on Sales
Labels	Number of Shirts	Production Costs	Income Based on Sales
Units	Shirts	$	$
Expression			
	0		
		165	
			206.25
		247.50	
	42		
			528

analysis

Making and Selling Shirts

Answer each of the following questions below in complete sentences.

9. Do the two lines intersect? If so, at what point?

10. In the context of this situation, how do you interpret the point of intersection?

11. In general, what does it mean to say that two lines intersect at a point?

12. In the context of this situation, what can you conclude about the portion of the graph to the left of the point of intersection?

    a. What can you conclude about the portion to the right of the intersection?

Connecting Algebraic to Graphical Representations

input

In the previous two problems about producing and selling markers and shirts, you studied the relationship between two related situations. From a mathematical perspective, you have examined two linear equations in two unknowns or a **system of two linear equations**.

A system of equations is said to have a **solution** if the graphs of the two equations intersect at a point (x, y). The value of the point of intersection, when substituted into each of the equations, represents a solution to *each* linear equation, so it is a solution to the *system* of equations.

Connecting Algebraic to Graphical Representations

process: model

Recall that in Making and Selling Shirts, the equation used to model your costs of $7.50 per shirt with a start-up fee of $22.50 was: $y_1 = 7.5x + 22.5$.

The equation used to model your income from selling shirts at $8.25 per shirt was:

$$y_2 = 8.25x.$$

At this selling price, no one seems to be buying your shirts. Your friend suggests that you lower your selling price to $7.50 per shirt.

1. At the reduced selling price, what is the new equation that will model income?

2. Create a graph showing both the equation that models your costs and the *new* equation that models your income.

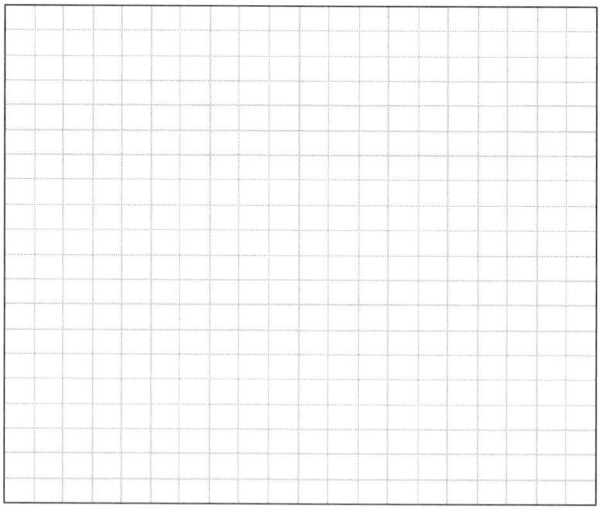

analysis

Connecting Algebraic to Graphical Representations

3. How do the graphs of the two equations relate to one another? Why do the graphs relate to one another in this manner?

4. Is there a solution to this system of equations? If so, what is the solution? If not, explain why there is no solution.

5. Would reducing your selling price per shirt to $7.50 be a wise business decision? Why or why not? Base your explanation on the graphs as well as on other factors.

process: model

Connecting Algebraic to Graphical Representations

Since you had difficulty selling your shirts at $8.25, and reducing the price will cause you to lose money, you decide to seek help from your old friends at U.S. Shirts.

In order to help you, U.S. Shirts decides that you will not be charged a set-up fee any longer. However, they will not reduce their charge of $7.50 per shirt.

6. Based upon U.S. Shirts dropping the set-up charge, what is the new equation that will model your costs?

7. Based upon a reduced selling cost of $7.50 per shirt, what is the equation to model your income?

8. Create a graph showing both the *new* equation that models your costs and the *new* equation that models your income.

analysis

Connecting Algebraic to Graphical Representations

9. How do the graphs of the two equations relate to one another? Why do the graphs relate to one another in this manner?

10. Is there a solution to this system of equations? If so, what is the solution? If not, explain why there is no solution.

11. If U.S. Shirts removes your set-up fee, is reducing your selling price per shirt to $7.50 a wise business decision? Why or why not? Base your explanation on the graphs as well as on other factors.

12. Based only upon the graph, how can you determine if there are zero solutions, one solution, or many solutions to a system of equations?

13. Based only on the problem scenario and the resulting equations, how can you determine if there are zero solutions, one solution, or many solutions to a system of equations?

Connecting Algebraic to Graphical Representations

process: practice

14. Graph the following system of equations:

$y = -x - 1$

$y = x - 3$

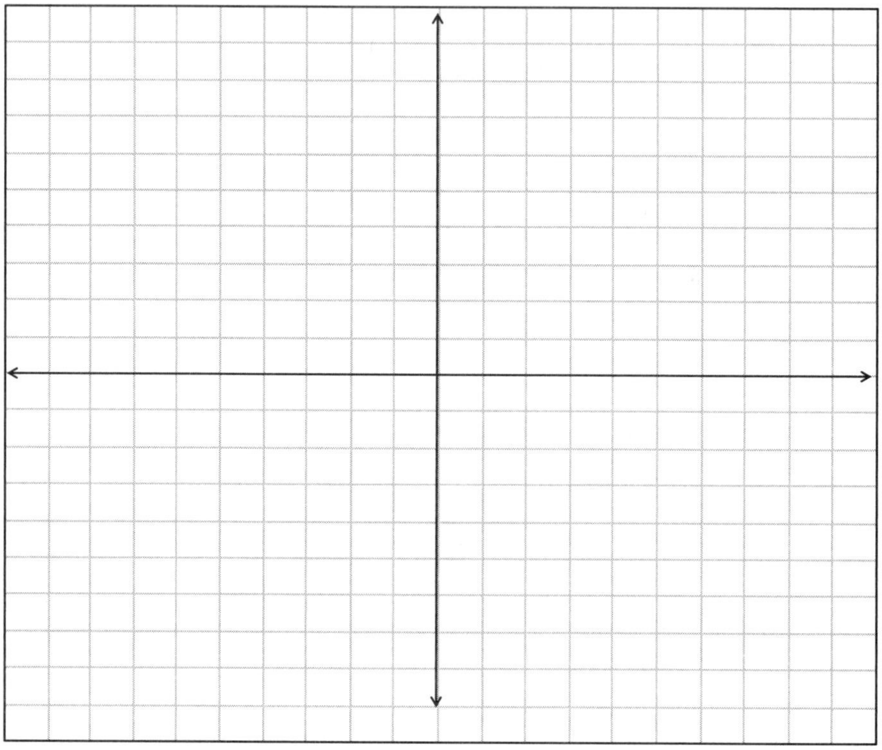

a. Based on the graph, describe the relationship between the two lines.

b. Is there a solution to the system? If so, what is the solution? Explain why it is a solution to the system.

Connecting Algebraic to Graphical Representations

process: practice

15. Graph the following system of equations: (Hint: You will want to write in the form $y = mx + b$.)

    $2y = 2x + 8$

    $3y = 3x + 6$

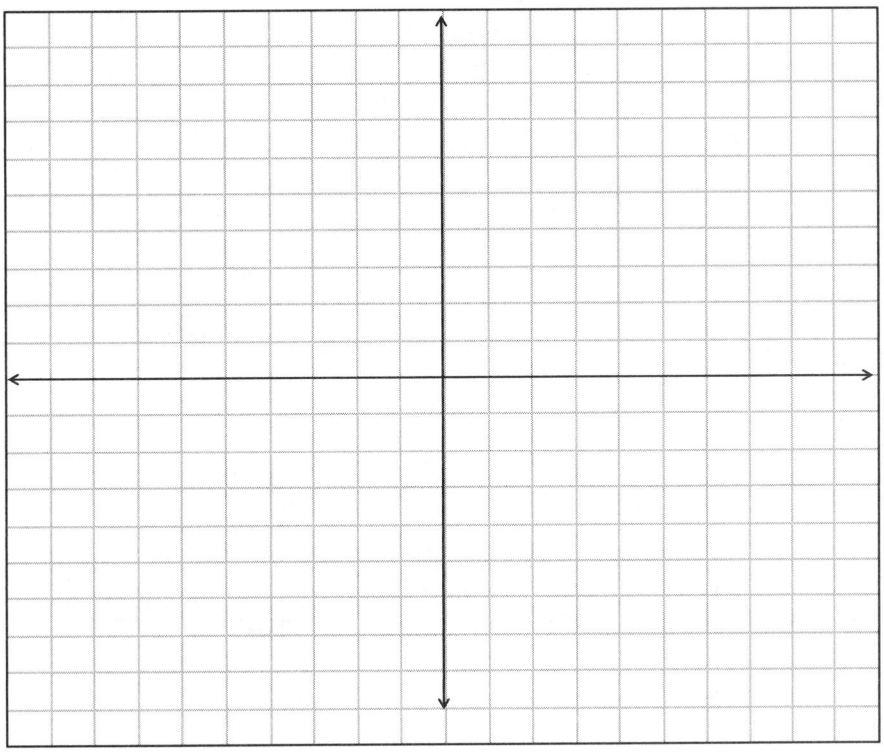

a. Based on the graph, describe the relationship between the two lines.

b. Is there a solution to the system? If so, what is the solution? Explain why it is a solution to the system.

process: practice

Connecting Algebraic to Graphical Representations

16. Graph the following systems of equations:

$y = 2x + 3$

$y = -x + 5$

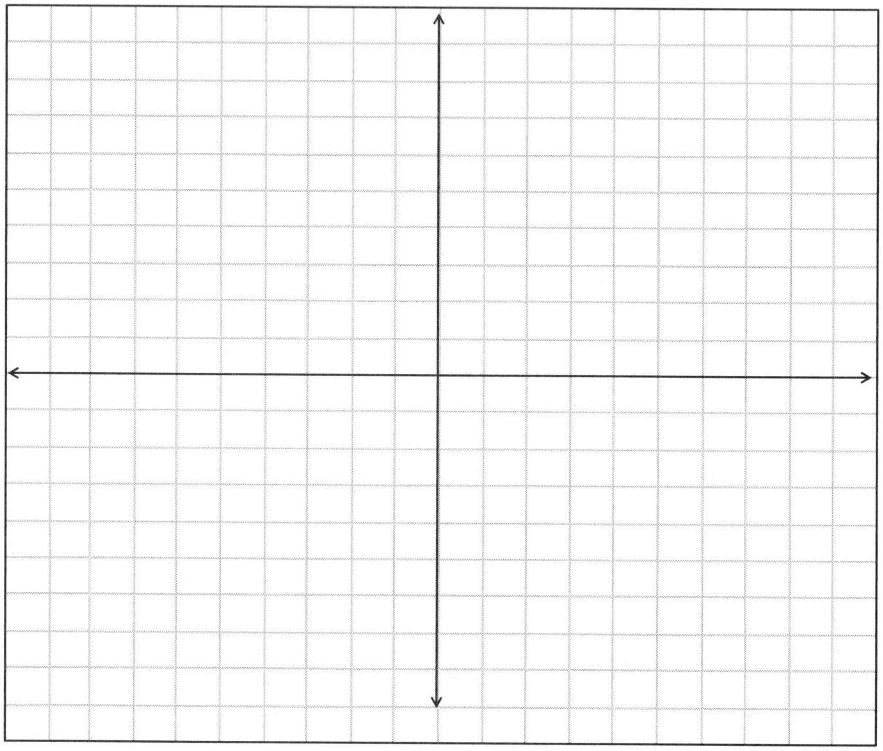

a. Based on the graph, describe the relationship between the two lines.

b. Is there a solution to the system? If so, what is the solution? Explain why it is a solution to the system.

analysis

Connecting Algebraic to Graphical Representations

17. What does it mean if a system has only one solution?

18. What does it mean if the system has no solutions?

19. What does it mean if every point is a solution for each equation?

20. What does it mean *graphically* if a point is <u>not</u> a solution to a system of equations?

21. What does it mean *algebraically* if a point is <u>not</u> a solution to a system of equations?

22. Can a point value satisfy one equation in a system and not the other? Explain.

process: summary

Connecting Algebraic to Graphical Representations

23. How can you use the equation to determine if lines are parallel (no points in common), intersecting (one point in common), or **coincident** (infinite number of points in common).

24. How can you tell from reading a problem scenario if lines are parallel, intersecting or coincident?

input

Finding the Better Paying Job

A friend of yours interviewed for two different sales positions in competing companies. The Stellar company pays $500 per week plus 10% commission on the total dollars from sales per week. The Lunar company pays $200 per week but offers a 20% commission on the total dollars from sales per week. Sales at both companies are seasonal. Your friend wants some help in determining which job option is best.

process: model

Finding the Better Paying Job

1. Write an equation that describes the salary model for Stellar. Indicate what the variables and constants mean in the equation.

2. Write an equation that describes the salary model for Lunar. Indicate what the variables and constants mean in the equation.

process: evaluate

Finding the Better Paying Job

To determine which option is best, gather some data by answering the following questions.

3. If the total sales are $1,200 for a particular week, what salary will your friend receive for that week from:

   a. Stellar?

   b. Lunar?

4. If the weekly salary from Lunar is $1200, what were the total sales for the week?

5. If the weekly salary for Stellar is $540, what will the salary be from Lunar?

6. Will the weekly salaries ever be equal? If so, under what conditions will it occur? What will be the amount of the salary?

process: model

Finding the Better Paying Job

7. Verify your answer to the previous question by graphing the two equations

process: model

Finding the Better Paying Job

8. Generate a table of values using the data gathered. Include at least four more values, where two sales totals generate a low weekly salary and two sales totals generate a high weekly salary.

Total Sales	Stellar Earnings	Lunar Earnings

Finding the Better Paying Job

analysis

9. Based on your findings, which company would you recommend to your friend and why?

   a. Are there any conditions under which you would change your recommendation? If so, what are they?

Finding the Better Paying Job

extension

10. Your friend had a third interview for a position with a company, called Solar. Solar pays a straight salary of $750 per week. Create the equation to represent this situation and graph it on the same set of axes where the salaries for Stellar and Lunar are graphed.

11. Under what conditions is Solar the best option?

12. You chose to take a job with Stellar and you want to earn at least $975 a week. Represent the situation algebraically. What is the solution set for the number of dollars in sales that you must earn weekly?

Solving Systems of Two Equations Algebraically

process: model

1. Graph the following system of two linear equations.

   $y = -4x + 11$

   $y = 3x - 7$

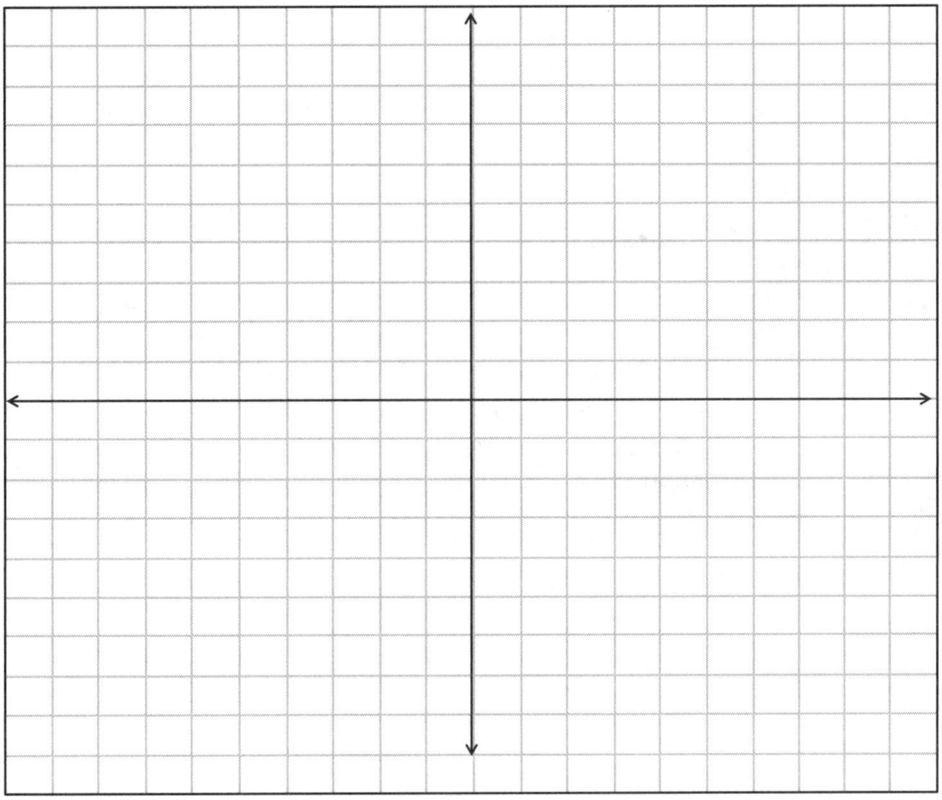

2. Determine the point of intersection.

3. Verify your solution algebraically by checking the point in the two equations.

Solving Systems of Two Equations Algebraically

analysis

4. Were you able to verify your solution algebraically? If not, explain why.

Solving Systems of Two Equations Algebraically

input

As you saw in the problem on the previous page, it can be difficult to estimate the point of intersection for a system of equations.

Why was it more difficult to find the solution in the last problem than in some of the previous problems you have examined?

Because estimating the point of intersection from the graph is not always very efficient, we have some other techniques we can use to solve systems and find the point of intersection. The first of these methods is called the **substitution method**. As you work in this section, try to decide why this method is called the substitution method.

In the problem on Making and Selling Shirts, the equations

$$y_1 = 7.50x + 22.50 \text{ and}$$

$$y_2 = 8.25x$$

were used to represent the cost to produce the shirts and the income from selling the shirts. The form of the equation used here is familiar to you. It is called the **slope-intercept form** of a linear equation.

Why is this form called the slope-intercept form?

Solving Systems of Two Equations Algebraically

input

Here is a system of equations:

$$y = 2x + 1$$

$$y = x + 3.$$

In what form are these equations written?

Notice that both equations use 'y equals."

Is it valid to then write:

$$2x + 1 = x + 3?$$

Why or why not?

Now, we have one equation instead of two and only one variable instead of two.

Solve this equation for x.

Now that we know the value of x, find y. Show all your work.

Using the information you just obtained, what is the point of intersection for this system of equations?

Check your solution by using the second equation to confirm that your point is actually on that line as well.

input

Solving Systems of Two Equations Algebraically

Another form that is commonly used to model a linear relationship is the **general form** for the equation of a straight line, which is

$$Ax + By = C$$

In this form, C represents a constant value or fixed output, A and B are non-zero constants, and x and y are variables.

Here is an example of a situation that can be modeled algebraically by an equation in this form:

> The snack bar is selling hot dogs (x) and chips (y). Hot dogs are $1.25 (A) and chips are $0.50 (B). The snack bar owners can determine the earnings (C) using the equation:
>
> $$Ax + By = C$$
>
> $$1.25x + 0.5y = C$$

extension

Solving Systems of Two Equations Algebraically

5. Given the following situations, which linear model do you think is most appropriate in this case, slope-intercept form or general form? Justify your choice.

   Your club has $550 to spend on tickets to a major sporting event. Since the club is a youth organization, the tickets are discounted. The cost of an adult ticket is $4 and the price of a student ticket is $3. You want to find out how many of each type of ticket you can buy and not exceed your spending limit.

input

Solving Systems of Two Equations Algebraically

We can use the substitution method to solve systems of equations no matter if the equations are in slope-intercept form or general form.

Here is a new system of equations:

$$2x + 3y = 7$$

$$y = x - 1$$

Replace the y in the first equation with the expression $x - 1$.

Why is it valid to replace y with $x - 1$?

Solve the resulting equation for x.

Use the x value you found in either of the original equations to find y.

What is the point of intersection for these two equations?

Verify the solution using the other equation.

process: evaluate

Solving Systems of Two Equations Algebraically

Here is a third system of two equations:

$$4x + 2y = 550$$

$$3x + y = 300$$

6. In which form is each of these two equations written?

7. Solve one of these two equations for y, so that it is written in slope-intercept form. (Hint: First think about which of the two equations will be simpler to solve for y.)

8. Now, use the substitution method to find the coordinates of the point of intersection for these two lines. Verify the solution.

process: practice

Solving Systems of Two Equations Algebraically

9. Solve the following systems of linear equations. Verify any solutions graphically by using your graphing calculator.

   a. $y = x - 2$ and $3x + 2y = 1$

   b. $y = -5x + 7$ and $y = x - 11$

   c. $y = 10(3x - 4)$ and $y = 10(-2 + 3x)$

process: practice

Solving Systems of Two Equations Algebraically

d. $x - 3y = 8$ and $y - 4x = 21$

e. $x + y = 8$ and $2x + 2y = 16$

analysis

Solving Systems of Two Equations Algebraically

10. Why is this method of solving a system of equations called the substitution method?

process: summary

Solving Systems of Two Equations Algebraically

11. Can a system of two linear equations always be solved by substitution? Explain.

12. Can a system of two linear equations always be solved graphically? Explain.

13. When and why might it be more useful to solve a system graphically rather than algebraically using substitution?

14. Which of these two methods of solving systems of linear equations do you prefer and why?

input

World Oil: Supply and Demand

Data on the world's crude oil reserves in 2001 showed that there were about 1 trillion (1,004.1 billion) barrels of oil worldwide. In 1970, there were 933 billion barrels in reserve. This difference represents an average rate of change of about 2.3 billion barrels annually. Assume that this rate of increase in reserves will continue in the future.

In 2001, the worldwide consumption of oil was 27.74 billion barrels per year. (Incidentally, this number was down from 33.5 billion barrels per year in 1970.) Assume that the rate of annual oil consumption is fixed at 27.74 billion barrels per year.

Under these conditions, study the relationship between the reserve supply of oil versus the demand for oil.

process: model

World Oil: Supply and Demand

1. Write the equation that models the worldwide supply of oil. Indicate what each variable and constant represents in the equation.

2. Write the equation that models the worldwide consumption of oil. Indicate what each variable and constant represents in the equation.

process: evaluate

World Oil: Supply and Demand

3. Based on the 2001 data, in how many years from 2001 will the amount of the oil in the reserves reach 1250 billion barrels? What year will that be?

4. Use the 2001 data to determine what year it would be if there were 1866 billion barrels of oil in reserve.

5. Based on the 2001 data, how much oil will the reserves contain i25 years from now?

6 At the 2001 rate of consumption, how much oil will be consumed by 2011?

7. At the 2001 rate of consumption, how long will it take to use 200 billion barrels of oil?

process: model

World Oil: Supply and Demand

8. Complete the following table using the data that you generated.

Labels	**Number of Years from 2001**	**Amount of Oil Consumed**	**Amount of Oil in Reserves**
Units			
Expression			

analysis

World Oil: Supply and Demand

10. Create a graph of each of the equations on the same grid. Do the lines intersect?

analysis

World Oil: Supply and Demand

11. Verify the result algebraically.

12. Interpret the meaning of the point of intersection in the context of this problem.

extension

World Oil: Supply and Demand

13. Is the comparison being made in this problem between supply and demand of oil a valid comparison to be making? Justify your response.

Solving Inequalities

input

Let's revisit Finding the Better Paying Job from page 2-40 for a moment. In solving that problem, you determined the sales needed to earn at least $975 per week from the Stellar company. Now determine the sales needed to earn at least $975 from the Lunar Company.

Solving Inequalities

process: model

1. Use and inequality to represent this situation algebraically. What is the solution set for the total number of dollars in sales per week?

Solving Inequalities

input

Suppose your friend learns that, at least initially, she cannot expect to earn more than $1600 per week in salary. You know she wants to earn at least $975 per week.

Solving Inequalities

process: model

2. Use an inequality to show that $1600 is the maximum expected weekly earnings. Solve the inequality to determine the expected sales for:

   a. The Stellar company

   b. The Lunar company

3. Use the inequality to show the sales requirements if you friend will earn between $975 and $1600 per week at:

   a. The Stellar company

   b. The Lunar company

Solving Inequalities

analysis

4. What are the implications of the above analysis for your friend if she works at

   a. Stellar

   b. Lunar

input

Solving Inequalities

A system of inequalities is much like a system of equations in that we are interested in finding the solution to the system. Recall for a line, every point on the line is a solution to the equation of the line. Therefore, the solution to a system of equations, if it exists, is the point that lies on both lines.

To find the solution set for an inequality graphically:

- Graph the inequality as if it were a line.
- Find any ordered pair that is *not* on the line.
- Substitute the ordered pair into the inequality.

If the ordered pair satisfies the inequality, then it is in the solution set. For an inequality, the solution set is the half-plane. The half-plane is the region of the plane above or below the line and may include the line. Every point in the regions represents a solution to the inequality. Therefore, the solution to a system of inequalities is the portion of the plane where the two half-planes intersect. Every point in this portion of the plane is a solution to both inequalities and therefore a solution to the system of inequalities.

To check that you have found intersecting regions, as before, choose any point in the solution set and test to make sure it satisfies both of the inequalities.

To solve any system of inequalities, graph each of the inequalities, shading the region that contains the individual solution sets. If the inequality is a strict inequality, then the line is not part of the solution set and will be represented by a dotted boundary line, then shade the area of overall a bit darker. This area is the solution set for the system. If both boundary lines are included in the solution, you will also want to note the point of intersection.

input

Solving Inequalities

You had a job throughout the school year that paid $5.00 per hour. As summer vacation is about to begin, you have begun looking for more work. You found a job that pays $15.00 per hour. You want to do both jobs but you cannot work more than 30 hours per week. You are saving for college and hope to make at least $300 per week.

process: evaluate

Solving Inequalities

5. Determine how you may divide your time between the two jobs. Write the inequalities representing the possible number of hours you can work between the two jobs and the possible earnings. Graph the two inequalities.

6. Choose at least four combinations of hours you will work at each of the two jobs, knowing that you cannot work more than 30 hours in all.

   a. Test the combinations to see which provides with you at least $300 in income. Show your work.

7. Use the method above to find three more combinations of hours that will fulfill both requirements.

analysis

Solving Inequalities

8. Of the choices you made, which helps you earn the most money?

process: evaluate

Solving Inequalities

9. Solve the following systems of inequalities graphically and shade the solution set. Name at least two points in each of the solution sets and two points outside the solution sets. Verify your choices algebraically.

   a. $y \geq -2$ and $y - x < 1$

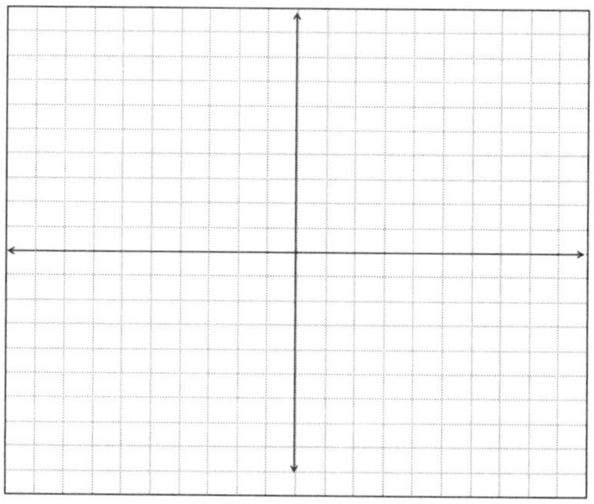

   b. $3y < 5x$ and $y \geq 3x$

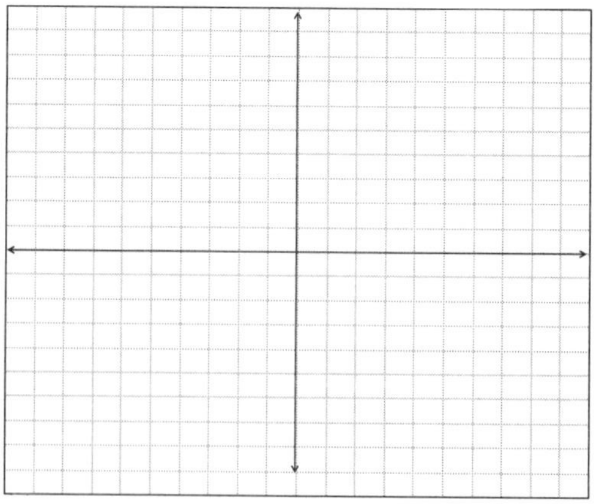

Solving Inequalities

process: evaluate

c. $3x - y \geq 6$ and $2x + y \leq 4$

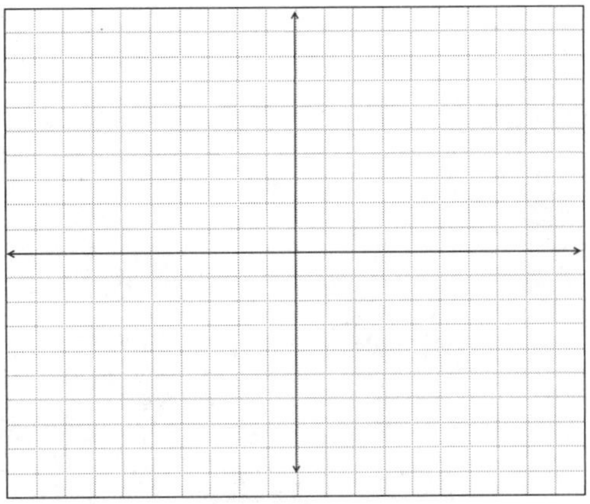

d. $y \leq 4 - 2x$ and $y \geq -2x - 1$

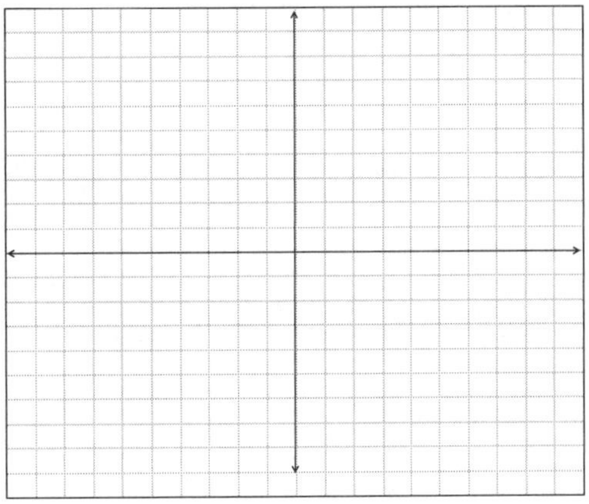

e. $x - y > 3$ and $x - y < -1$

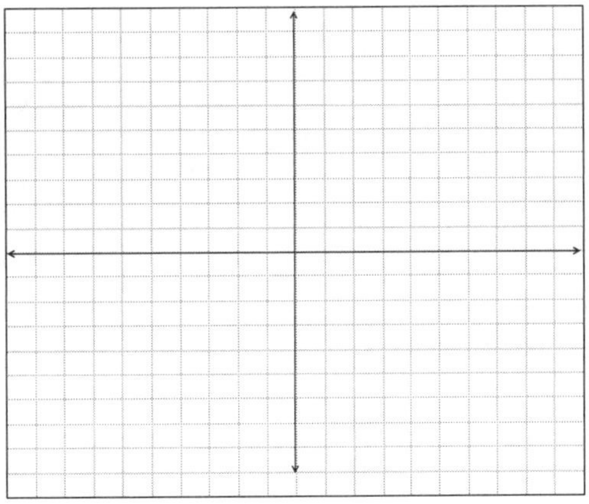

Solving Inequalities

input

You are graduating from high school and will be attending a local college. The firm that sponsored your internship wants you to continue to work for them and is willing to raise your salary to $9.50 per hour. You can also continue working at the job that pays $15.00 per hour that you started in the summer. You still want to earn at least $300 per week, but you need to decrease your hours to 20 per week at the most in order to have enough time to study.

Solving Inequalities

process: evaluate

10. Determine some options for dividing your time between the two jobs. To make this analysis, model the situation algebraically and graphically.

process: evaluate

Solving Inequalities

11. Name at least 4 different combinations of hours that you could work at each job.

12. Of the choices you made, which combination will allow you to earn at least $300?

13. Which combination helps you earn the most money?

Systems of More than Two Linear Equations and Inequalities

input

When helping your friend to pick between Stellar and Lunar, you compared the following relationships:

Stellar: $0.1x + 500 = y_1$

Lunar: $0.2x + 200 = y_2$

In each case, x represents the total sales in dollars and y is the earnings. Suppose now your friend has alternative opportunity from Solar. Solar is offering a weekly base salary of $350.00 with 15% commission on the total income from weekly sales.

Systems of More than Two Linear Equations and Inequalities

process: model

1. Write an algebraic equation representing the new salary option offered by Solar.

2. Graph the equations representing each of the three salary options.

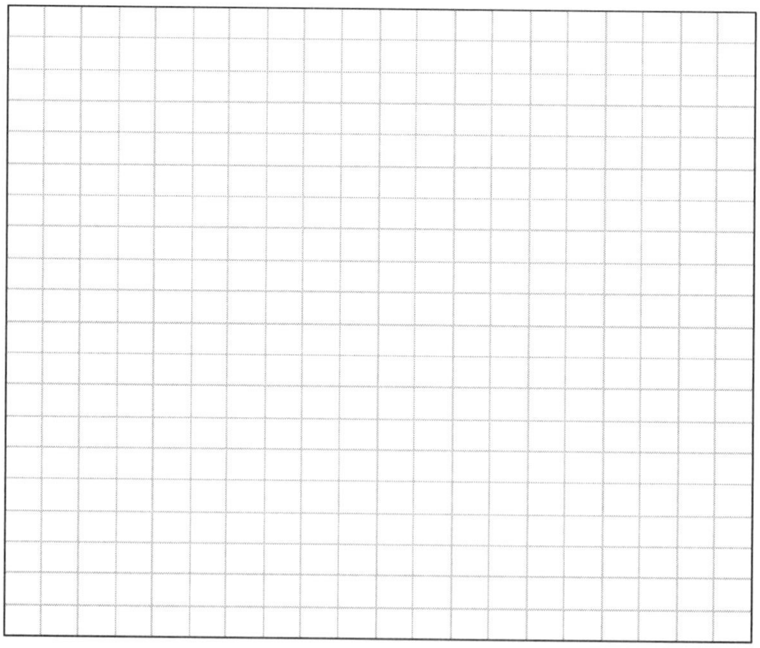

process: evaluate

Systems of More than Two Linear Equations and Inequalities

3. Looking at Stellar, Lunar, and this new offering from Solar, determine if the three options ever equal each other. If the options are all equal, when does this occur?

analysis

Systems of More than Two Linear Equations and Inequalities

4. When is Stellar's salary option the best?

5. When is Lunar's salary option the best?

6. When is Solar's salary option the best?

7. Write a short report summarizing all your findings.

process: model

Systems of More than Two Linear Equations and Inequalities

8. Graph the following three equations:

   $y = 2x + 1$

   $y = 3x$

   $y = -x - 1$

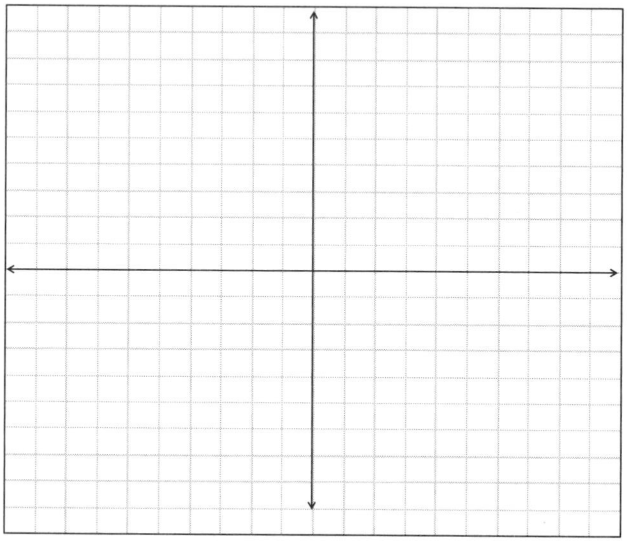

analysis

Systems of More than Two Linear Equations and Inequalities

9. Describe the relationship between the three lines.

10. How many points of intersection are there among the three lines? What are they?

11. Do any of these points represent a solution to the system of all three equations? Why or why not

process: model

Systems of More than Two Linear Equations and Inequalities

12. Graph the following three inequalities:

$y \leq 2x$

$y \geq x + 1$

$y \leq 3x - 1$

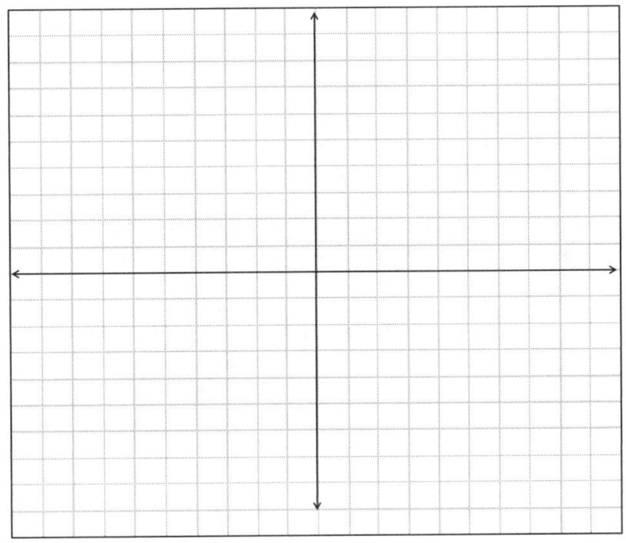

a. Is there a point where all three boundary lines intersect? If so, at what point?

b. Shade each of the half planes that fulfill the given inequalities.

c. Shade the region where all three half-planes intersect, if they do intersect.

input

Systems of More than Two Linear Equations and Inequalities

As you saw in the previous three problems, systems of equations and inequalities can contain more than two equations or inequalities. Sometimes the best method for solving a system is by graphing, while other times algebraic methods such as substitution will be best.

In addition to the substitution method for solving systems of linear equations, there is a method called **elimination.** This method allows you to transform a system, say of two equations in two unknowns, into one equation in one unknown and solve for the remaining variable. The elimination method is also called the addition/subtraction method.

input

Systems of More than Two Linear Equations and Inequalities

For example, you can solve systems of two linear equations using the elimination method.

$$3x + y = 5$$

$$4x - y = 2$$

Begin by choosing a variable to eliminate. For this system, choose to eliminate y.

Now that you have decided which variable, how do you think you will eliminate it?

Add the two equations to eliminate the variable y.

Now that you have one equation in one unknown, solve for it.

Now find the value of the variable you eliminated initially using one of the two equations.

What is the solution to the system?

Verify your result.

Systems of More than Two Linear Equations and Inequalities

input

Let's look at another example.

$$2x - 3y = 6$$
$$3x + y = 4$$

On inspection, how is this system different from the previous one?

What has to be done to one (or more) of the equations so that the equations can be added (or subtracted) to eliminate a variable?

Determine which variable you want to eliminate and rewrite the system in order to perform the elimination procedure.

Eliminate the variable and solve for the remaining unknown.

Solve for the variable you eliminated, by using one of the two original equations.

What is the solution to this system?

Verify the solution using the other one of the original equations.

process: summary

Systems of More than Two Linear Equations and Inequalities

To summarize the process for eliminating variables:

Step 1: Choose a variable to eliminate.

Step 2: Determine if the coefficients of that variable are identical or vary only by a sign, (e.g., 3 and –3) in the two equations. If this is the case, skip to Step 4.

Step 3: If the coefficients vary by more than just the sign, you must write equivalent equations for one or both, so that the process of adding or subtracting the equations will eliminate the variable.

Step 4: Eliminate the variable by adding or subtracting the equations.

Step 5: Solve for the remaining variable.

Step 6: Solve for the eliminated unknown using one of the original equations.

Step 7: Check the solution by using the other of the original equations

process: evaluate

Systems of More than Two Linear Equations and Inequalities

13. Let's apply this procedure to a system of three equations:

$2x + 3y = 1$

$x - 2y = 4$

$2x + y = 3$

The process for solving a system of three equations will be the same with an added step. After you choose the variable to eliminate, choose two equations to eliminate the variable and solve for the unknown. To find the remaining unknown, you will often find it easier, although not mandatory, to use the equation that was not selected for the elimination process. Recall that for a point to be a solution, it must satisfy all the equations.

Solve the system shown above and indicate each step in your process.

extension

Systems of More than Two Linear Equations and Inequalities

One final method for solving systems of equations involves matrices. A **matrix** is a rectangular arrangement of numbers situated in rows and columns. Each number in a matrix is called an **element**. The dimensions of a matrix have the form row x column. For example, the following matrix A is an example of a 2 x 3 or square matrix with 2 rows and 3 columns.

$$A = \begin{bmatrix} 2 & 3 & 4 \\ 4 & 2 & 3 \end{bmatrix}$$

To solve a system of equations using matrices, you must understand how to multiply two matrices and also how to find the inverse of a matrix. First, let us look at **scalar multiplication**, which is the multiplication of a constant times each element of the matrix. Scalar multiplication is essentially an application of the distributive property.

$$3\begin{bmatrix} 2 & 3 & 4 \\ 4 & 2 & 3 \end{bmatrix} = \begin{bmatrix} 6 & 9 & 12 \\ 12 & 6 & 9 \end{bmatrix}$$

To add two matrices A and B, look at the following example:

$$A = \begin{bmatrix} 2 & -3 & 5 & 0 \\ 4 & 1 & -2 & 2 \\ 2 & 2 & -1 & 1 \end{bmatrix} \quad B = \begin{bmatrix} 0 & 1 & 1 & 4 \\ 2 & 3 & 3 & 1 \\ 3 & 2 & 3 & 0 \end{bmatrix}$$

$$A + B = \begin{bmatrix} 2 & -3 & 5 & 0 \\ 4 & 1 & -2 & 2 \\ 2 & 2 & -1 & 1 \end{bmatrix} + \begin{bmatrix} 0 & 1 & 1 & 4 \\ 2 & 3 & 3 & 1 \\ 3 & 2 & 3 & 0 \end{bmatrix}$$

$$= \begin{bmatrix} 2 & -2 & 6 & 4 \\ 6 & 4 & 1 & 3 \\ 5 & 4 & 2 & 1 \end{bmatrix}$$

Based on this example, what must be true about the sizes of the matrices to be added or subtracted?

Systems of More than Two Linear Equations and Inequalities

input

To multiply two matrices A and B

$$A = \begin{bmatrix} 2 & 3 \\ 4 & 5 \end{bmatrix} \quad B = \begin{bmatrix} -2 & 1 \\ 3 & -3 \end{bmatrix}$$

look at the example below:

$$\begin{bmatrix} 2 & 3 \\ 4 & 5 \end{bmatrix}\begin{bmatrix} -2 & 1 \\ 3 & -3 \end{bmatrix} = \begin{bmatrix} 2(-2)+3(3) & 2(1)+3(-3) \\ 4(-2)+5(3) & 4(1)+5(-3) \end{bmatrix}$$

$$= \begin{bmatrix} 5 & -7 \\ 7 & -11 \end{bmatrix}$$

Generalizing from the example above, to find the product of two rectangular matrices, multiply the entries of the:

First row of A and the first column of B. Add them together. This is the entry for row 1, column 1.

First row of A and the second column of B. Add them together. This is the entry for row 1, column 2.

Second row of A and first column of B. Add them together. This is the entry for row 2, column 1.

Second row of A and the second column of B. Add them together. This is the entry for row 2, column 2.

When applying this method to find the solution to a system of two equations in two unknowns, you will need to know how to write a matrix equation defined as AX = B. Let us do that by looking at the following system and constructing the matrix equation.

$$x + 3y = 3$$

$$2x + y = 6$$

In this matrix equation, A is the matrix of coefficients of the variables of each of the terms in the system, X is a matrix of the variables, and B is a matrix of the constants. For the system above, the matrix equation AX = B is:

$$A \bullet X = B$$

$$\begin{bmatrix} 1 & 3 \\ 2 & 1 \end{bmatrix}\begin{bmatrix} x \\ y \end{bmatrix} = \begin{bmatrix} 3 \\ 6 \end{bmatrix}$$

The solution to a matrix equation, namely solving for X, is $X = A^{-1}B$. We will now look at how to find the inverse matrix.

Systems of More than Two Linear Equations and Inequalities

extension

For the matrix $A = \begin{bmatrix} a & b \\ c & d \end{bmatrix}$, the inverse of the matrix is defined as:

$$A^{-1} = \frac{1}{ad - bc}\begin{bmatrix} d & -b \\ -c & a \end{bmatrix}$$

The inverse matrix exists if and only if $ad - bc \neq 0$.

Now using the matrix $A = \begin{bmatrix} 1 & 3 \\ 2 & 1 \end{bmatrix}$, determine if the inverse exists, calculate it.

Having found the inverse, you can now find $X = A^{-1}B$ and the solution to the system. Multiply A^{-1} by B.

Maximizing or Minimizing Functions (Optional)

process: model

1. Graph the following system of inequalities:

   $y \geq 2x$

   $y \geq 4x + 12$

   $y \leq -2x + 6$

   $y \leq -4x - 6$

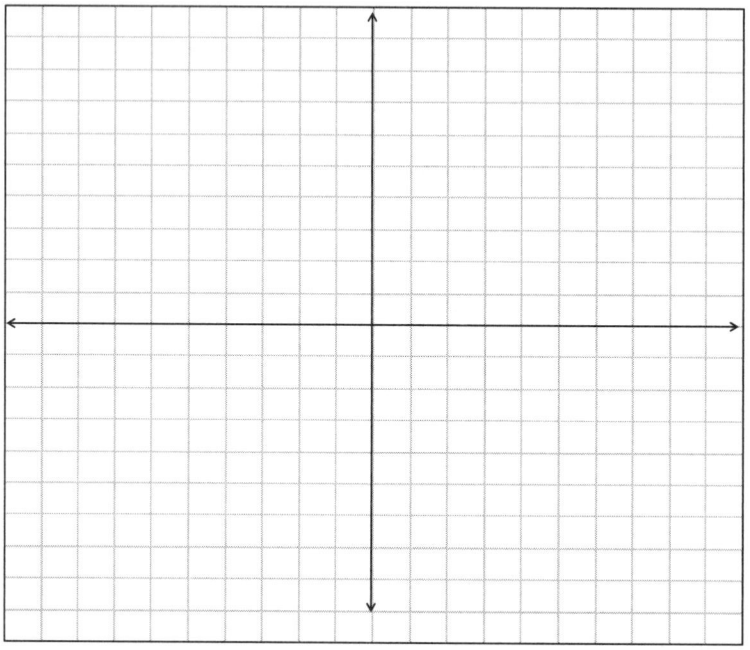

   a. Indicate the boundary equations for each member in the system.

   b. Determine the points of intersection between the boundary lines.

process: model

Maximizing or Minimizing Functions (Optional)

2. Graph the following system of inequalities:

$y \geq 0$

$x \geq 0$

$5x+6y \geq 30$

$y \geq -6x + 24$

$x + y = 9$

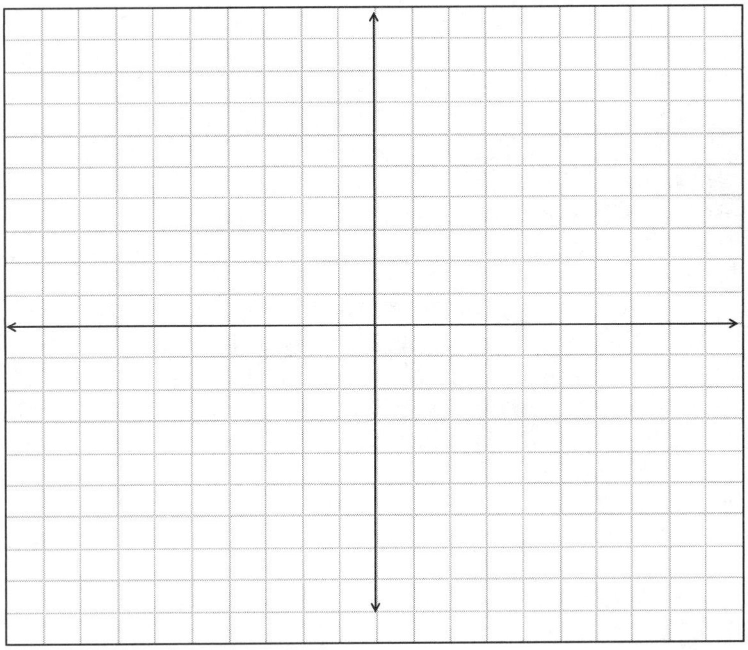

a. Indicate the boundary equations for each member in the system.

b. Determine the points of intersection between the boundary lines.

input

Maximizing or Minimizing Functions (Optional)

You have been looking at the solution set of systems involving inequalities and equations. The ability to solve such systems is important not only in mathematics, but in economics and business as well. The area of mathematics that uses systems of this type to solve problems is called Linear Programming. **Linear Programming** is applied to problems in which it is desirable to maximize or minimize the value of a linear function subject to given conditions, called constraints, on the variables. The constraints are usually expressed as the linear equalities.

Problems involving production and profit are examples where linear programming models would be applied. Here is one example:

A small local company produces two different models of windows. One model takes a single worker 2 hours to produce while the other model takes a single worker 3 hours to produce. The company has 24 employees who work 8 hours per day for a total of 192 hours per day. The first model window is sold for $250 per window and the second model is $375 per window. The company needs to determine how many of each type should be produced to maximize its profits.

process: model

Maximizing or Minimizing Functions (Optional)

3. Identify the variables in the scenario above.

4. Identify the variable quantity that you want to maximize.

5. Identify each of the constraints and express it as an inequality. In identifying the constraints, consider the fact that some constraints are not stated explicitly, such as you cannot produce a negative number of windows.

Maximizing or Minimizing Functions (Optional)

process: model

6. Graph the inequalities representing the constraints.

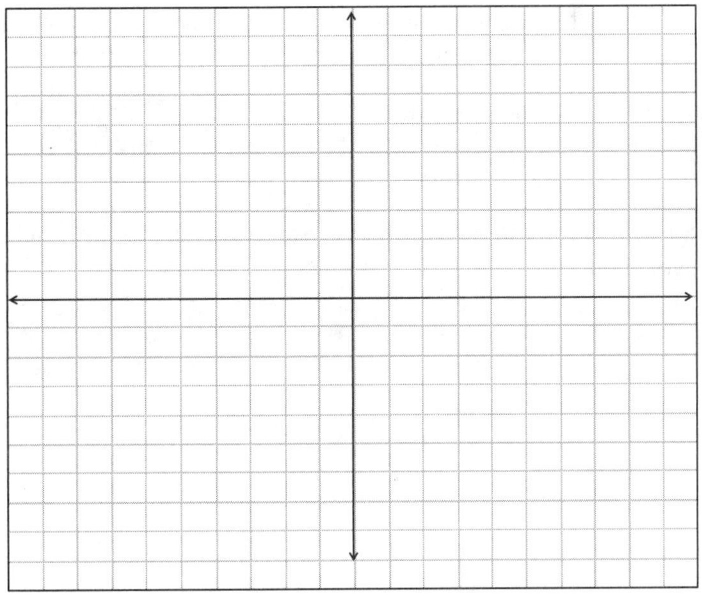

Maximizing or Minimizing Functions (Optional)

analysis

7. Looking at your graph, can you conjecture which points will maximize your profit?

8. Test your conjecture.

input

Maximizing or Minimizing Functions (Optional)

By graphing each inequality and finding the solution set, you have found the set of all ordered pairs that satisfy all the constraints. The points, of which there are infinitely many, all lie in the polygonal region formed by the boundary equations. Evaluating the function at all of these points to determine which value will maximize it is impossible.

Fortunately for us, mathematicians have shown that the maximum and minimum values of a function are always represented by vertex points of the region. We can use this fact to find the maximum or minimum value.

In general, to find the maximum or minimum value, you evaluate the function at each of the vertices.

input

Maximizing or Minimizing Functions (Optional)

Your aunt and uncle raise show dogs. The dogs require a food that contains special nutrients. To make sure there are no mistakes, your relatives made a table that shows amounts of nutrients in each of the foods.

Type of Food	Amount of Calcium	Amount of Iron
Doggie Dog	1 unit per pound	$\frac{1}{2}$ unit per pound
Dog Yums	$\frac{1}{4}$ unit per pound	1 unit per pound

To maintain their proper weight, the dogs must eat a certain amount. Overall, your relatives calculate that for all the dogs they are raising, the total amount of food needed per day is 50 pounds. The dogs eat a mixture of both brands. For a healthy coat, the dog's daily food intake must consist of at least 10 units of calcium and not more than 40 units of iron. In total, all the dogs combined may not consume more than 125 pounds of food per day.

process: evaluate

Maximizing or Minimizing Functions (Optional)

9. If Doggie Dog costs \$0.55 per pound and Dog Yums costs \$0.70 per pound, what is the least possible cost for food for the dogs per day?

   a. Graph the situation as defined.

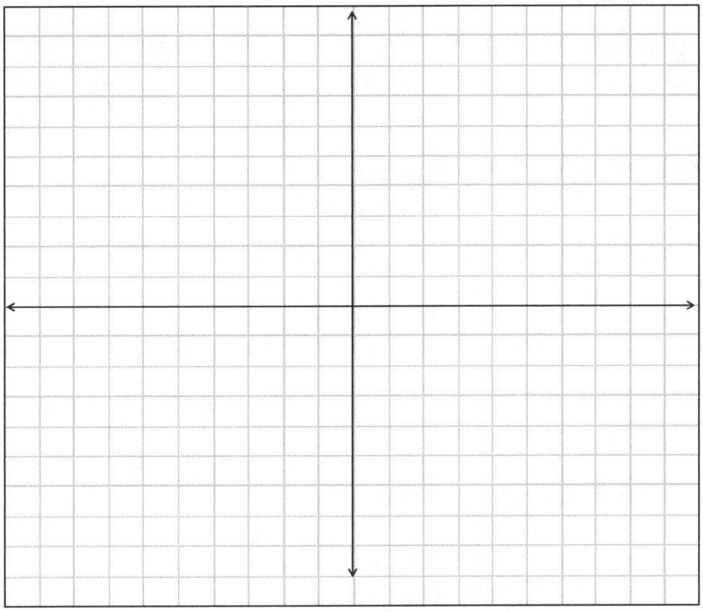

10. If the cost of Doggie Dog were raised to \$0.80 per pound, would they have to change the combination of foods in order to keep the daily cost for food to a minimum?

input

Arithmetic Sequences

Your cousin, who is in college, earns extra money by tutoring middle school students in science. The charge is $28 per hour for one student plus $8 per hour for each additional student. Your cousin said the size of the group varies. How much money will your cousin charge if *n* students are participating in the one-hour tutorial session?

process: model

Arithmetic Sequences

1. Define the set of input and output values based on this situation.

2. Generate a table of input and output values.

Labels		
Units		
Expression		

process: model

Arithmetic Sequences

3. Describe how to find your cousin's earnings based on the number of students in the tutor group?

4. Describe how to find the next number in the earnings column knowing only the previous number.

input

Arithmetic Sequences

The list of outputs you generated, (28, 36, 44, 52, 60, 68, 76, 84), is called a sequence. A **sequence** is an ordered list of numbers. Each number in the sequence is called a **term**.

The input values (domain) for a sequence are consecutive integers usually beginning at 0 or 1. The output values (range) are the terms.

process: evaluate

Arithmetic Sequences

5. Look at the following sequences of numbers. For each sequence, list the next three terms and describe how each new term was generated.

   a. 5, 9 13, 17,

   b. 12, 32, 52, 72, 92,

   c. 1, 2.5, 4, 5.5,

   d. 12, 9, 6, 3, 0,

analysis

Arithmetic Sequences

6. Is there a pattern for finding the next term given the previous term? Describe the pattern.

process: evaluate

Arithmetic Sequences

7. Look at the following sequences of numbers. For each sequence, list the next three terms and describe how each new term was generated.

   a. 1, 2, 4, 8, 16,

   b. 1, 4, 9, 16, 25,

   c. $1, \frac{1}{3}, \frac{1}{9}, \frac{1}{27},$

analysis

Arithmetic Sequences

8. How are the patterns you described for the sequences in question 5 different than the patterns you described for the sequences in question 7?

input

Arithmetic Sequences

Each of the sequences you examined in question 5 is called an **arithmetic sequence.** An arithmetic sequence is a sequence of numbers having a **common difference,** *d,* between each two successive terms.

The terms of the sequence are represented by a_n where *a* is any number and *n* is the position of the term in the sequence.

process: model

Arithmetic Sequences

9. Based on what you have found in the previous questions, generate an arithmetic sequence using the following criteria:

   a. The first term is 2.

      The difference between the terms is 3.

      Write the first six terms.

   b. The first term is 0.

      The difference between the terms is 0.5.

      Write the first 8 terms.

   c. The first term is –1.

      The difference between the terms is –2.

      Write the first five terms.

Arithmetic Sequences

analysis

10. Determine which of the following sequences is an arithmetic sequence. If a sequence is arithmetic, indicate the common difference.

    a. 4, 12, 20,. . .44

    b. 3, 9, 27, 81

    c. 5, 2, 6, 3, 6

    d. $\frac{1}{2}, \frac{1}{4}, \frac{1}{16}$

    e. 1, -7, -15,

analysis

Arithmetic Sequences

11. Generate the arithmetic sequence representing the outputs of f(x) = 2x + 5 when the inputs are integers from 0 to 8.

    a. Is this sequence arithmetic? How do you know?

    b. What is a_1?

    c. What is the common difference *d*?

    d. What is the relationship between the equation for the function above and a_1?

    e. What is the relationship between the equation above and *d*?

    f. How do you find a_2 based on a_1? What is a_2?

    g. What is a_4? How can you find a_4 directly from a_1 without finding a_3?

    h. What is a_{10}? How can you find a_{10} without having to find all the terms before it? Check the result by finding a_{10} using the term before it.

    i. Find a50 without finding the terms preceding it.

    j. How did the equation f(x) = 2x + 5 help you answer these questions?

input

Arithmetic Sequences

A formula such as $f(x) = 2x + 5$ represents a rule for finding a particular term in the sequence. This is called an **explicit formula.** Many times, explicit formulas are like the equations you have previously used. The only difference is that in this case, to find the 5th term ($a_5 = 13$), you would not substitute 5 into the equation. The first term ($a_1 = 5$) is the output from substituting 0 into the equation. Based on this, what number would you substitute into the equation to find the 5th term?

Therefore, the actual explicit formula for this sequence is written as

$$2(n - 1) + 5 = a_n$$

What does 5 represent in the arithmetic sequence?

What does 2 represent in the arithmetic sequence?

Another type of formula used with sequences is called a **recursive formula**, sometimes referred to as the **NOW NEXT formula.** A recursive formula expresses each new term based on the preceding term. In this example, the recursive formula would read:

$$a_{n+1} = a_n + 2$$

(next = now + 2)

process: model

Arithmetic Sequences

12. Given the arithmetic sequence 4, 6, 8, 10:
    a. Determine the common difference. Given the first term $a_1 = 4$, show how each term is generated with respect to the preceding term. Generate the recursive formula.
    b. Using the recursive formula, find a_7.
    c. Write the explicit formula for this sequence.
    d. Use the explicit formula to find a_{20}.

process: model

Arithmetic Sequences

13. If a_n is the nth term and *d* is the common difference, write a recursive formula for this sequence.

    a. Write the explicit formula for this sequence.

    b. Test the explicit formula by finding the 5th term of a sequence whose first term is 3 with a common difference of 6.

    c. Use your recursive formula to confirm that you are correct.

process: summary

Arithmetic Sequences

The equation that represents the pattern for the arithmetic sequence is called a recursive equation because each term is determined by applying the expression to the preceding term.

To summarize, if the equation for a sequence can be written as:

$$a_n = a_1 + d(n - 1) \text{ (Explicit formula)}$$

$$a_{n+1} = a_1 + d \text{ (Recursive formula)}$$

where n is the number of the term, a_1 is the first term, and d is the common difference, then the sequence is an arithmetic sequence.

input

Arithmetic Sequences

The U.S. Postal Service increased the cost of mailing a first-class letter weighing up to one ounce to $0.37 in 2002. The cost for mailing a letter increases $0.23 for each additional ounce up to 13 ounces.

analysis

Arithmetic Sequences

14. Is this situation best modeled by an arithmetic sequence or a linear function? Justify your answer.

process: model

Arithmetic Sequences

15. Using the notation for an arithmetic sequence, let a_n represent the cost in dollars for first class postage for a letter that weighs n ounces and write an explicit equation for a_n.

    a. Write a recursive formula to model the same situation.

process: evaluate

Arithmetic Sequences

16. What is the domain of the arithmetic sequence?

17. Based on the equation you have written, what is the cost of a letter that weighs 9 ounces?

18. If the postage is $3.13, what is the weight of the letter?

extension

Arithmetic Sequences

19. In general, when do you think a sequence model might be more appropriate than a linear model?

20. Is it reasonable to interpret the common difference of the arithmetic sequence as the slope? Explain.

21. The graph of a function with positive slope is increasing and the graph of a negative function is decreasing. Under what condition do you think an arithmetic sequence is increasing and under what condition do you think it will be decreasing?

process: practice

Arithmetic Sequences

22. Given the following explicit formulas for arithmetic sequences, find the first 6 terms and write the recursive formula of each sequence.

    a. $a_n = 5n + 3$

    b. $a_n = 3n - 4$

    c. $a_n = 6(n - 1) + 5$

    d. $a_n = -2(n - 1) + 5$

23. Write an explicit and recursive formula representing the nth term of each arithmetic sequence.

    a. 20, 15, 10 , 5, 0,...

    b. 2, 18, 34, 50, 66, ...

    c. 10, 14, 18, 22,...

    d. 100, 72, 44, 16,...

Arithmetic Series

input

In the previous section of this unit, you learned about a particular type of number sequence called an arithmetic sequence.

What is an arithmetic sequence of numbers?

In this section of the unit, you will learn about series and in particular about arithmetic series.

A **series** is the sum of the terms in a sequence. The sum of terms in an arithmetic sequence is an **arithmetic series**.

For example, the first problem in the last section concerned your cousin's tutoring business. The arithmetic sequence of numbers generated from that problem was:

28, 36, 44, 52, 60, ...

What is the sum of the first five numbers in the sequence?

Using mathematical notation, you have found the fifth sum, which is denoted as S_5. This sum is found by adding a_1, a_2, a_3, a_4, and a_5, which you just did.

In general, the sum of the first n numbers in a sequence is denoted as S_n. S_n is found by adding the first n terms in the sequence and is the nth number in the series.

$$S_n = a_1 + a_2 + a_3 + \ldots + a_{n-1} + a_n$$

You can also write any term in a series using a special summation notation symbol called sigma, represented by $\sum$. Sigma is the Greek letter for S – think Sum. For example, the expression below represents the sum of the first five values in the above arithmetic sequence.

$$\sum_{n=1}^{5}(20+8n)$$

To evaluate this expression, simply substitute the numbers from 1 to 5 into the expression 20 + 8n and add them together.

Substitute 1 for n in the expression 20 + 8n. What value do you calculate?

Substitute 2 for n in the same expression. Now, substitute 3. Now, 4. And lastly, 5.

Add these 5 numbers together. You just evaluated $\sum_{n=1}^{5}(20+8n)$.

process: evaluate

Arithmetic Series

1. Expand each of the following by showing the numbers that must be added together. Then, find the sum of the numbers.

   a. $\sum_{n=1}^{6} 1200n =$

   b. $\sum_{n=1}^{8} (n+4) =$

   c. $\sum_{k=1}^{4} (3k)-2 =$ (Note that only the 3k is indicated as part of the sigma-notation.)

2. Use the sigma notation to rewrite each of the following and then find the indicated sum.

   a. S_3 for the sequence of numbers generated by $a_n = 2^n$.

   b. S_7 for the arithmetic sequence indicated by $a_n = \frac{1}{2}n$.

   c. S_{10} for the sequence of odd numbers beginning at 11.

process: model

Arithmetic Series

One way to visualize a series is to use a grid made up of square units. To visualize the arithmetic series for the sequence 3, 7, 11, 15, create a set of three square units, a set of seven square units, a set of 11 square units, and a set of 15 square units.

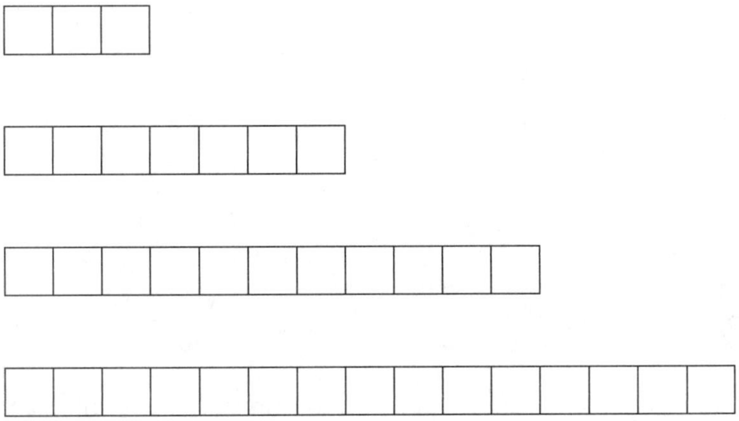

Create a model for the sum of 3, 7, 11, and 15 by stacking the grids on top of one another.

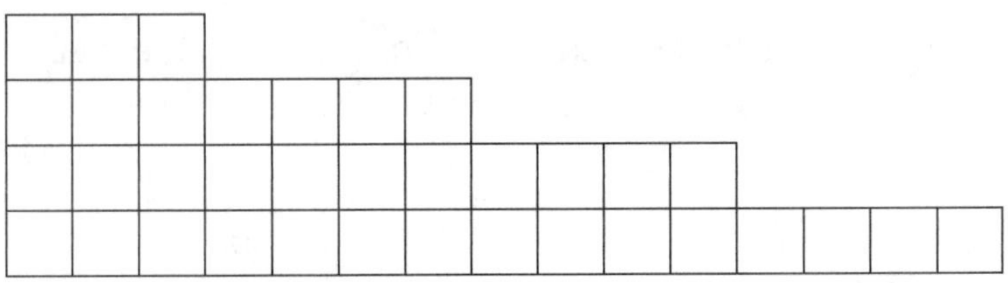

Make a second copy of the model above and put it together with the first model to form a rectangle, like so.

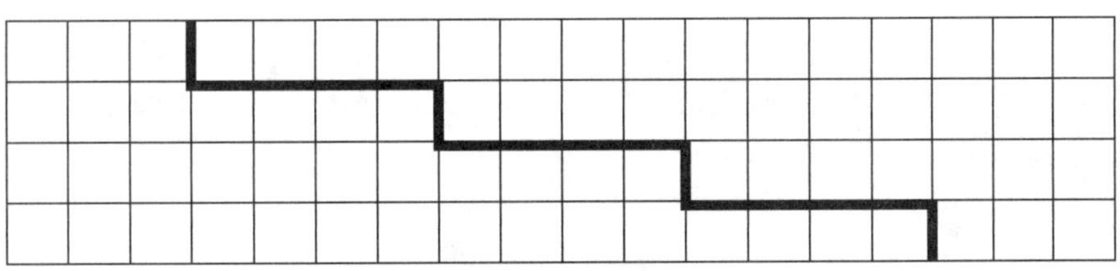

3. What is the area of the rectangle above? Show the numbers you multiplied.

4. How is the sum of the four terms related to the area of the rectangle?

5. Based upon this area, what is the sum $\sum_{n=1}^{4}[4(n-1)+3]$?

Arithmetic Series

process: model

6. Use the square units on graph paper to model 2 + 4 + 6 + 8 + 10. Tape your model into your text in the space below.

   a. Create a copy of your model and attach it to the first model to create a rectangle.

   b. What is the area of your rectangle? Write the numbers you multiplied to find the area.

   c. How is the sum of the five terms in this problem related to the area of the rectangle?

   d. Represent this sum using sigma notation. What is the sum?

7. Use the square units on graph paper to model the first three terms in the sequence $a_n = 4 + 5(n - 1)$. Tape your model into your text in the space below.

   a. Create a copy of your model and attach it to the first model to create a rectangle.

   b. What is the area of your rectangle? Write the numbers you multiplied to find the area.

process: model

Arithmetic Series

c. How is the sum of the terms in this problem related to the area of the rectangle?

d. Represent this sum using sigma notation. What is the sum?

8. Complete the following table based upon the three arithmetic series problems you just finished.

	Rectangle Width	Rectangle Length	Area Written as Length • Width	Sum of Series in Terms of Area
3 + 7 + 11 + 15				
2 + 4 + 6 + 8 + 10				
S_3 for 4 + 5(n – 1)				

analysis

Arithmetic Series

9. How is the width of your rectangle related to the sum of the series?

10. How is the length of the rectangle related to the sum of the series?

11. How is the area of the rectangle related to the sum of the series?

Arithmetic Series

analysis

12. Write a sentence or two describing how you can find the sum or the first n terms in an arithmetic series using an area model.

13. Generalize your findings from this activity into a formula for the nth sum in a series (S_n).

Arithmetic Series

process: practice

14. Find the 7th sum of the arithmetic series (S_7) based on the sequence: 2 + 5 + 8 + ...

15. What is the sum of the first 100 positive integers?

16. In the auditorium, the first row has 24 seats and each row thereafter has three more seats. If the last row has 48 seats, how many seats are in the auditorium?

process: practice

Arithmetic Series

17. Use what you have developed to find the following sums.

a. $\sum_{w=1}^{5}(2w-1)$

b. $\sum_{n=1}^{5}(2n)-1$

18. A woman borrowed $6000 for a period of 5 years. Her monthly payments are $145, $144.75, $144.50, $144.25, $144, ... How much did she pay over the entire loan?

a. How much interest did she pay over the life of the loan?

process: summary

Arithmetic Series

19. Compare and contrast an arithmetic sequence with an arithmetic series.

analysis

Functions and Inverse Functions

1. The data tables below represent two functions and g(x). Study the tables and describe what you observe about the relationship between the variables.

Table for the first function

Input	Output
-2	-3
-1	-1
0	1
1	3
2	5
3	7

Table for the second function

Input	Output
-3	-2
-1	-1
1	0
3	1
5	2
7	3

2. Generate a data table as above for the following two functions.

   a. $y = x - 3$

   b. $y = x + 3$

Table for the first function

Input	Output

Table for the second function

Input	Output

analysis

Functions and Inverse Functions

3. What do you observe about the relationship between the inputs and outputs in this situation?

4. Using your graphing calculator, graph $y = x$ and the two equations in question 2. What do you observe about their graphs?

input

Functions and Inverse Functions

Each of the examples above represents a function and its inverse. In both cases, the input and output of the first function were reversed for the second function. This says that for any function $f(x)$ consisting of ordered pairs (x, y) or $(x, f(x))$, the inverse consists of all ordered pairs (y, x). The roles of independent and dependent variables have been switched. This was evident in the input and output tables as well as from the graphs. The inverse of a function is denoted by $f^{-1}(x)$, read as f inverse of x.

Switching the role of the variables can be achieved through a series of algebraic manipulations. Given $f(x) = 4x - 3$, find the inverse by:

1. Replacing $f(x)$ with y

2. Interchanging the x and y variables in the equation

3. Solving for y

4. Replacing y with $f^{-1}(x)$

process: practice

Functions and Inverse Functions

5. Find the inverse of the following linear functions:
   a. $f(x) = 3x - 2$

   b. $y = -x + 4$

   c. $f(x) = 2x + 2$

   d. $y = \frac{x}{2} + 4$

process: model

Functions and Inverse Functions

6. Using your graphing calculator, graph y = x, and then graph, one pair at a time, each function and its inverse from question 5. Draw a sketch of each pair of graphs below.

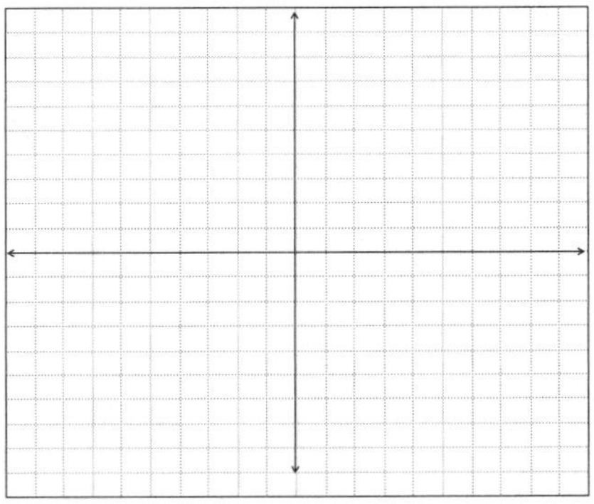

a.

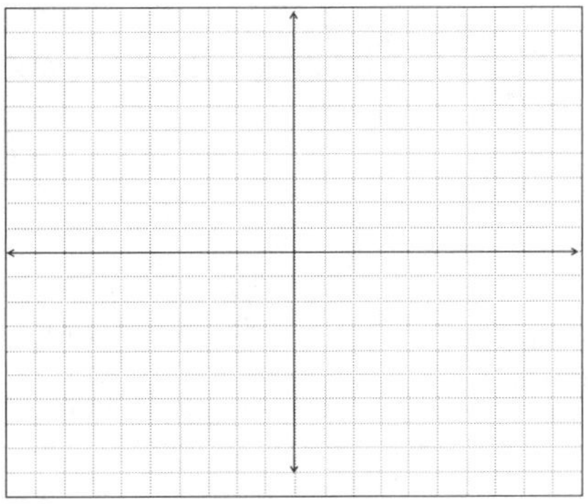

b.

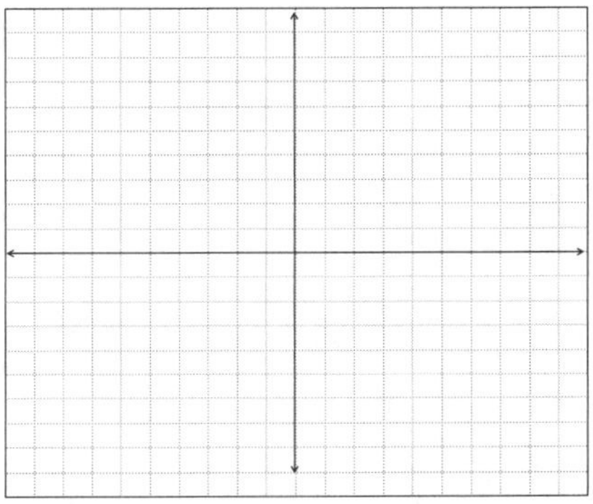

c.

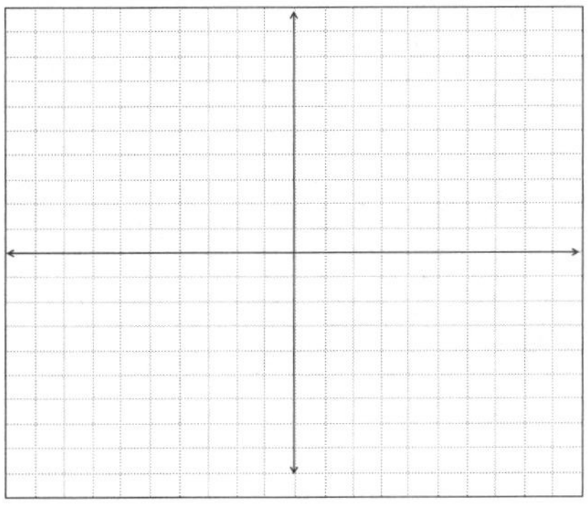

d.

analysis

Functions and Inverse Functions

7. What observation can you make about the graphs with respect to the line y = x?

8. How could you represent the inverse function graphically without doing any algebraic manipulations first?

Unit 3: Parallel and Perpendicular Lines

Contents

preview

Erathosthenes of Syene

About 220 BC, Erathosthenes estimated the circumference of the Earth.

He assumed that the Earth was a sphere and that the sun's rays were parallel. He determined that on the longest day of the year the sun was directly over the town of Syene (now called Aswan), because the sun's light shone directly down a deep well in the town. On the longest day of the year, he measured the angle of the shadow cast by a vertical pole in Alexandria, which is 500 miles north of Syene. The measure of the angle cast by the shadow of the vertical pole angle was 7.2°. (This angle is labeled $\angle 2$ in the figure.)

1. Use this angle measure to compute the circumference of the Earth. Show all of your work.

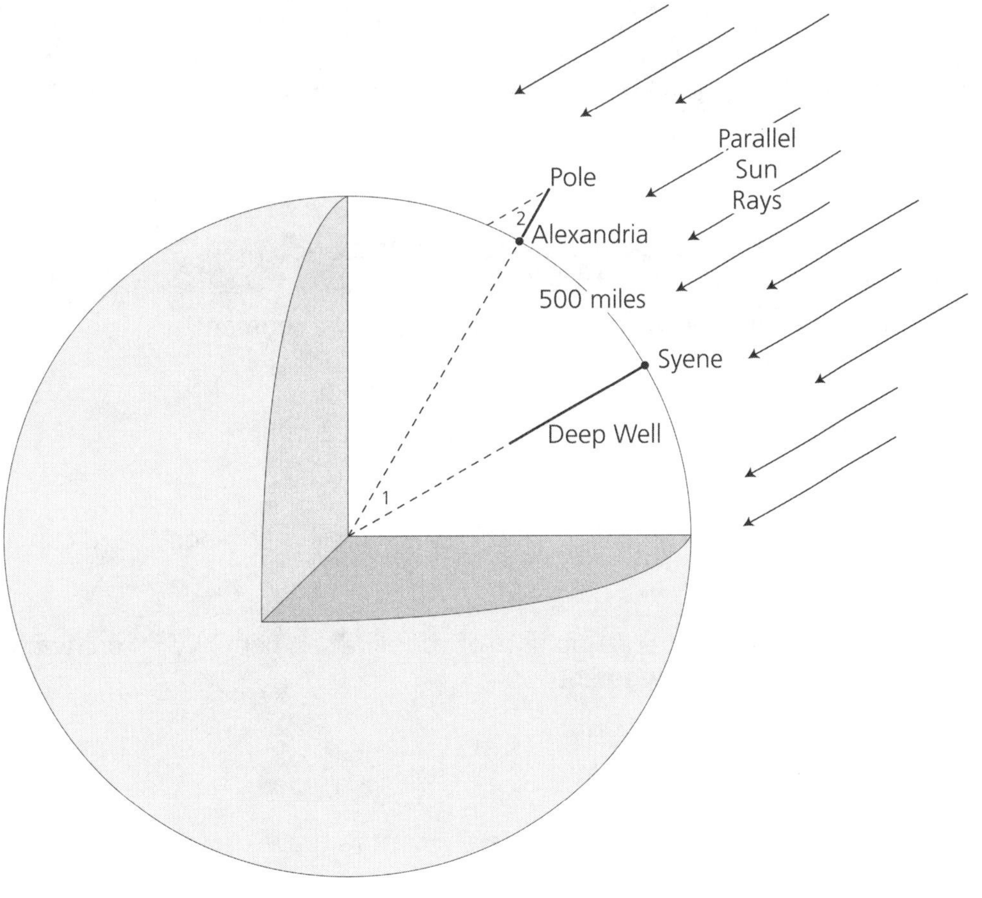

input

Symbols and Geometric Objects

Mathematicians use symbols as a shorthand notation for many things in geometry. To symbolize points, capital letters are used. For example, the letter A is used to symbolize this point.

By connecting two points, you form a **line segment**. The two points that are connected are called the **endpoints**. The endpoints of this segment are H and I. They determine this line segment. A line segment is named according to its endpoints.

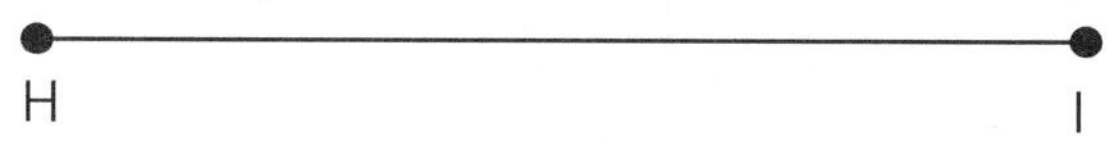

The segment contains every point between the endpoints as well as the endpoints.

process: model

Symbols and Geometric Objects

1. What are two ways you can name this line segment?

2. How many points does the line segment contain?

3. Form a line segment by locating and labeling two points and connecting them.

4. What are the two endpoints of your line segment? What are two ways you can name this line segment?

5. How many line segments are required to draw:
   a. A square
   b. A rectangle
   c. A pentagon
   d. An octagon

input

Symbols and Geometric Objects

Instead of writing out the words "line segment" every time you want to talk about a segment, you can use a shorthand or symbolic mathematical notation. Line segment *HI* is usually written as: $\overline{HI}$ or $\overline{IH}$. To indicate a line segment, simply write the names of the two endpoints in capital letters and draw a dash above the two letters.

process: model

Symbols and Geometric Objects

6. Use this shorthand or symbolic mathematical notation to write the possible names of the line segment you created on the previous page.

process: practice

Symbols and Geometric Objects

7. Name the three sides of this triangle, and use shorthand notation to symbolize the three sides of this triangle as line segments.

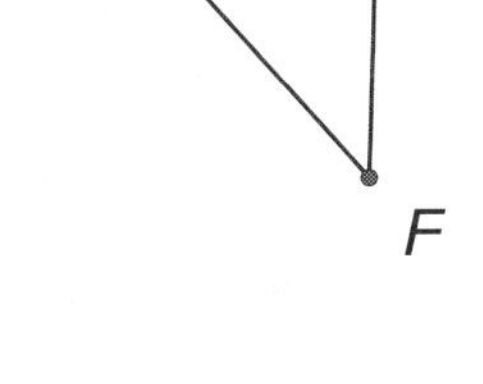

process: model

Symbols and Geometric Objects

8. Use symbolic notation to write a possible name for this segment.

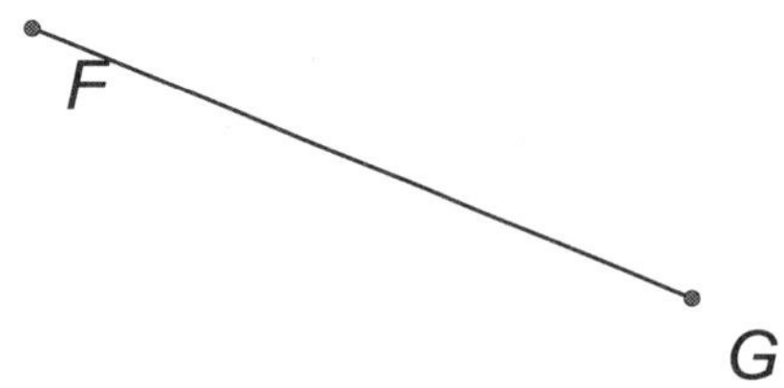

   a. Measure this segment using your centimeter ruler. What is its length?

Instead of writing the words "The length of segment FG is ___ cm," you can use the following symbolic or shorthand notation: *FG* = _____cm. Note that if we write $\overline{FG}$, we mean segment FG, but if we write FG, we mean the length of segment FG.

9. Use a straightedge to draw a triangle in the space below. Name the vertices of your triangle R, O, and L.

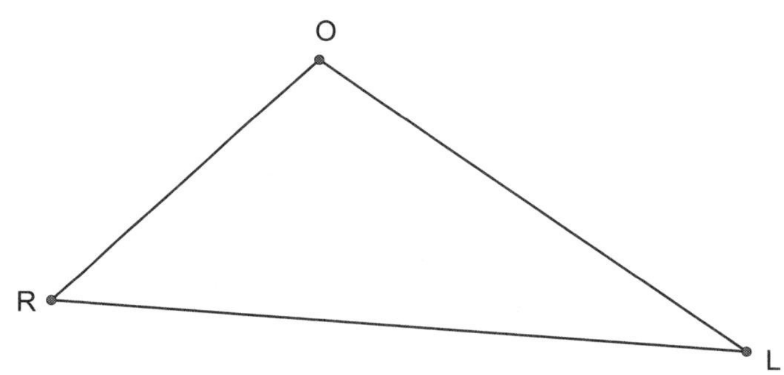

   a. Measure each side of your triangle in centimeters. Record your findings using shorthand notation to symbolize the length for each segment.

Symbols and Geometric Objects

input

If a line segment is extended indefinitely in one direction, it becomes a **ray**. For example, if extended through endpoint I, the object becomes ray HI.

If extended through point H, the object becomes ray IH.

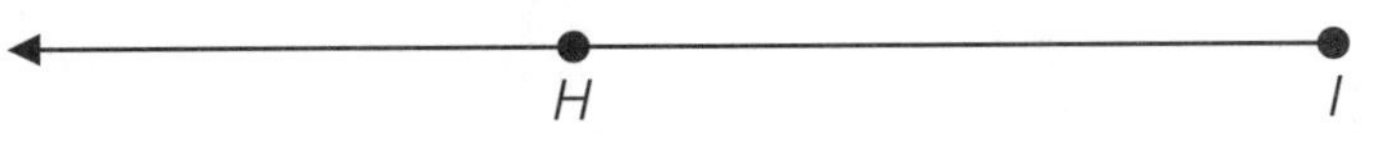

An arrow on one end of a ray indicates it has infinite length in that direction.

The shorthand or symbolic notation used to indicate rays is very similar to the notation for line segments, except an arrow is placed over the point through which the ray extends. For example, ray HI becomes $\overrightarrow{HI}$ or $\overleftarrow{IH}$ and ray IH becomes $\overrightarrow{IH}$ or $\overleftarrow{HI}$.

Symbols and Geometric Objects

analysis

10. How many endpoints does a ray have?

11. A ray is sometimes called a half-line. Why is this so?

process: practice

Symbols and Geometric Objects

12. Create a ray in the space below.

  a. What is the endpoint of this ray?

  b. Through which point does this ray extend?

  c. What is the length of this ray?

  d. Write one possible shorthand notation for this ray?

input

Symbols and Geometric Objects

If a line segment is extended infinitely in both directions, it becomes a **line**. For example, if $\overline{HI}$ is extended through endpoints H and I, it becomes line HI.

The arrows indicate that this line extends infinitely in both directions.

analysis

Symbols and Geometric Objects

13. How many endpoints does a line have?

14. Based upon the shorthand or symbolic notation you already learned for segments and rays, how will you symbolize line HI using this notation?

process: practice

Symbols and Geometric Objects

15. Create a line in the space below and write both possible shorthand notations for this line.

analysis

Relationships Between Lines

1. Given two lines that are **coplanar** (both in the same flat surface), what are the ways in which these lines can relate to one another?

2. Draw sets of lines to illustrate the relationships you cited above.

In the previous unit and in Integrated Math I, you saw many examples of **intersecting lines**.

3. If two lines or segments intersect, what does that mean?

4. If two lines intersect, what is the maximum number of points they have in common? Explain.

5. What is the minimum number of points two intersecting lines have in common? Explain.

6. What is unique or special about the point of intersection for two lines?

process: model

Relationships Between Lines

Cats-R-Us sells cat toys. In order to produce its new squealing mouse, Cats-R-Us has to pay its production company a $200 set-up fee and $10 per toy produced. Cats-R-Us sells each toy for $12.

7. Write the algebraic equation to represent the *cost* to produce the toys.

8. Write the algebraic equation that represents the *income* form selling the toys.

process: model

Relationships Between Lines

9. Graph the two equations that model the cost and the income from the squealing mouse toys on the same set of axes.

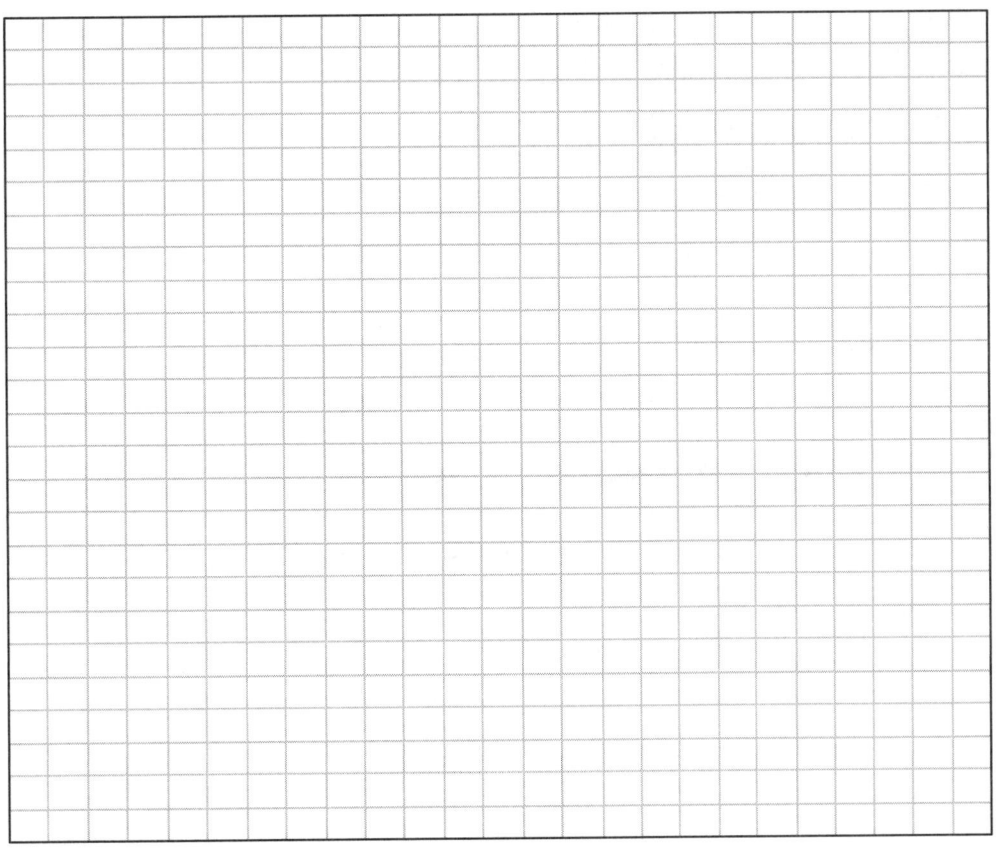

analysis	**Relationships Between Lines**
	10. Will the two lines that represent the income and the cost equations ever intersect? How do you know?

analysis	**Relationships Between Lines**
	11. If the lines that represent the cost and income to Cats-R-Us do intersect, what will the point of intersection represent? 12. Do lines always intersect? Explain.

process: model	**Relationships Between Lines**
	The squealing mouse toys were not exactly jumping off the store shelves, so Cats-R-Us decided to lower the selling price to $10. 13. Write the new algebraic equation that represents the *income* from selling the toys at $10. 14. Graph the line that represents the new income at the reduced selling price on the set of axes on the previous page.

analysis

Relationships Between Lines

15. Will the new line that represents the income ever intersect the line that represents the cost to produce the squealing mouse toys? How do you know?

16. If lines do not intersect, but are graphed on the same set of axes, what other possible relationships could they have to one another?

input

Relationships Between Lines

As you saw in the previous section on Symbols and Geometric Objects, two points can be used to name and represent a line. To symbolize lines, mathematicians also use a single lowercase letter. This lowercase letter represents all of the points on the line at once. For example the letter g represents the following line.

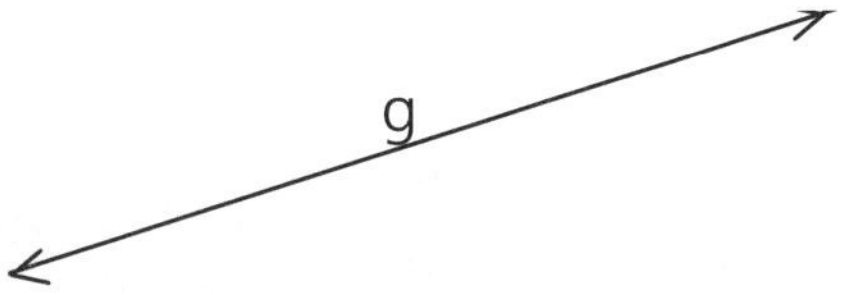

To symbolize parallel lines, use a small set of parallel lines (//). If you want to tell someone that line g is parallel to line m, you would write g // m. On a sketch, an equal number of arrowheads on lines show that lines are parallel.

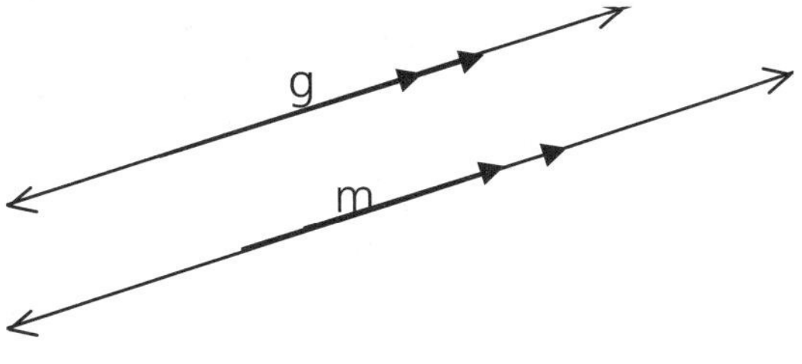

process: model

Relationships Between Lines

17. Use symbolic notation to show that line AB is parallel to line CD.

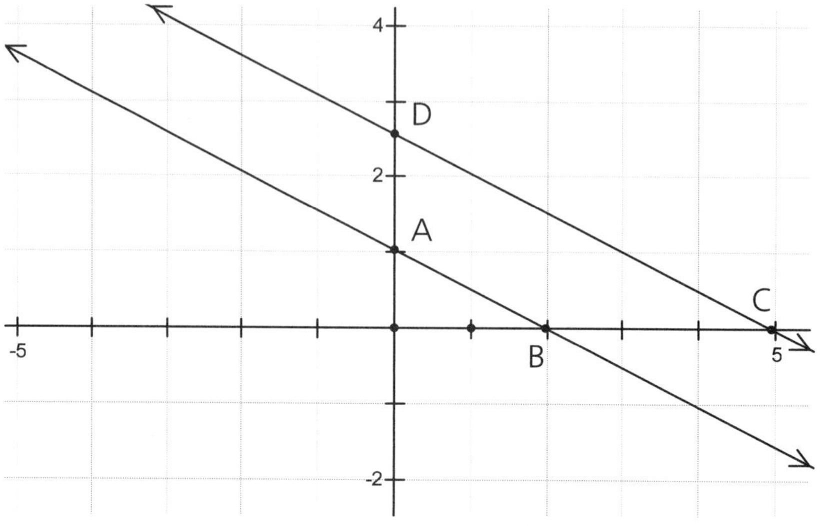

analysis

Relationships Between Lines

18. If lines or line segments are parallel, what does that mean?

analysis

Relationships Between Lines

19. Are there any other ways in which two lines can relate to one another if we do not stipulate that the lines are coplanar? If so, use two pencils or pens to illustrate this relationship for your other group members. If not, explain why not.

analysis

Relationships Between Lines

In fact, lines that are not parallel do *not* necessarily intersect.

20. Try holding two pencils or straight objects so that they will never intersect, even if indefinitely extended. If the straight objects that you are holding are still parallel to one another, turn one of the objects and move it up or down so that it will not intersect the other. Sketch what you viewed.

Two lines are said to be **skew** if they do not intersect, but are also not parallel.

21. Are skew lines coplanar? Explain.

process: model

Relationships Between Parallel Lines

1. Graph two parallel lines in the coordinate plane.

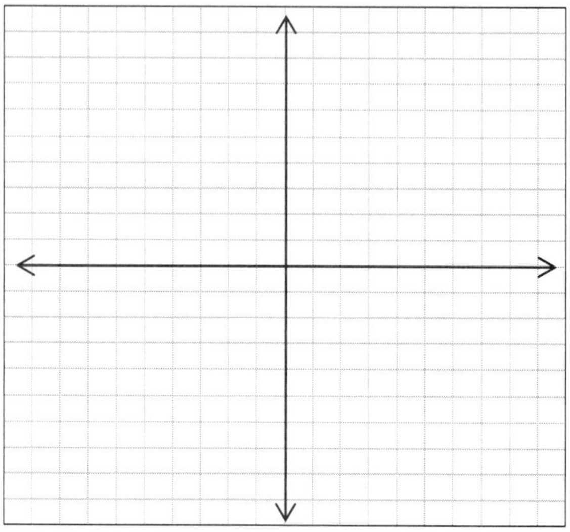

2. Use what you know to create the equations of the lines you just graphed.

analysis

Relationships Between Parallel Lines

3. Compare the equations of the two lines. What is the same from one equation to the other? What is different from one equation to the other?

process: model

Relationships Between Parallel Lines

4. Graph two different parallel lines in the coordinate plane. If you graphed lines that rose from left to right, now graph lines that fall from left to right. If you graphed lines that were steep, now graph lines that are not steep.

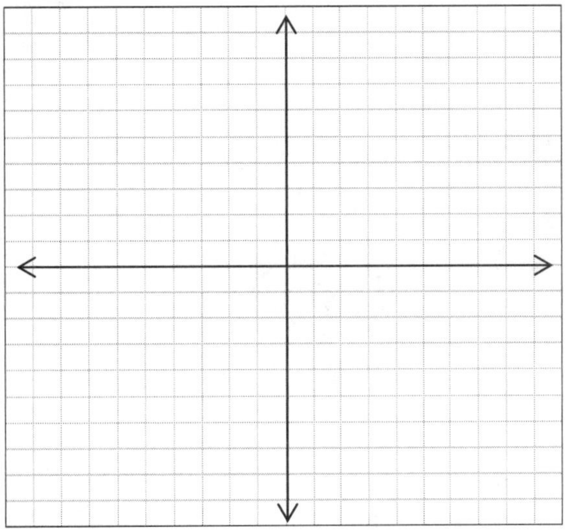

5. Create the equations of the lines you just graphed.

analysis

Relationships Between Parallel Lines

6. Compare the equations of the two lines. What is the same from one equation to the other? What is different from one equation to the other?

7. How is this set of lines the same as the previous set? How is the set of lines different than the previous set?

process: model

Relationships Between Parallel Lines

8. Graph each of the following equations on the same grid.

$y = 2x$

$y = 2x + 3$

$y = 2x - 5$

$y = 2x + 5$

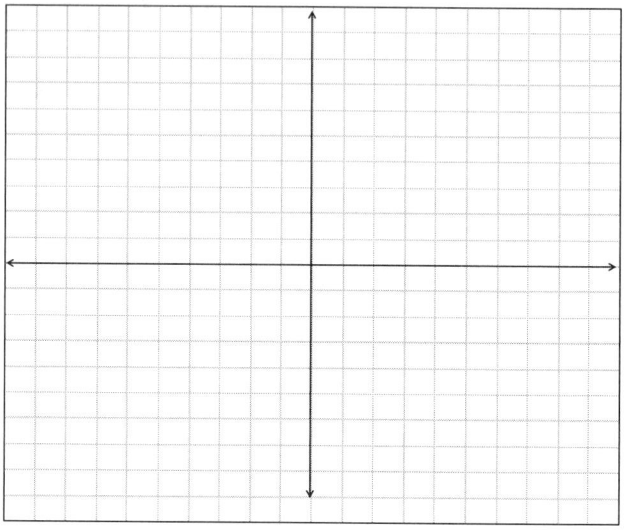

analysis

Relationships Between Parallel Lines

9. What do the four equations have in common? What do the graphs of the four equations have in common?

10. Based solely on the equation, how can you determine that these lines are parallel?

11. Generalize the relationship you just found.

process: evaluate

Relationships Between Parallel Lines

12. What is the *slope* of a line parallel to the line $y = 200x + 93$?

13. What is the *slope* of a line parallel to the line $y = -7x + 20$?

process: summary

Relationships Between Parallel Lines

14. If two lines are parallel, then tell what you know about the equations of these lines.

15. If coplanar lines have equal slopes, then how will the graphs of these lines relate to one another?

16. Write the equations of 3 different lines parallel to the line $y = 17 + 5x$.

$y =$ ____________ ____________

$y =$ ____________ ____________

$y =$ ____________ ____________

17. Write the equation of three different lines that intersect the line $y = 17 + 5x$.

$y =$ ____________ ____________

$y =$ ____________ ____________

$y =$ ____________ ____________

process: model

Angles

1. In the figure below, you could point to angle *W* or angle *C* with little difficulty. Ask another group member to point to angle *A*, and shade the angle to which he or she pointed.

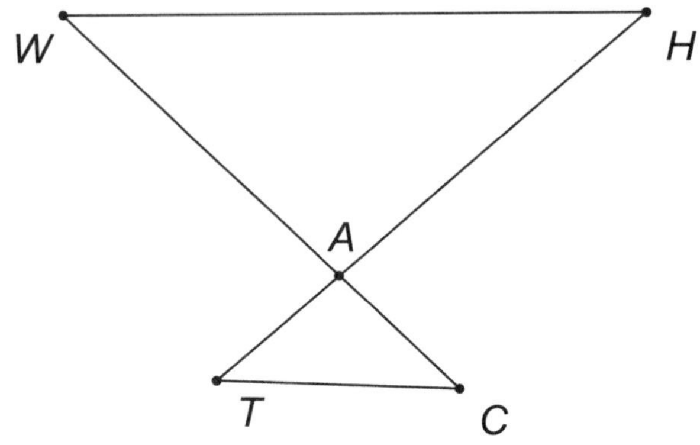

2. In the following figure, ask a different group member to point to angle A, and shade the angle to which he or she pointed.

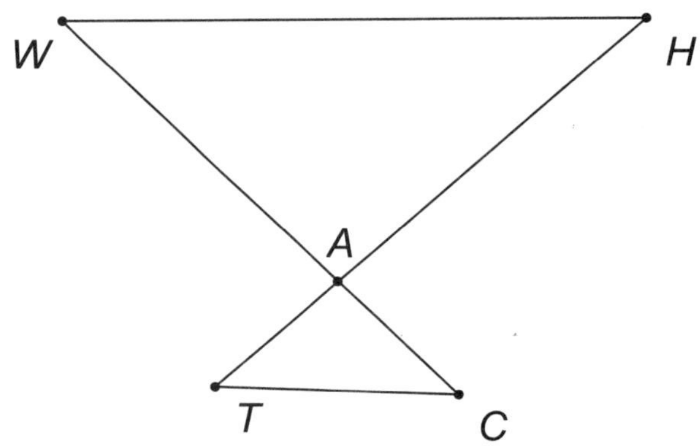

3. In the following figure, ask a third group member to point to angle A, and shade the angle to which he or she pointed.

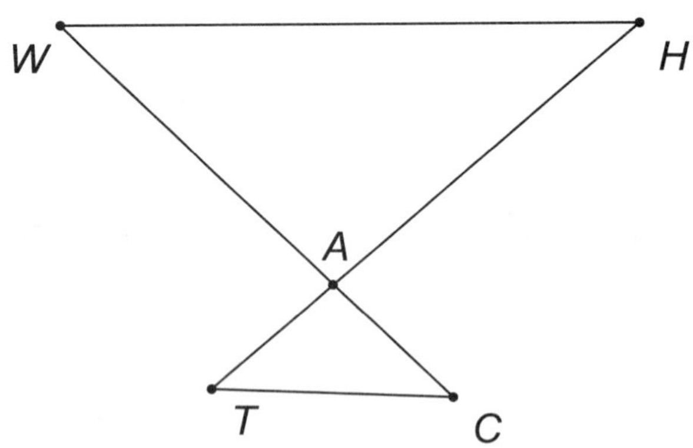

analysis

Angles

In a certain class, Albert shaded the angle where the letter A appears because he thought that the angle name should appear in the angle being named. Betsy shaded the angle directly below where the letter A appears because she thought it made sense that the label for the angle should appear directly across from the angle being named. Carlos shaded the angle to the right of where the letter A appears; saying that he thought picking the angle to the right or to the left would be correct.

4. Do you agree with your group members' choices? Do you agree with Albert, Betsy, or Carlos? Explain.

5. If you want everyone to shade the same region to indicate angle *A*, how would you accomplish this? Can you suggest other ways to name angle *A* that would avoid confusion, while accurately describing its exact location? (Do not add other letters to the figure. Use only what is there.)

input

Angles

Notice that each angle is formed by the intersection of two sides.

Here's how to use symbols to indicate the angles in triangle TWO:

Angle T can be written as: $\angle T$.

Angle W can be written as: $\angle W$.

Angle O can be written as: $\angle O$.

In order to avoid confusion, angles are often named using three letters:

$\angle T$ can also be written as: $\angle WTO$ or $\angle OTW$.

$\angle W$ can also be written as: $\angle TWO$ or $\angle OWT$.

$\angle O$ can also be written as: $\angle TOW$ or $\angle WOT$.

	Angles
analysis	6. Do you notice any pattern or logic when using three letters to name an angle? Explain. 7. Why can't we use two letters to name an angle?

	Angles
input	Another way to avoid confusion when naming angles is to use numbers that are positioned near the vertex of the angle.

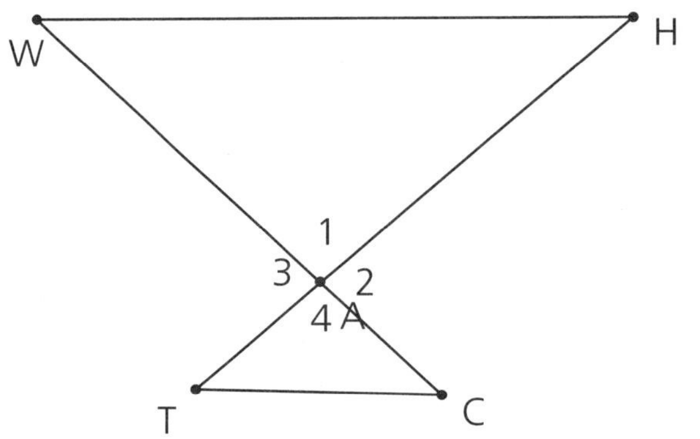

In the figure above, ∠WAH can now be called ∠1.

8. Use the numbers in the figure to name ∠TAC.

9. Name ∠HAC in two other ways.

10. Name ∠WAT in as many other ways as possible.

Angles

process: summary

11. What are the three different ways to name the angle below?

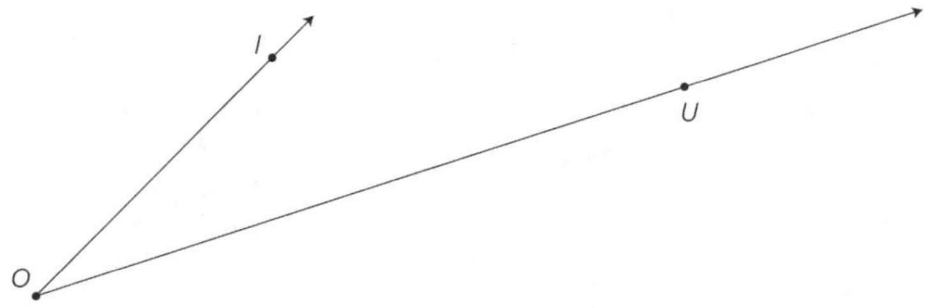

12. Why is there a greater possibility for confusion when naming angles than when naming lines or segments?

Angles

analysis

13. How many points do you need to draw an angle?

14. How many sides does an angle have?

15. Draw an angle that is greater in measure than $\angle IOU$.

16. If you were asked to demonstrate that the angle you just drew is greater in measure than $\angle IOU$, how would you do it? Explain.

Measuring Angles

input

We use the word *measure* to refer to the size of an angle. *(Note that the letter m preceding the angle symbol means "measure of.")* Angles are measured in units called *degrees*. The smaller the size of the angle, the fewer the number of degrees. The larger the size of the angle, the greater the number of degrees. For example, an angle that measures 74 degrees is larger than an angle that measures 41 degrees.

Here's how you can write angle measures using the symbol for degrees:

74 degrees can be written as: 74°.

2 degrees can be written as: 2°.

Measuring Angles

process: evaluate

1. How many degrees are in an angle that forms the corner of a square?

2. If you are facing forward and turn around to face backward, how many degrees have you turned?

3. How many degrees are in a circle (one full revolution)?

4 Circle your best estimate for the measure of ∠NOT ($m\angle$NOT).

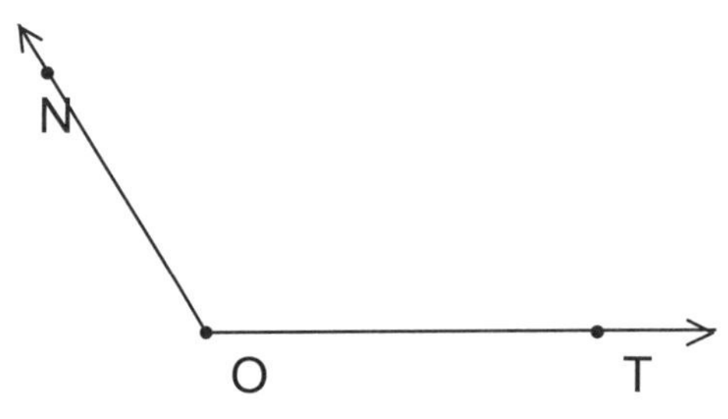

a. 20°
b. 90°
c. 120°
d. 170°

process: practice

Measuring Angles

5. Circle your best estimate for the $m\angle DOY$ (measure of $\angle DOY$).

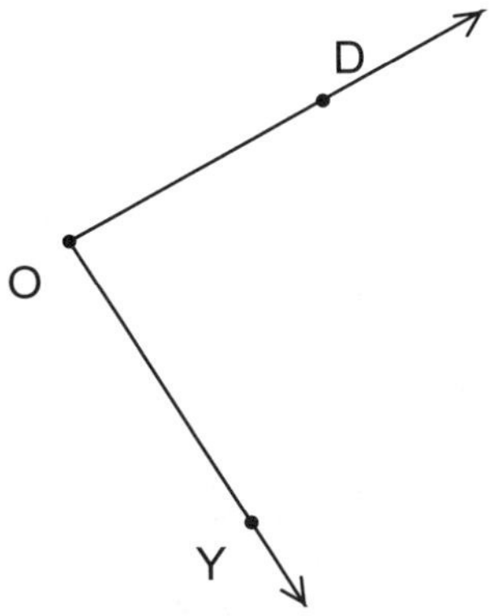

a. 10°
b. 150°
c. 45°
d. 90°

6. Circle your best estimate for the $m\angle HEY$:

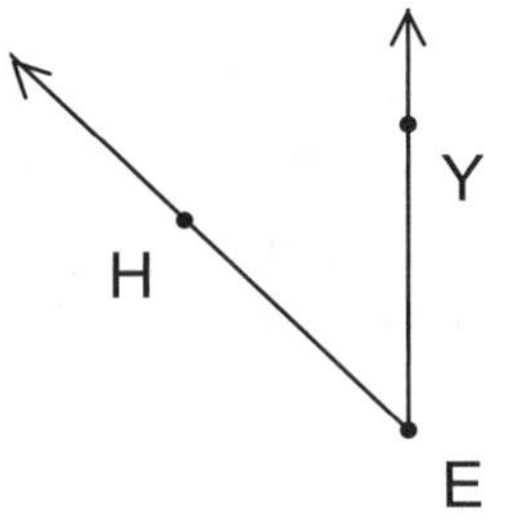

a. 90°
b. 45°
c. 22°
d. 1°

analysis

Measuring Angles

7. Consider this statement: the measure of $\angle TOW = 45°$. Draw a diagram illustrating this statement.

input

Measuring Angles

A **protractor** is a tool that is used to measure angles. Here is what a protractor looks like.

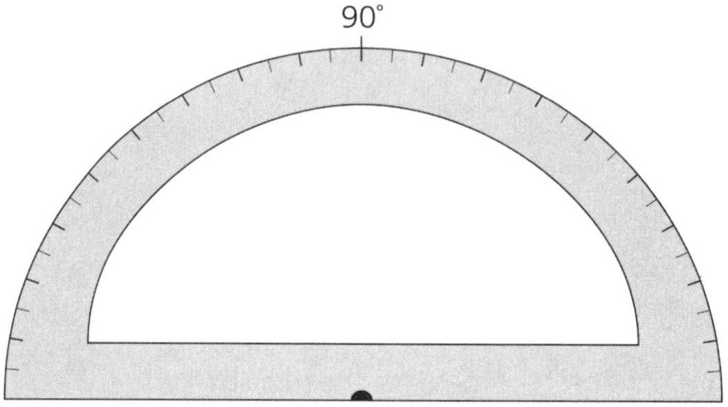

The first step in learning to use a protractor is to place the protractor in the correct position on the angle.

Notice the notch in the middle along the bottom of the protractor. (On some protractors this may be a diamond, cross hairs, or some other design.) Place the notch on the vertex of your angle. Align the bottom of the protractor (or 0°, if 0° is not at the bottom of the protractor) with a side of the angle.

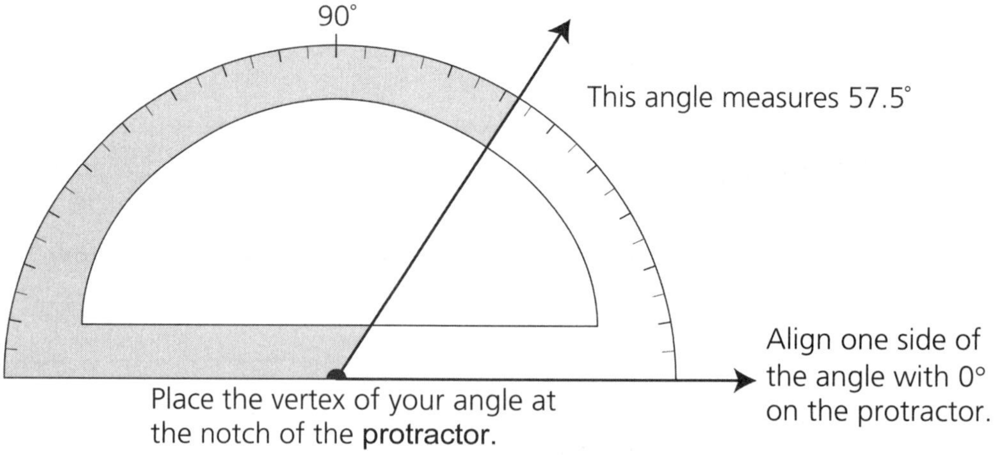

input

Measuring Angles

The second step in learning to use a protractor is to notice that it has two scales. Both scales measure angles between 0 and 180 degrees. Deciding which scale to use involves examining the angle you wish to measure. Decide if the angle is smaller or larger than a 90° angle.

If the angle appears to be less than a 90° angle, use the scale that measures the angle to be less than 90°.

If the angle appears to be larger than a 90° angle, use the scale that measures the angle to be greater than 90°.

If you are unsure about whether the angle is smaller or larger than 90°, start counting from the 0° mark at the first side of your angle. Count along the edge of the protractor until you reach the point where it interests the second side of the angle. From there, read the measure of the angle.

Note that when you align one side of your angle with the protractor, you may use either side of your angle, and you may use either 0° mark on your protractor. Just be sure that the interior of the angle is contained within the protractor. (The interior of the angle is white in the example below.)

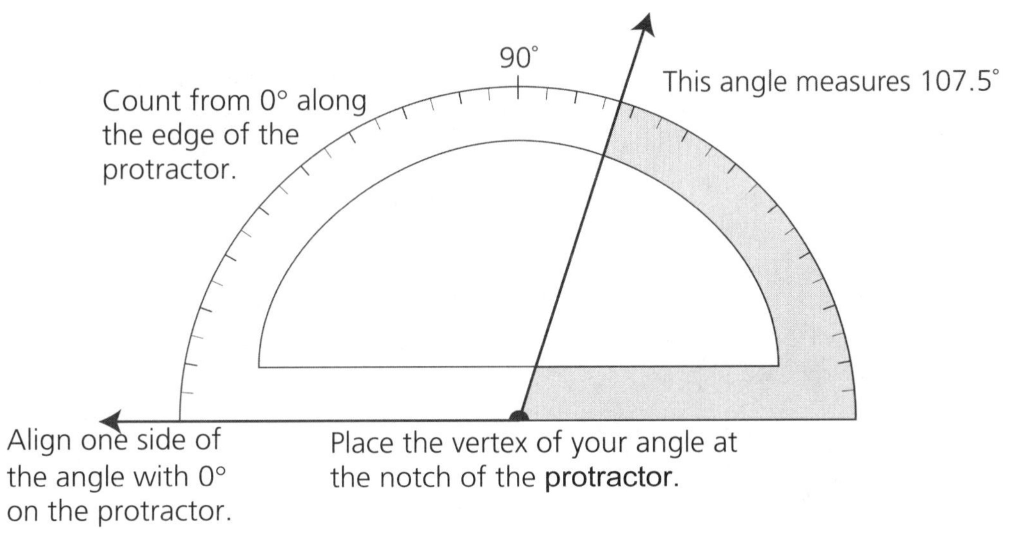

Measuring Angles

process: evaluate

8. The marked angle measures 90°. Note that a right angle has a "¬" symbol drawn at its vertex to show that its measure is 90°. Look at each of the other angles and decide whether it measures more than 90°, less than 90°, or equal to 90°. Write the word *more*, *less,* or *equal* next to each angle.

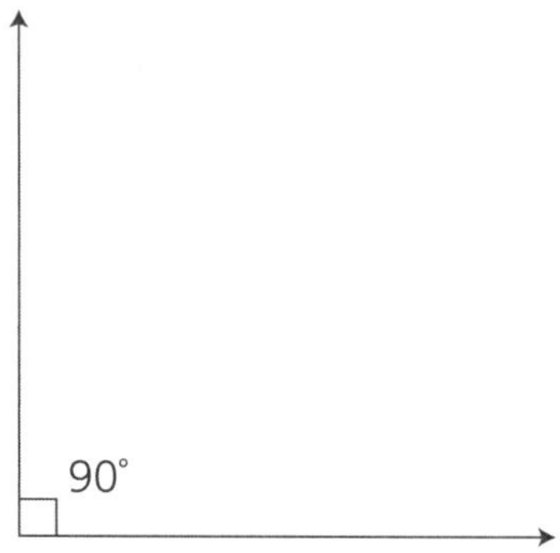

a. 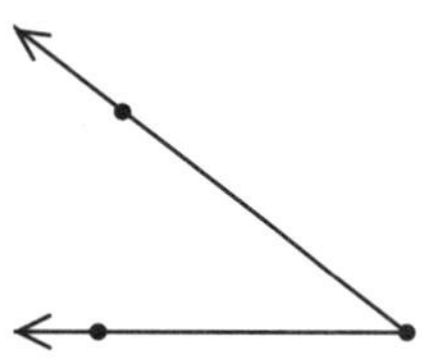

b. 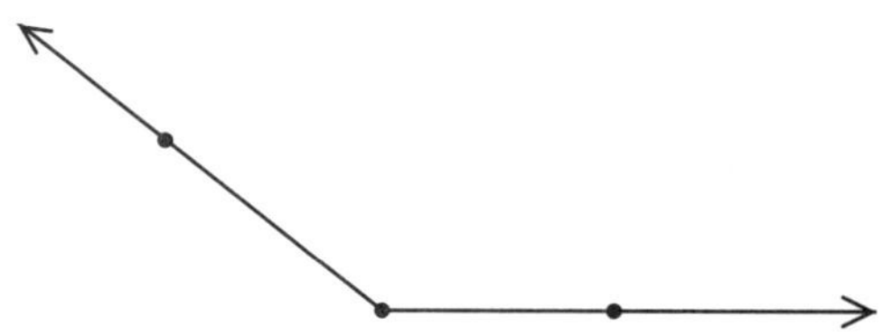

c. 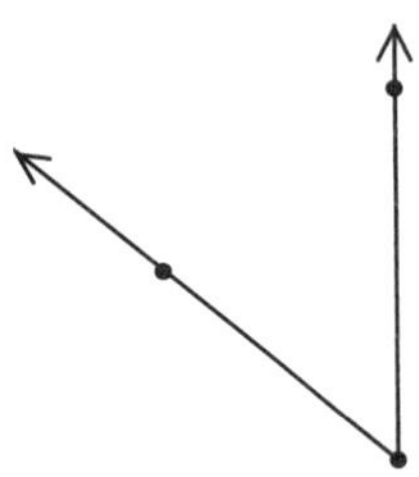

d. 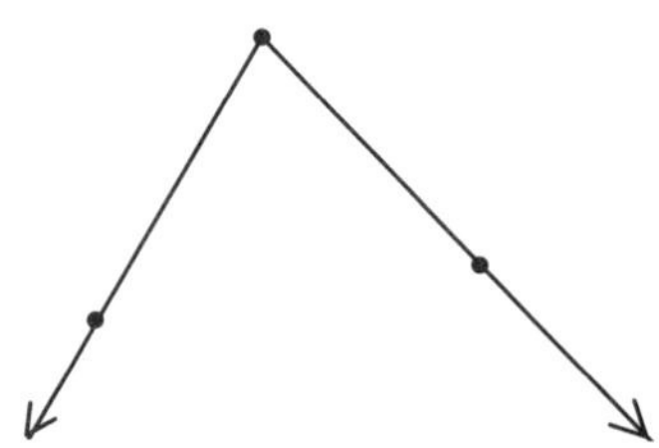

e. 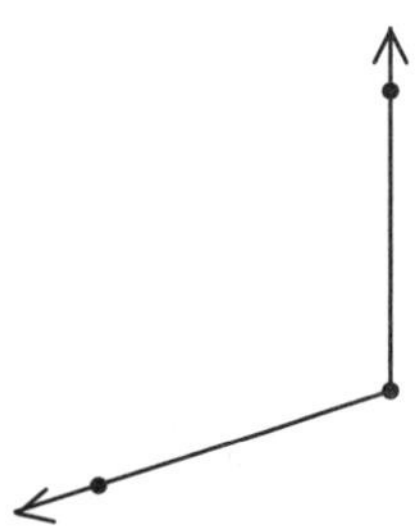

f. 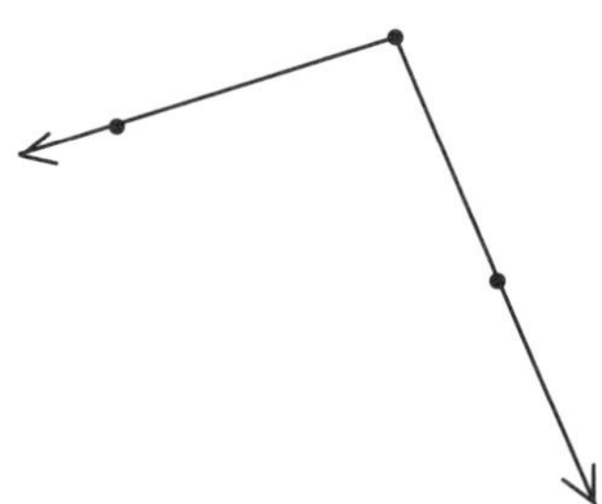

input

Measuring Angles

If an angle measures more than 0° and less than 90°, it is called an **acute angle**. If an angle measures exactly 90°, it is called a **right angle.** If an angle measures more than 90° and less than 180°, it is called an **obtuse angle**. If an angle measures exactly 180°, it is called a **straight angle**.

process: evaluate

Measuring Angles

9. Write the word *acute*, *right, obtuse*, or *straight* near each angle.

a. 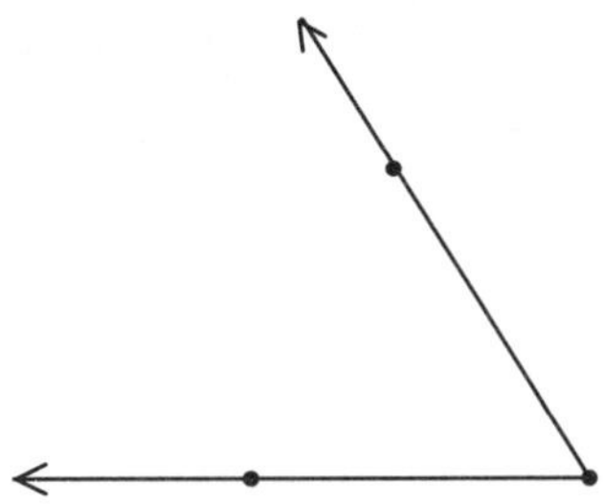

b. 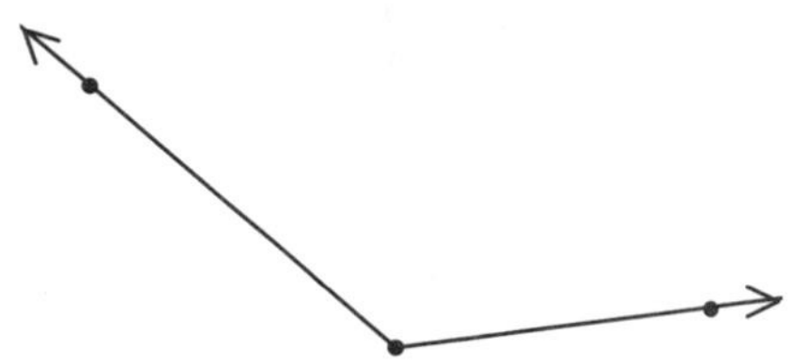

c. 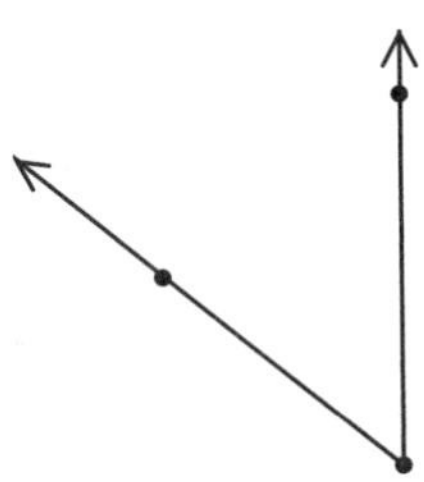

d. 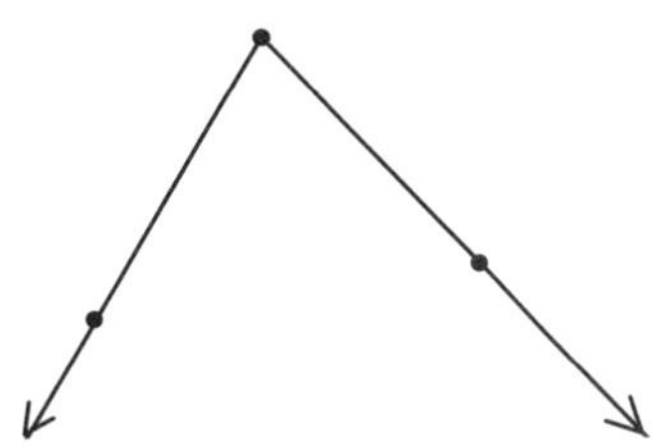

e. 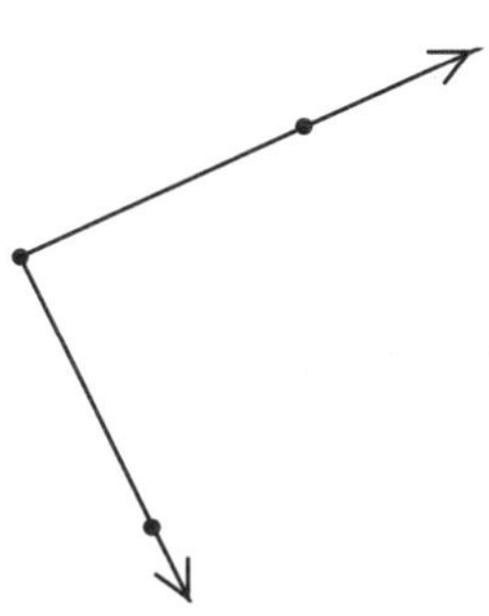

f. 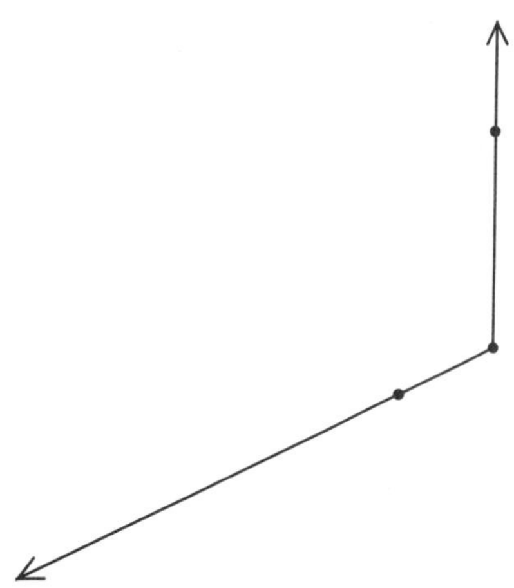

Measuring Angles

process: evaluate

g. 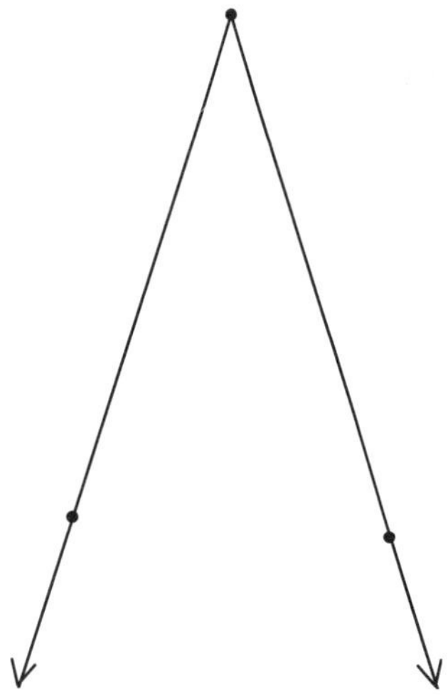

h. 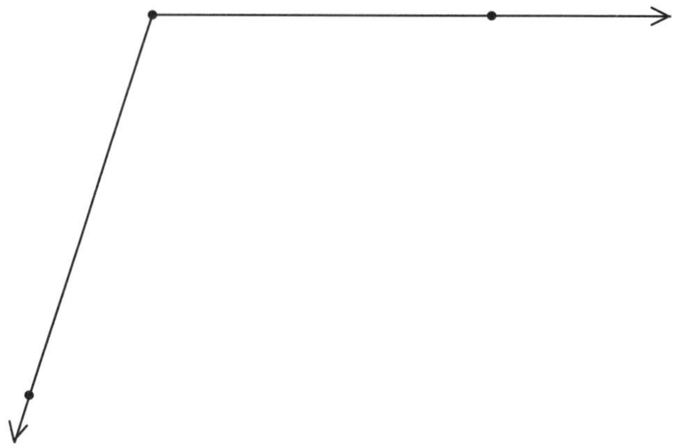

i. 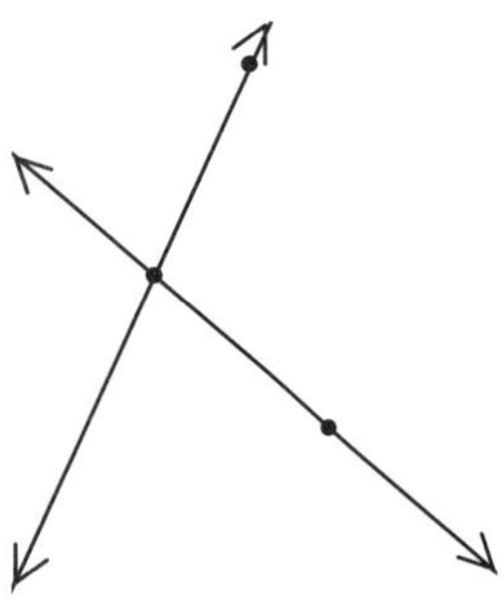

j. 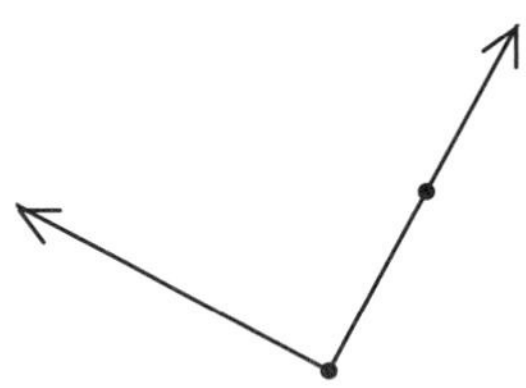

k. 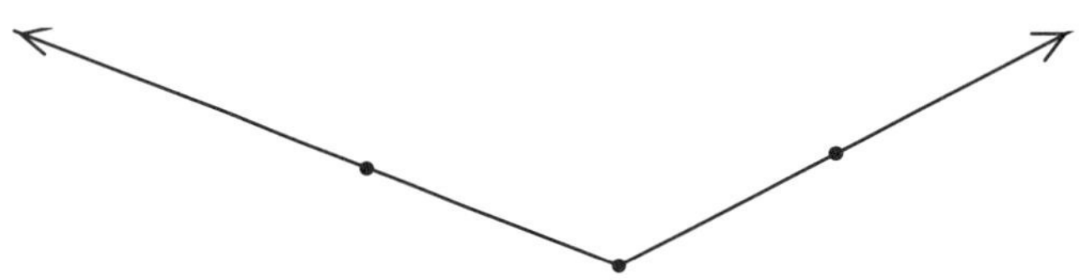

l. 

10. Measure these angles and confirm that you were correct about whether the angle was acute, right, obtuse, or straight. Record your measures inside each angle.

analysis

Measuring Angles

11. Which angle do you think has the greater measure, the one on the left or the one on the right? Explain.

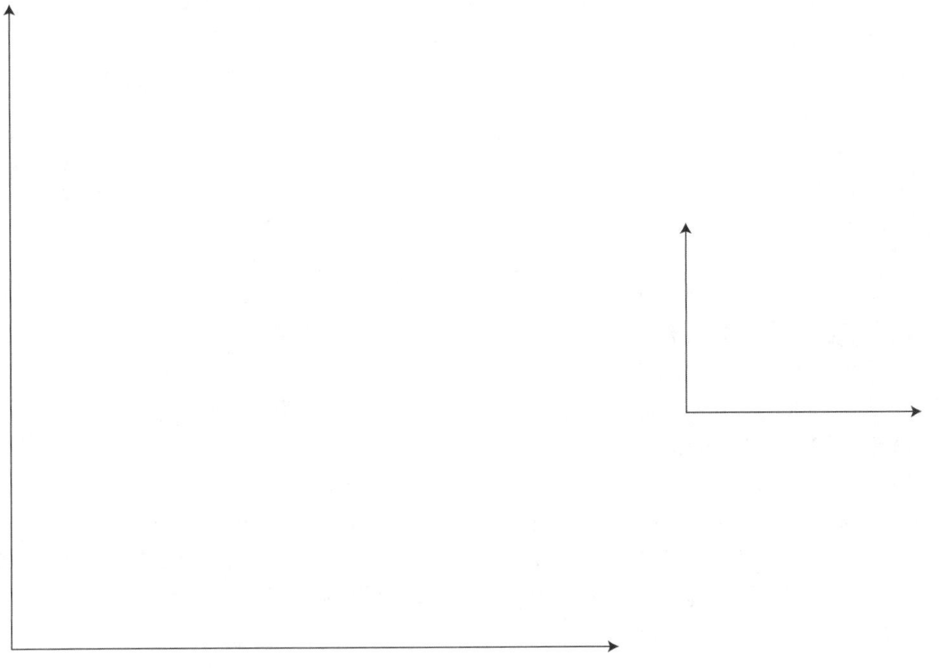

a. Measure each angle to confirm your prediction. Record your measures on the angles.

b. How do the pictured lengths of the sides of an angle influence its measure?

c. The sides of an angle are usually rays. What does that tell you about the lengths of the sides of angles?

process: evaluate

Measuring Angles

12. If the minute hand on a clock moves from the 2 to the 6, how many degrees has it traveled? Draw a picture and write complete sentences explaining your answer.

extension

Measuring Angles

13. If the time on a clock is 1:50, what is the degree measure of the angle between the minute hand and the hour hand? Draw a picture and write complete sentences explaining your answer. (Hint: The answer is not 120 degrees.)

process: summary

Measuring Angles

14. Explain in your own words how to use a protractor to measure the following angle.

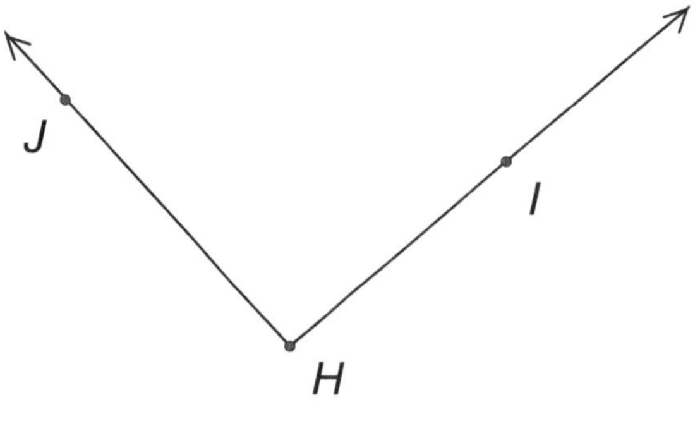

input

Relationships Between Perpendicular Lines

Any two lines (or segments) that intersect and form right angles are called **perpendicular lines** (or **perpendicular segments**).

Use a straightedge and a protractor to draw two lines that intersect and form a right angle.

analysis

Relationships Between Perpendicular Lines

1. Decide if the lines in this figure are perpendicular. Explain how you arrived at your conclusion.

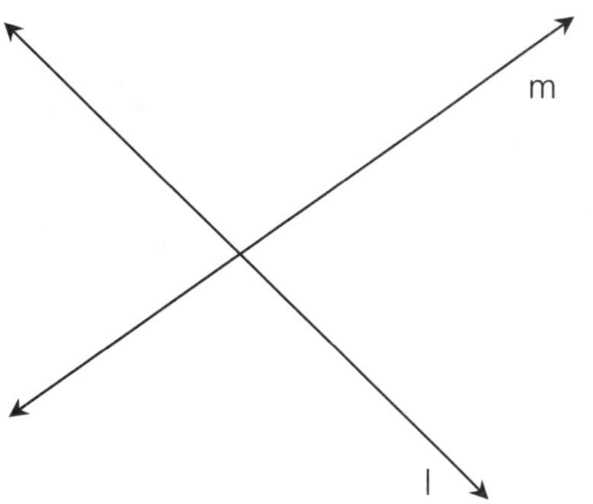

Relationships Between Perpendicular Lines

process: model

2. Graph each of the following pairs of lines, and measure the angle between the lines. Record the measure on the graph.

   a. $y = \frac{2}{3}x$

      $y = -\frac{3}{2}x$

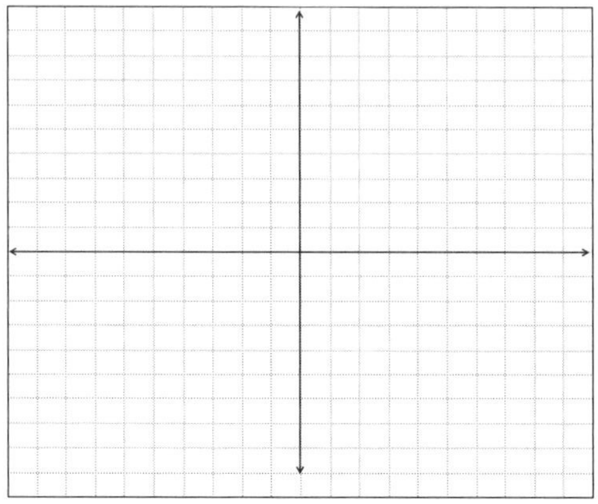

   b. $y = \frac{4}{3}x + 1$

      $y = -\frac{3}{4}x + 4$

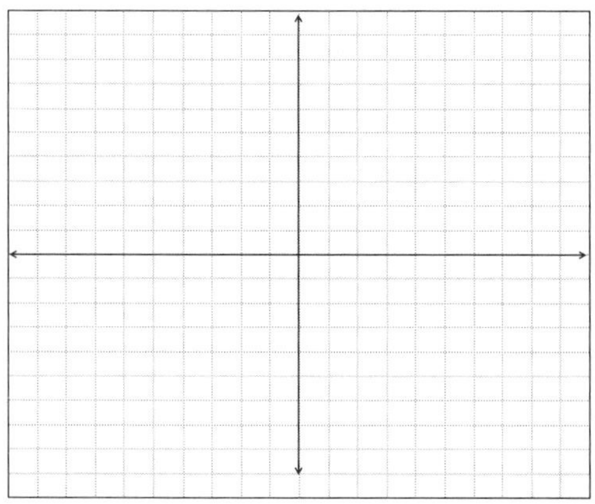

   c. $y = -\frac{5}{2}x - 2$

      $y = \frac{2}{5}x + 3$

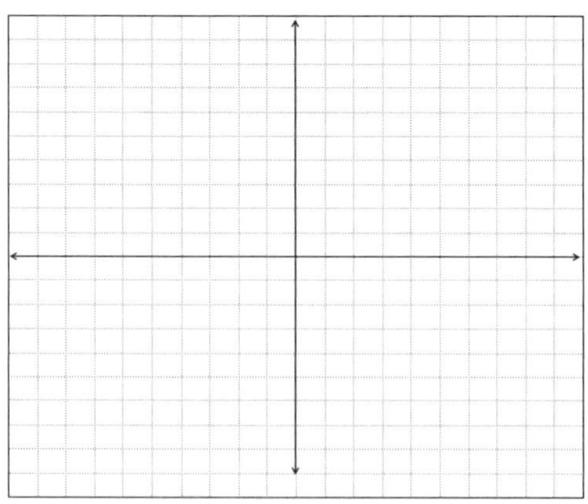

Relationships Between Perpendicular Lines

analysis

3. What relationship exists between the pairs of lines that you graphed in question 2?

4. What relationship do you see between the two *equations* for the lines graphed in question 2?

5. Based on your responses to the previous questions, what is the slope of a line that is perpendicular to $y = \frac{1}{2}x + 5$?

   a. Create an equation for a line that is perpendicular to $y = \frac{1}{2}x + 5$.

   b. Graph both lines on the following grid and verify that the lines are perpendicular.

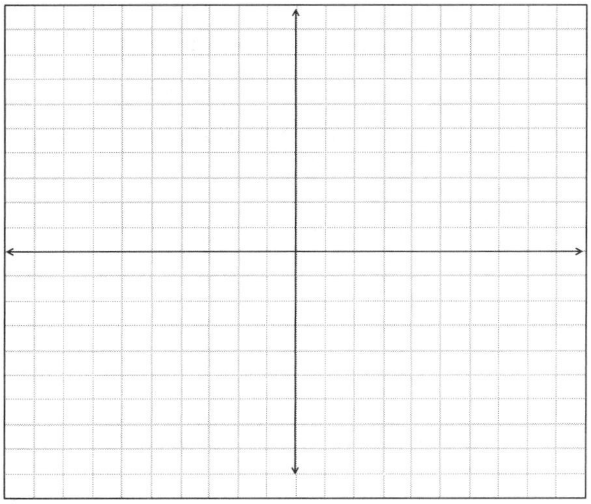

process: model

Relationships Between Perpendicular Lines

6. Graph two perpendicular lines in the coordinate plane.

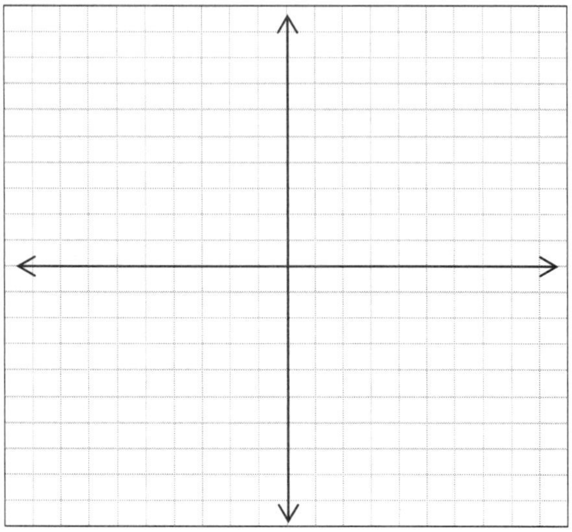

7. What are the equations of the lines you just graphed?

8. Graph two other perpendicular lines in the coordinate plane.

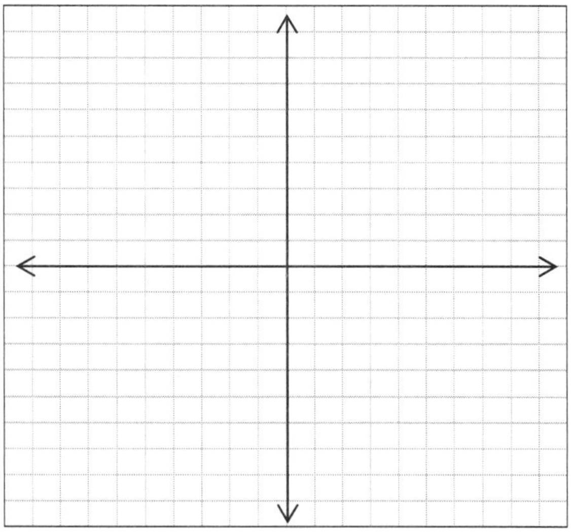

9. What are the equations of the lines you just graphed?

10. Does the relationship you found for the slopes of perpendicular lines hold in your two examples? If the relationship does not hold, why is that the case?

analysis

Relationships Between Perpendicular Lines

11. The equation of a line is $y = 3x + 4$. Determine an equation of a line that is perpendicular to the given line and explain how you determined this equation.

    a. Graph both lines on the grid and verify that the lines are perpendicular.

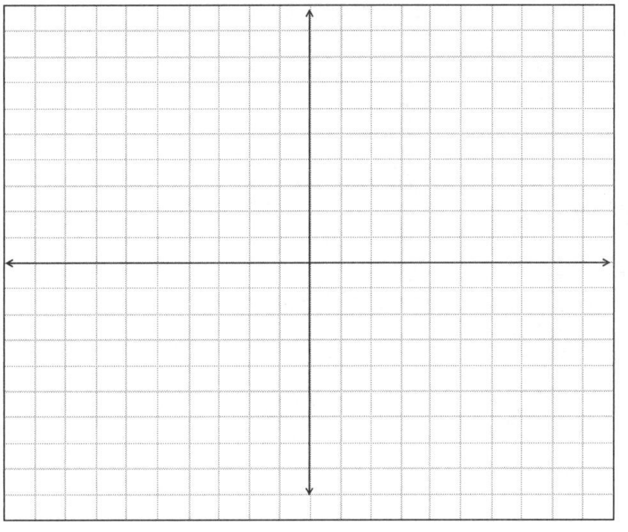

12. In general, how can you manipulate the slope of the first line in a pair of perpendicular lines to obtain the slope of the second line in each pair? What relationship exists between the slopes of perpendicular lines?

Relationships Between Perpendicular Lines

analysis

13. Why do the slopes of two perpendicular lines relate to one another in this particular way?

14. Find the product of the slopes for the lines in:

    a. Question 2a

    b. Question 2b

    c. Question 2c

    d. Question 5

    e. Question 7

    f. Question 9

    g. Question 11

15. What do you observe about the product of the slopes of perpendicular lines? Why is this so?

process: practice

Relationships Between Perpendicular Lines

16. What is the slope of a line perpendicular to the line $y = -7x + 20$?

17. Write the equations of 2 different lines perpendicular to the line $y = -\frac{3}{2}x$.

$y =$ ______________________________

$y =$ ______________________________

18. Verify that the lines for your equations in question 17 are perpendicular by graphing them on the following grid.

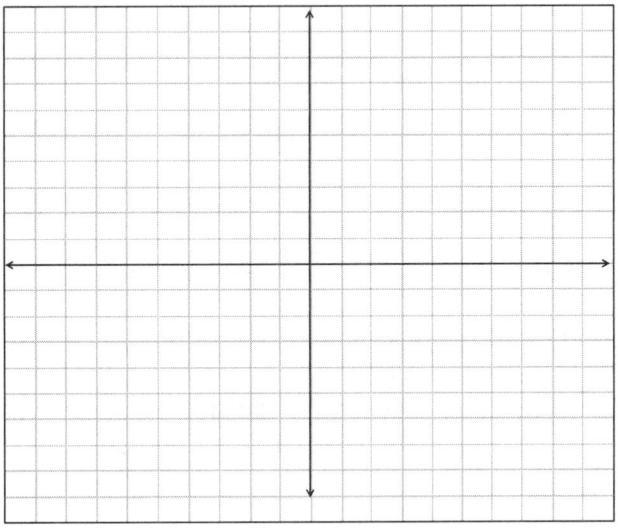

Relationships Between Perpendicular Lines

analysis

19. Graph a vertical line and a line perpendicular to it on the grid.

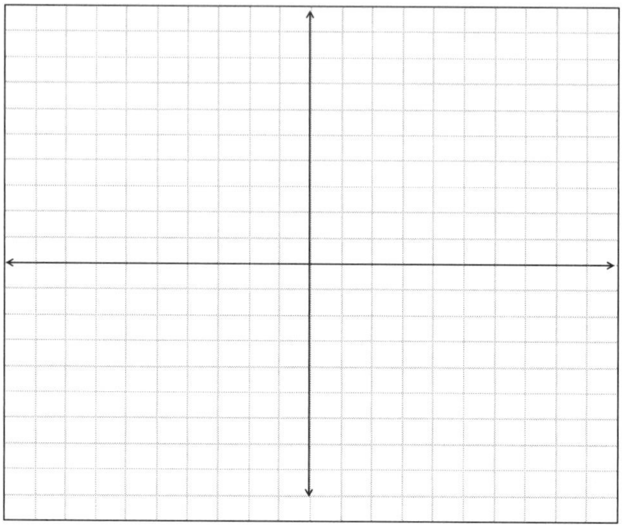

a. What are the equations of these two lines?

b. What is the slope of the vertical line?

c. What is the slope of the horizontal line?

d. Does the relationship you found on page 3-38 for the slopes of perpendicular lines still hold? Explain your response.

process: summary

Relationships Between Perpendicular Lines

23. By recognizing the fact that the slopes for the equations $y = -2.4x$ and $y = -2.4x + 3$ are both -2.4, what do you know about the relationship between these lines?

24. Based only on the equations of the lines, how can you tell if two lines are *parallel*? Based only on the equations of the lines, how can you tell if two lines are *perpendicular*?

25. If two lines are perpendicular, then how are the slopes of the two lines related to one another?

26. Two lines intersect at the origin. One of the lines has a slope of $-\frac{1}{2}$. Provide the equations of these two lines.

27. Summarize your findings about parallel and perpendicular lines and the equations for these lines.

Perpendicular Lines and Segments

input

Earlier in this unit, you learned shorthand or symbolic ways to represent lines, segments, rays, and length. There is a symbolic or shorthand way of representing the fact that lines or segments are perpendicular. You use a symbol that looks like an upside-down letter T. It is $\perp$. If you want to symbolically show that line *l* is perpendicular to line *m*, you write $l \perp m$.

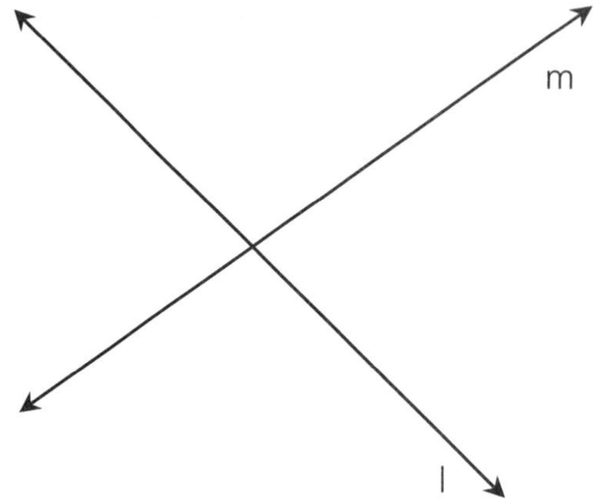

Recall that the **altitude** of a triangle is the height of the triangle.

The **base** of a triangle is a special name for the side of the triangle to which an altitude is drawn.

Perpendicular Lines and Segments

process: practice

1. What relationship exists between the base and the altitude of a triangle?

2. Draw the altitude of ΔGEO, using $\overline{EO}$ as the base of the triangle. Then, use the new symbol you just learned to show the relationship between the base and the altitude.

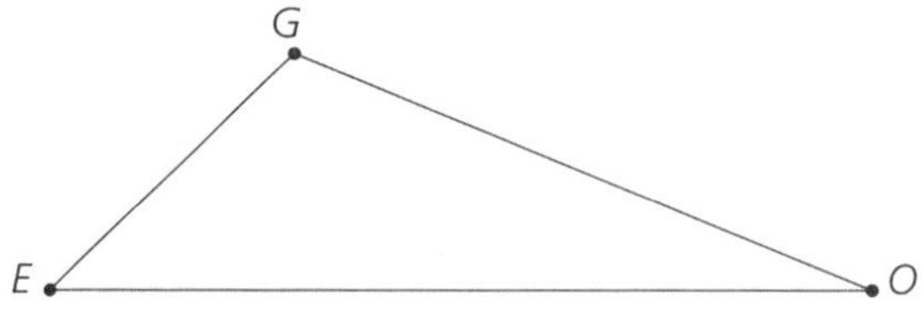

process: practice

Perpendicular Lines and Segments

3. Draw the altitude of ΔGEO, using $\overline{EG}$ as the base of the triangle. Then, show the relationship between the base and the altitude, using the new symbol you just learned.

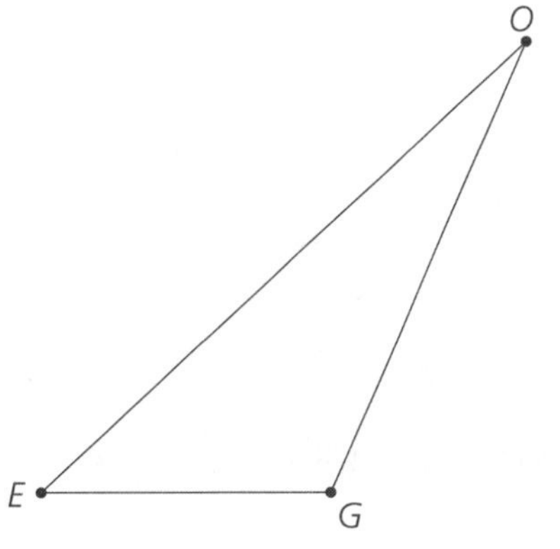

4. Draw the altitude of ΔGEO, using $\overline{OG}$ as the base of the triangle. Then, symbolize the relationship.

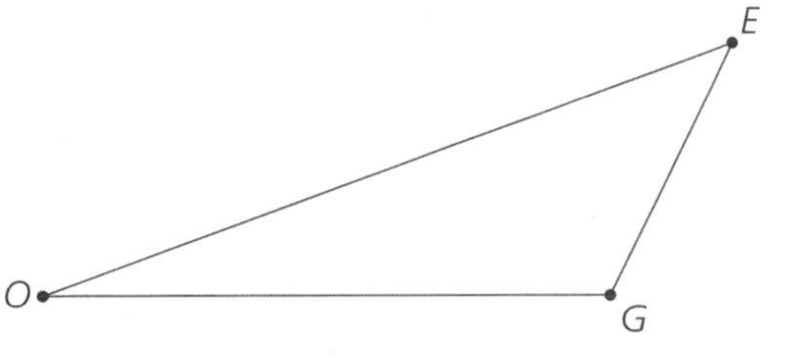

analysis

Perpendicular Lines and Segments

5. The triangles in problems 2 - 4 are all the same triangle.

   a. Explain why it might not be immediately obvious that the triangles in problems 2 – 4 are the same.

Perpendicular Lines and Segments

analysis

b. How many different altitudes can a triangle have?

c. Are all of the altitudes of a triangle equal in length?

d. What are the lengths of the three altitudes in ΔGEO?

6. What property of a triangle makes an altitude appear exterior to the triangle?

7. What property of a triangle makes an altitude appear interior to the triangle?

8. If an altitude is neither interior to the triangle nor exterior to the triangle, what is the third option? What property of a triangle makes this happen?

   a. Create a triangle in which at least one of the possible altitudes is neither exterior nor interior to the triangle.

process: model

Perpendicular Lines and Segments

9. Is it possible to draw more than one line perpendicular to a given line? Attempt to create at least two lines that are perpendicular to the line in this figure.

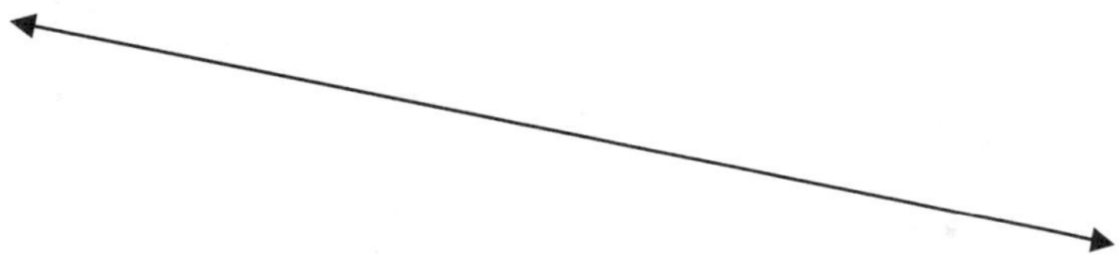

   a. How many lines can be created that are perpendicular to the line above?

10. Here is a line and a point on the line.

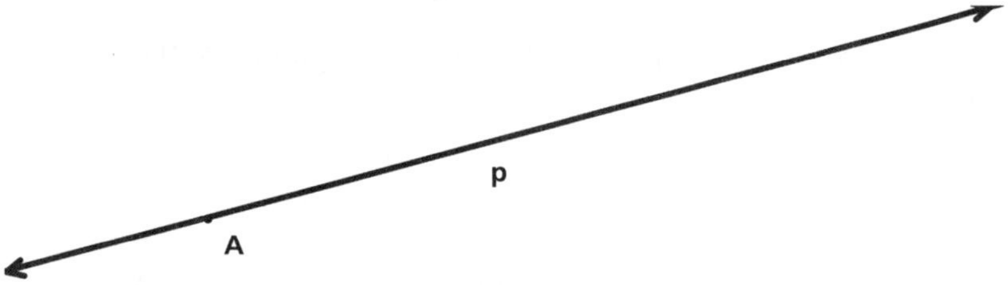

   a. Draw as many lines as possible *through the given point* that are perpendicular to the line. How many lines did you draw?

   b. Use symbolic notation to describe the relationship between line p and your newly created line.

   c. How many lines can be perpendicular to a given line *at a certain point on that line*?

process: model

Perpendicular Lines and Segments

11. Here is a line and a point *not* on the line.

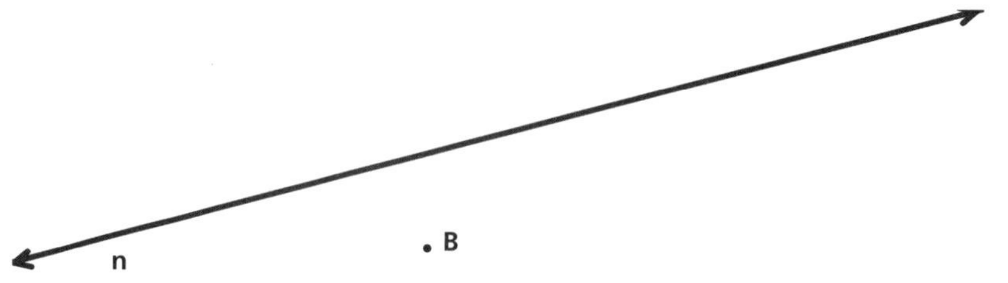

a. Draw as many lines as possible *through the given point* that are perpendicular to the line. How many lines did you draw?

b. Use symbolic notation to describe the relationship between line n and your newly created line.

c. How many lines *that pass through a point not on the line* can be perpendicular to a given line?

12. Using the diagram below, determine if it is possible to draw a line that is *perpendicular* to a given line segment and also *bisects* the line segment?

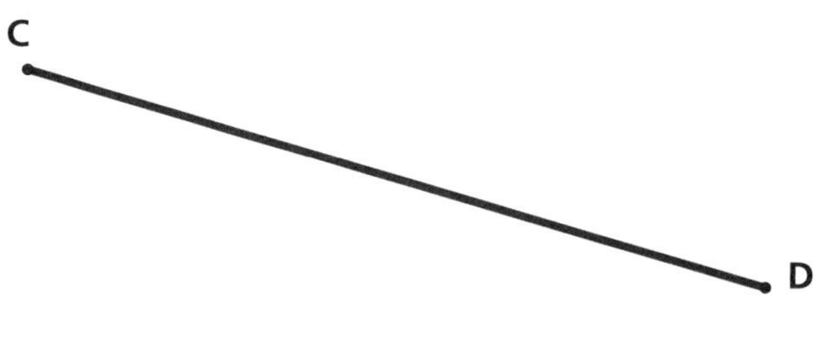

a. How will you prove that your line is perpendicular to and bisects the given line segment?

process: model

Perpendicular Lines and Segments

b. Use symbols to represent the two relationships that exist between $\overline{CD}$ and your newly created line or segment. (Hint: Label the midpoint.)

input

Perpendicular Lines and Segments

A line that is perpendicular to a line segment and also bisects it is called a **perpendicular bisector**. Every line segment has a perpendicular bisector.

process: summary

Perpendicular Lines and Segments

13. Create the perpendicular bisector of the following line segment.

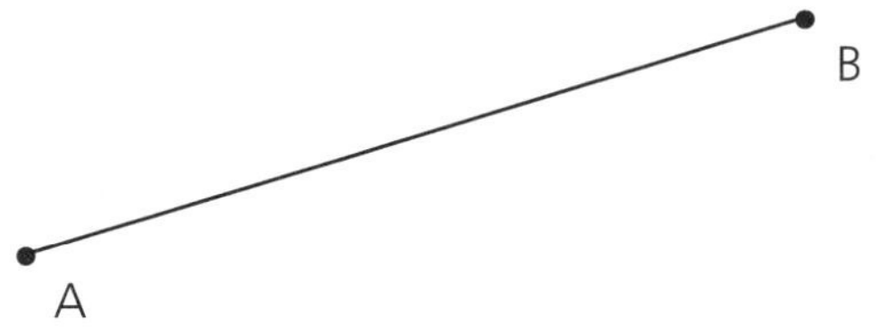

Vertical Angle

input

Just as certain angles have special names like right angles or straight angles, certain angle pairs have special names.

In this figure, ∠1 and ∠3 are an example of a pair of **vertical angles**.

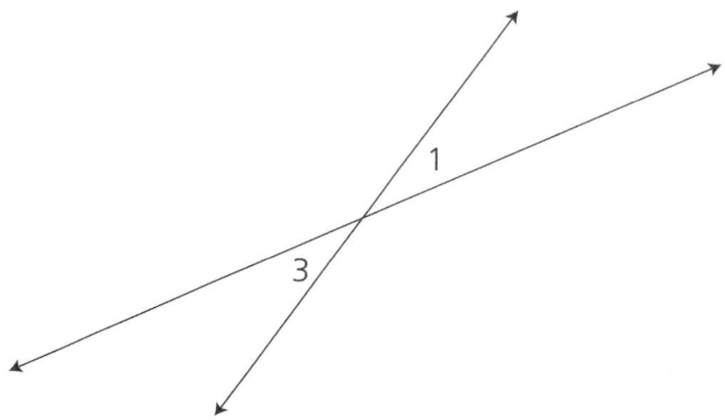

Vertical angles can be seen in the hinged pieces of a baby gate that guards the top of the stairs. When the gate is folded or open, the measures of the vertical angles 1 and 3 are at their minimums.

Vertical Angles

process: evaluate

1. Measure angles 1 and 3 and record their measures on the figure.

Gate is open.

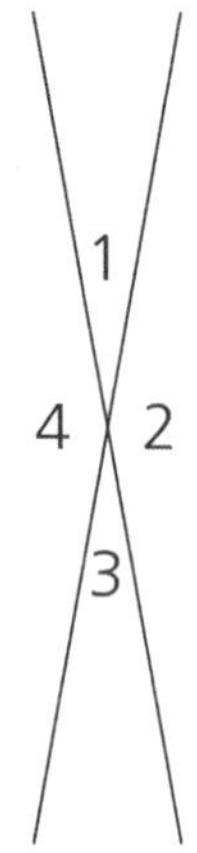

process: evaluate

Vertical Angles

When the gate is stretched across the doorway or closed, the measures of the vertical angles change.

2. Measure angles 1 and 3 in the picture of the closed gate and record their measures on the figure.

Gate is closed.

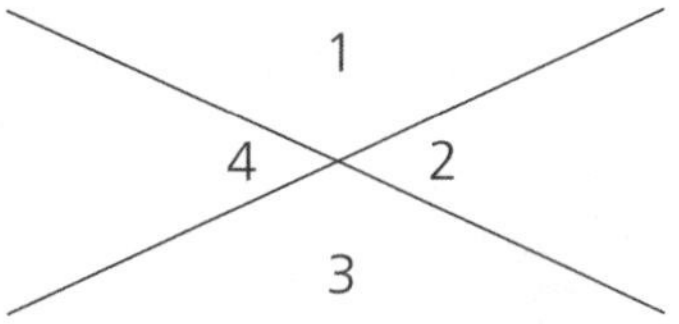

analysis

Vertical Angles

3. The following figures are all examples of angle pairs that are NOT vertical angles.

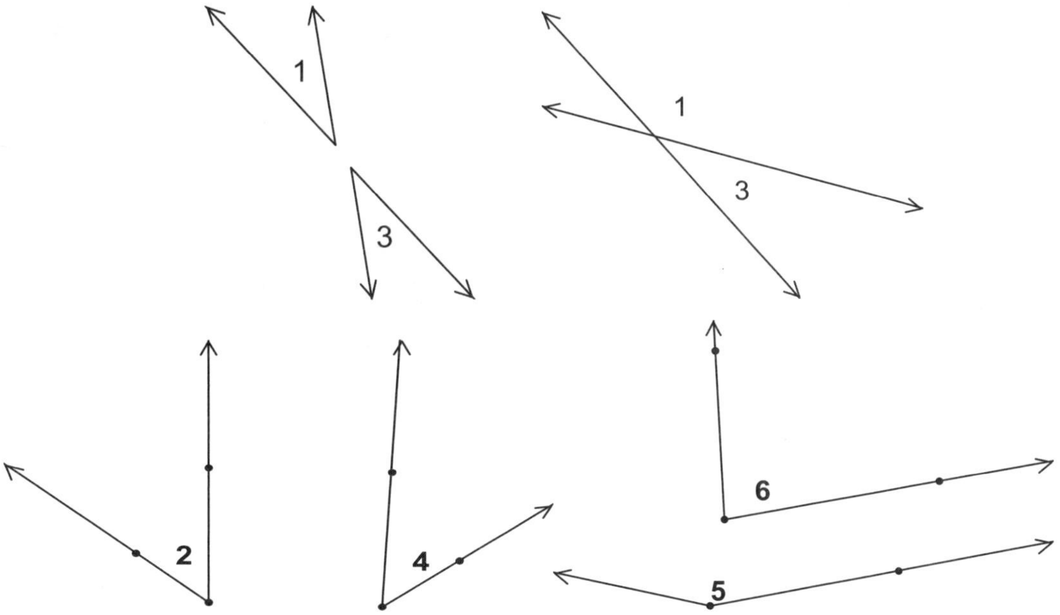

Describe what is visually different about pairs of angles that are vertical angles and pairs of angles that are not vertical angles.

process: model

Vertical Angles

By definition, **vertical angles** are the pairs of non-adjacent angles formed when two lines intersect.

4. Draw two examples of pairs of vertical angles.

   a. Measure these angles and record their measurements on the drawing.

5. Draw two more pairs of angles that are not vertical angles.

   a. Measure these angles and record their measurements on the drawing.

process: practice

Vertical Angles

6. List all of the pairs of vertical angles you see in this figure.

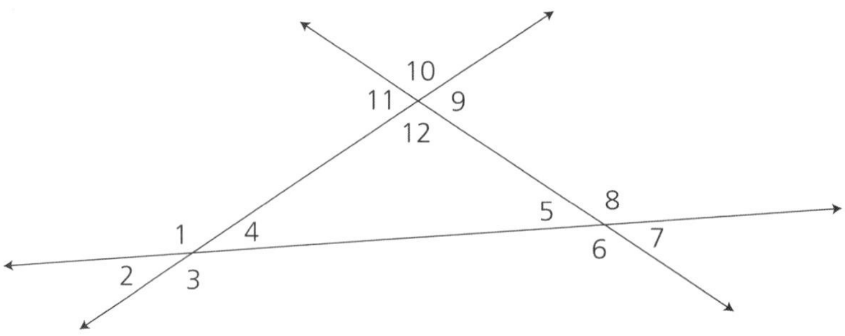

Vertical Angles

analysis

7. Measure each angle in the six pairs of vertical angles in the previous figure. Record these measurements on the figure.
   a. What observations can you make about the measures of the two angles in a vertical pair of angles?
   b. Will your observations about vertical angles always hold true?

8. In your own words, describe what *defines* a pair of vertical angles.

9. What additional properties hold true with all pairs of vertical angles?

Angle Pairs

input

As you saw in the previous section, some pairs of angles are given special names. In this section, you will learn about other pairs of angles that have special names.

Two angles whose measures sum to 90° are **complementary angles**. For example, the following pairs of angles are complementary.

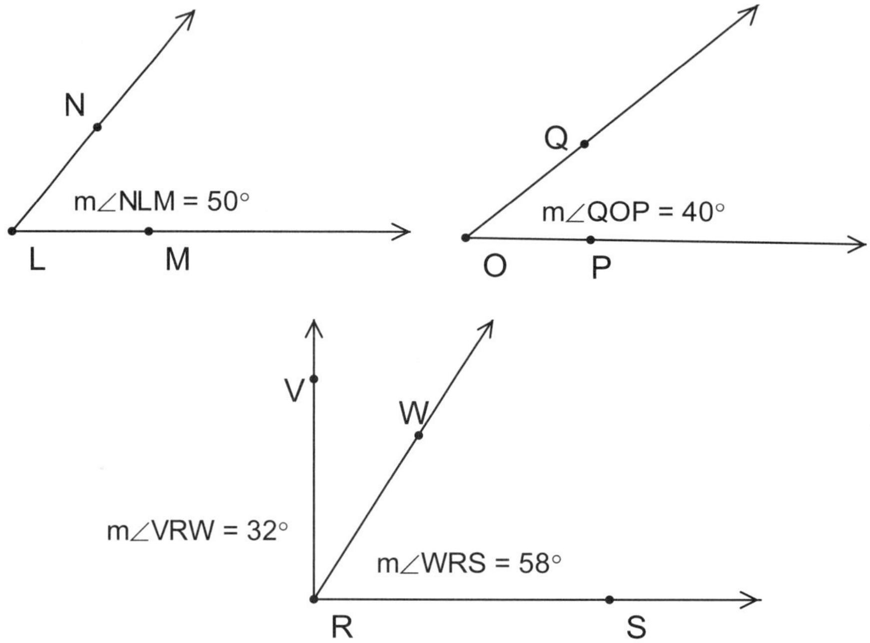

These angles, however, are NOT complementary to one another.

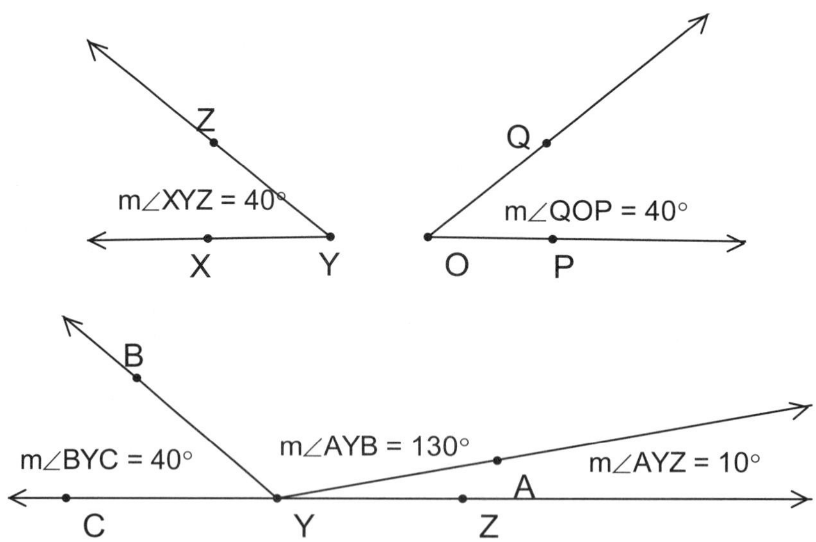

process: practice

Angle Pairs

1. Use the following angles to choose at least two pairs of complementary angles. More than two pairs are pictured.

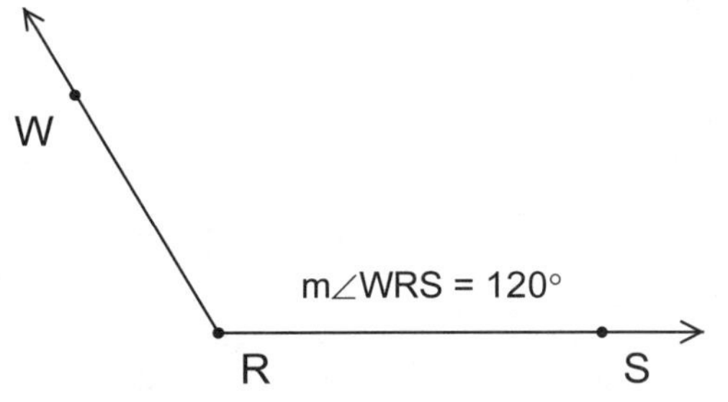

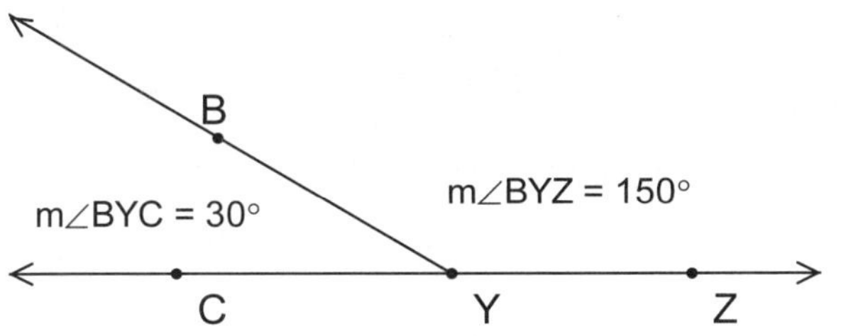

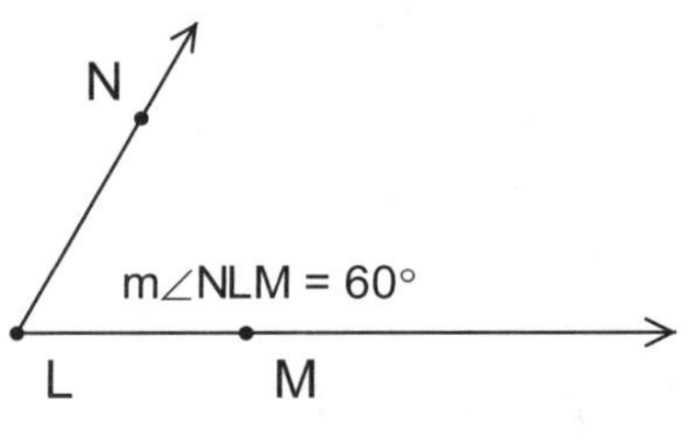

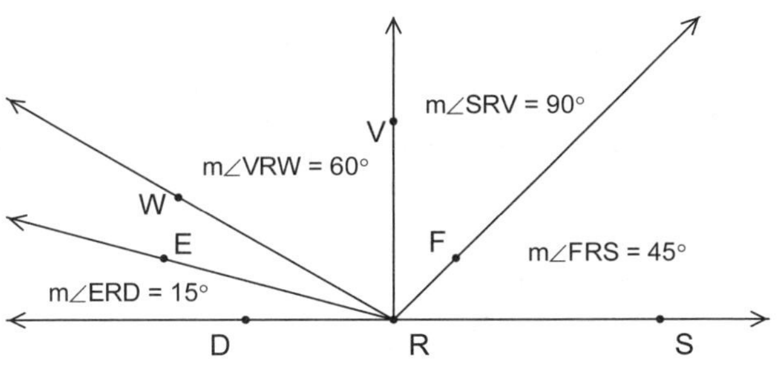

2. Draw your own pair of complementary angles.

Angle Pairs

input

Two angles whose sum is 180° are **supplementary angles**.

For example, the following pairs of angles are supplementary angles.

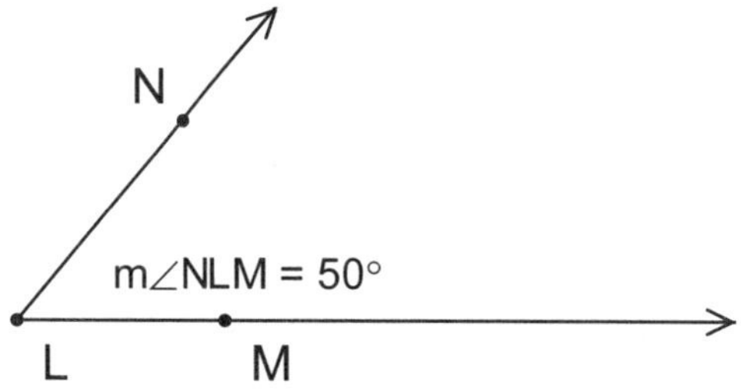

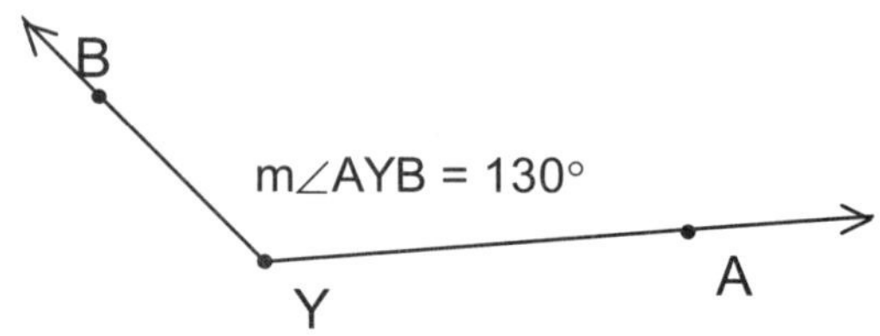

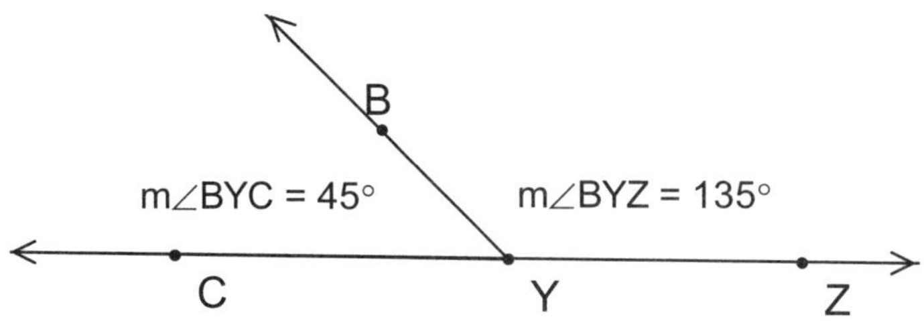

These angles, however, are NOT supplementary.

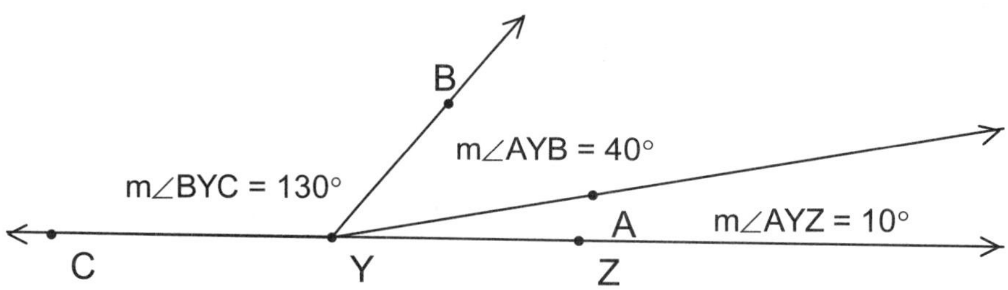

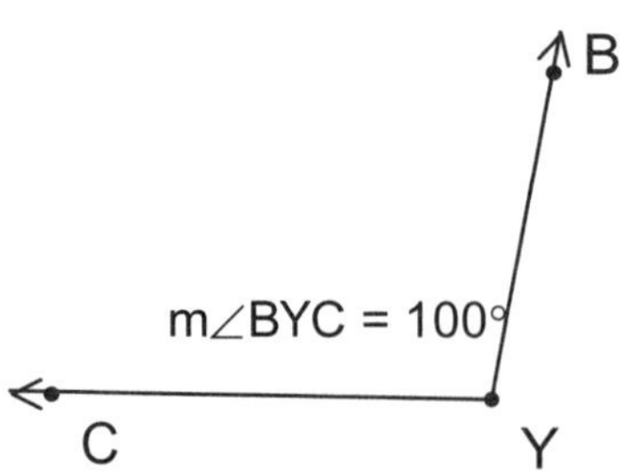

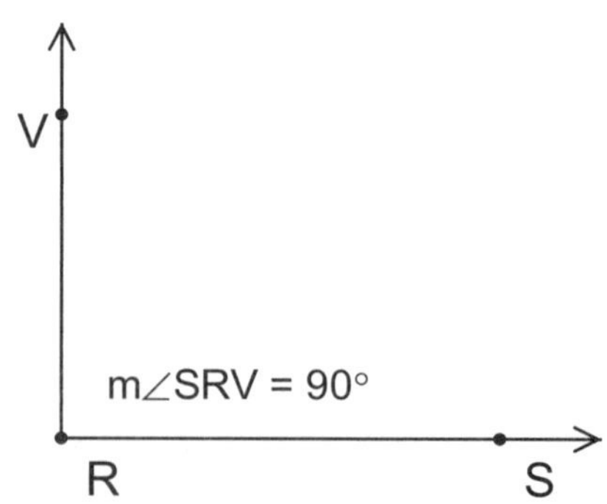

process: practice

Angle Pairs

3. Refer back to the set of angles used for question 1. Using these angles, choose at least two pairs of supplementary angles.

4. Draw your own pair of supplementary angles.

analysis

Angle Pairs

5. Are any pairs of angles in the following figure complementary? If so, which angles? If not, why not?

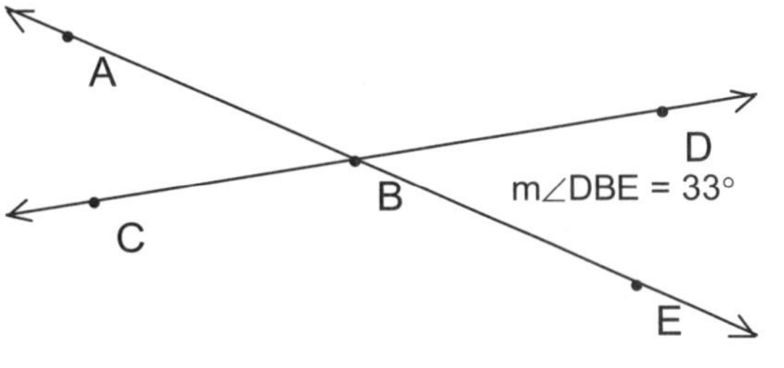

6. Are any pairs of angles in the above figure supplementary? If so, which angles? If not, why not?

preview

More Angle Pairs

1. What does the following statement mean:

In our school, the cafeteria is adjacent to the gymnasium.

input

More Angle Pairs

Two angles are **adjacent angles** if and only if they share a vertex and a common side between them. For example, ∠ADB and ∠BDC are adjacent. They share vertex D and ray DB between them.

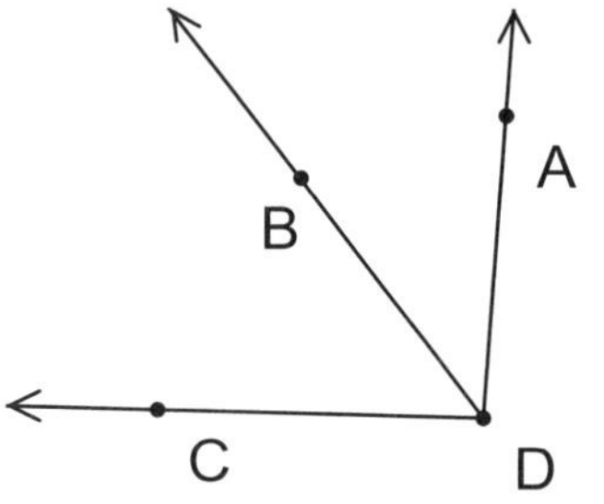

analysis

More Angle Pairs

2. Based upon the definition above, why is ∠ADB NOT adjacent to ∠ADC?

process: practice

More Angle Pairs

3. Return to the angles pictured following question 1 on page 3-53. Choose at least three pairs of adjacent angles.

4. Draw a pair of angles that are adjacent.

5. Is ∠K adjacent to ∠M? Why or why not?

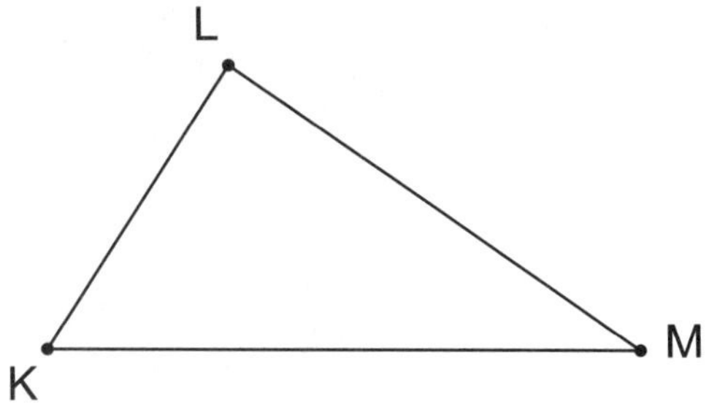

6. Are ∠'s ABD and CBE adjacent? Why or why not?

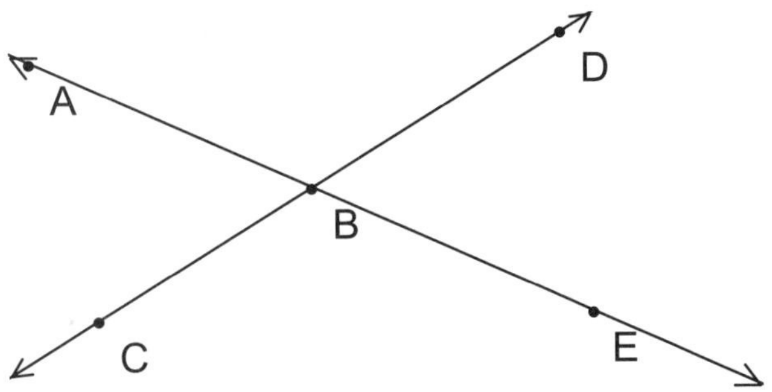

7. Are ∠'s ABD and ABC adjacent in the above figure? Why or why not?

input

More Angle Pairs

Not only are ∠'s ABD and ABC adjacent angles, they are a special type of adjacent angles. Angles ABD and ABC form a **linear pair**.

Here is another example of a linear pair of angles.

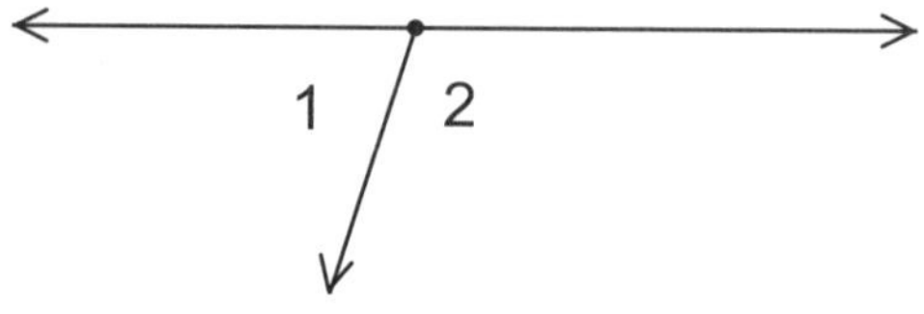

Here is an example of angles that are *not* a linear pair.

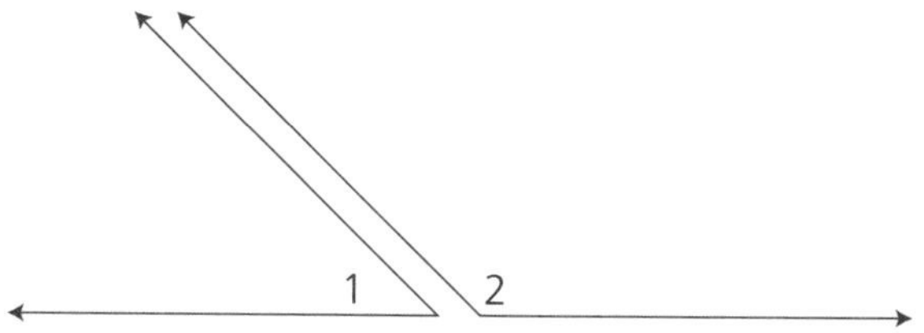

analysis

More Angle Pairs

8. What is visually different about the angles that form a linear pair and the angles that don't?

9. Write complete sentences describing a linear pair of angles. What conditions have to be met in order for a pair of angles to be a linear pair?

process: model

More Angle Pairs

10. Draw two more examples of angles that are linear pairs and two more examples of angles that are not linear pairs.

analysis

More Angle Pairs

11. Choose the correct word or words to complete the following statement.

    A linear pair of angles can also be classified as __________.

    a. complementary b. supplementary

    c. both of the above d. neither of the above

12. Explain your choice in the previous question.

13. This diagram is a gauge for the gas tank in an automobile. The tank has a 14-gallon capacity. Measure the angles in this linear pair to find the number of gallons remaining in the tank and the number of gallons that have been used.

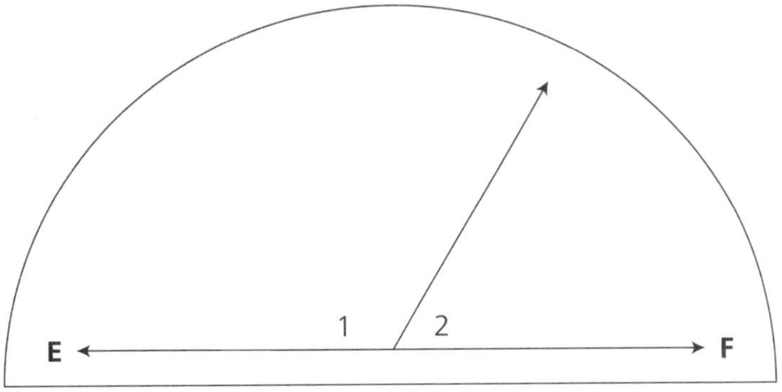

process: practice

More Angle Pairs

14. List all of the linear pairs of angles you see in this figure.

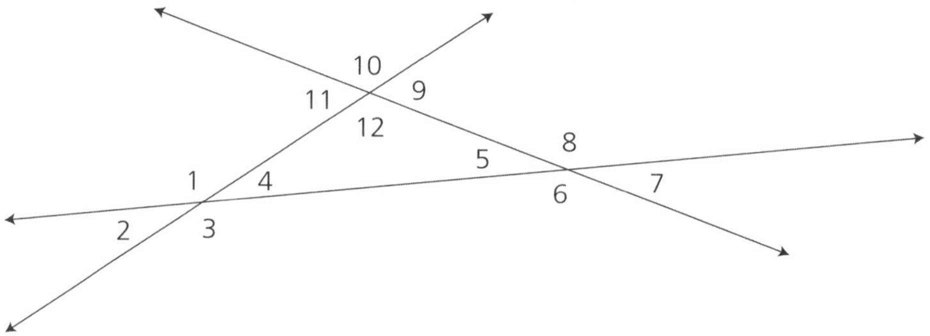

process: summary

More Angle Pairs

15. Based upon the examples and non-examples in this section, write a complete definition of **linear pair**.

16. How does the mathematical definition of **adjacent angles** agree with the everyday definition of the word *adjacent*?

More Angle Pairs

process: summary

17. Write a summary of the information you have learned about angle pairs in the last two sections of text.

18. Tell whether each statement is true or false. If the statement is false, correct it to make it true. If the statement is true, show an example to illustrate it.

    a. Vertical angles can be supplementary.

    b. Vertical angles can be complementary.

    c. Vertical angles can be adjacent.

    d. A linear pair is complementary.

process: model

Angles Formed by Transversals

1. In the space below, use a straightedge to create two coplanar lines that are *not* parallel.

2. Now create one line that intersects both of the non-parallel lines at separate points.

input

Angles Formed by Transversals

A line that intersects at least two other lines in distinct points is called a **transversal**. Mathematicians often refer to a transversal as "cutting" the lines it intersects.

process: evaluate

Angles Formed by Transversals

3. How many angles are created in your figure above (in which the transversal intersects two non-parallel lines)?

4. Find the measure of all the angles formed by the transversal where it intersects the other two lines. Record the measurements on the picture you created in question 1.

process: evaluate

Angles Formed by Transversals

5. Harvey misplaced his protractor in the process of completing the last question. He had measured only two of the angles in his figure. Find the measures of the remaining angles without using a protractor.

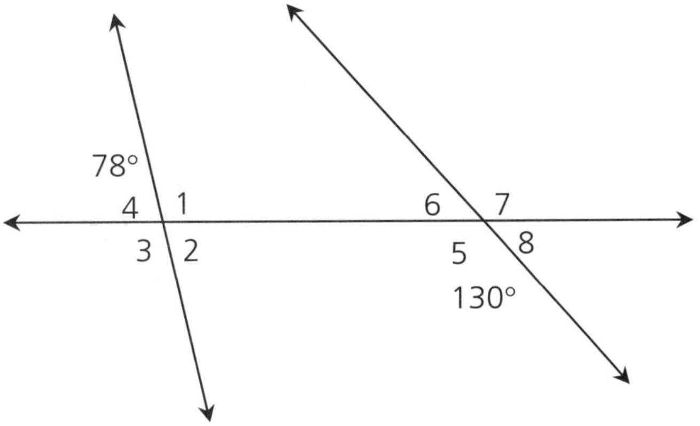

analysis

Angles Formed by Transversals

6. Explain the reasoning you used to find m∠1.

7. Explain the reasoning you used to find m∠2.

8. Explain the reasoning you used to find m∠3.

9. Explain the reasoning you used to find m∠6.

10. Explain the reasoning you used to find m∠7.

11. Explain the reasoning you used to find m∠8.

process: model

Angles Formed by Transversals

12. How many transversals do you see in each figure?

a.

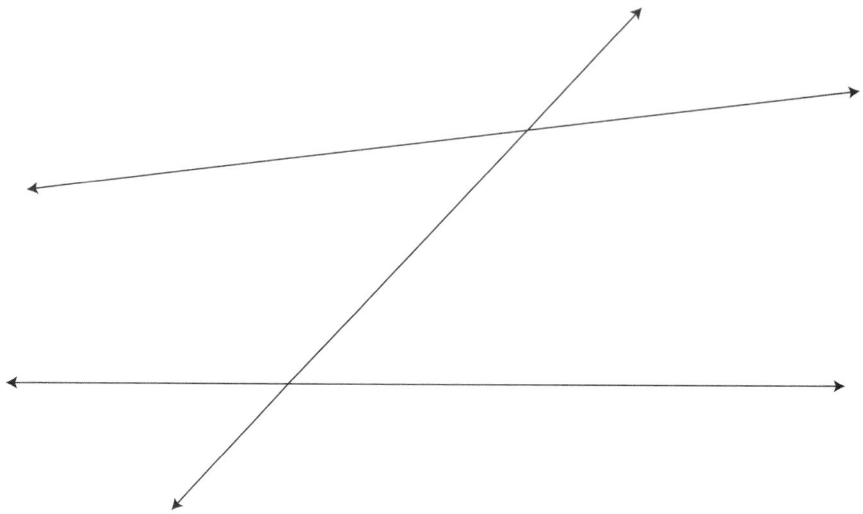

b.

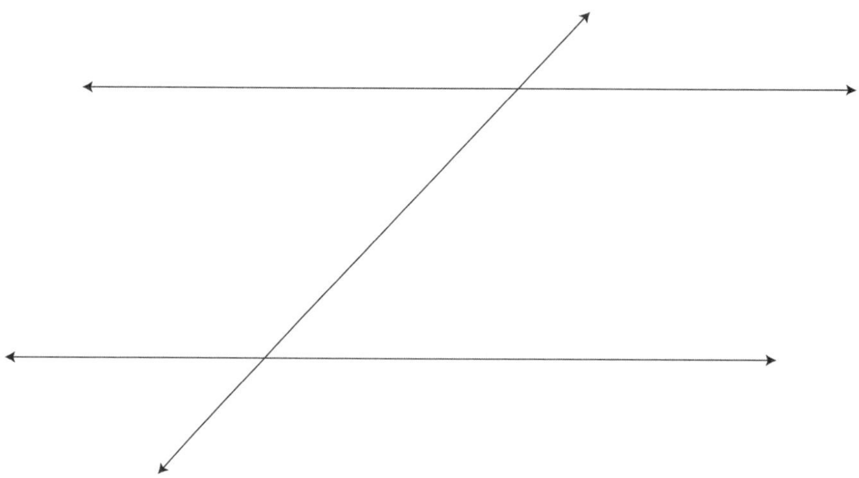

analysis

Angles Formed by Transversals

13. Transversals form many pairs of angles. You have already learned about certain pairs of angles. Fill in the second column of this chart by providing the reason for each relationship in the first column.

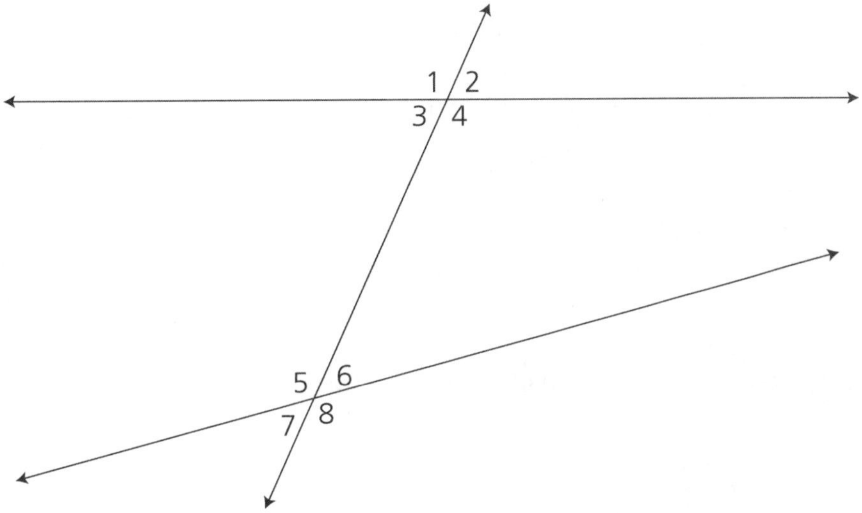

Relationship Between a Pair of Angles	Why is this statement true?
$m\angle 1 = m\angle 4$	
$m\angle 6 = m\angle 7$	
$m\angle 1 + m\angle 2 = 180°$	
$m\angle 3 + m\angle 4 = 180°$	
$m\angle 6 + m\angle 8 = 180°$	
$m\angle 5 + m\angle 7 = 180°$	

a. Can you state a relationship about any other angle pairs? If so, add them to the blank spaces in the table.

Angles Formed by Transversals

input

In addition to the vertical angles and the linear pairs of angles, five other types of angle pairs are formed when two lines are intersected by a transversal. These angle pairs are:

- Alternate interior angles
- Alternate exterior angles
- Interior angles on the same side of the transversal (same-side interior angles)
- Exterior angles on the same side of the transversal (same-side exterior angles)
- Corresponding angles

Angles Formed by Transversals

analysis

Notice that four of these five types of angle pairs use either the designation of "interior" or "exterior." Of the eight angles pictured on the previous page, four of them are interior angles and four of them are exterior angles.

14. Conjecture about which four angles are interior angles and which four angles are exterior angles. Provide your reasoning.

Angles Formed by Transversals

input

Alternate interior angles are pairs of angles formed by the intersection of a transversal with two other lines. All pairs of alternate interior angles have two features: they are located on alternate sides of the transversal (line 3) and between the other two lines (line 1 and line 2).

One example of alternate interior angles is:

$\angle 3$ and $\angle 6$.

Provide another example of alternate interior angles.

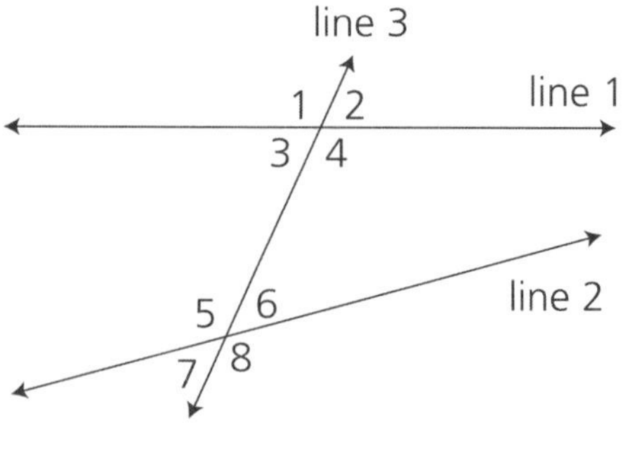

input

Angles Formed by Transversals

Alternate exterior angles are pairs of angles formed by the intersection of a transversal with two other lines. All pairs of alternate exterior angles have two features: they are located on alternate sides of the transversal (line 3) and outside the other two lines (line 1 and line 2).

One example of alternate exterior angles is:

$\angle 1$ and $\angle 8$.

Provide another example of alternate exterior angles.

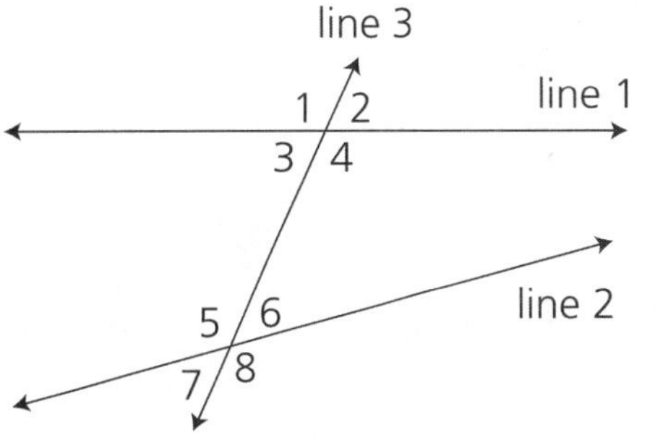

analysis

Angles Formed by Transversals

15. Based on what you've learned about alternate interior angles, what two features must **same-side interior angles** have?

16. Provide an example of same-side interior angles from this figure.

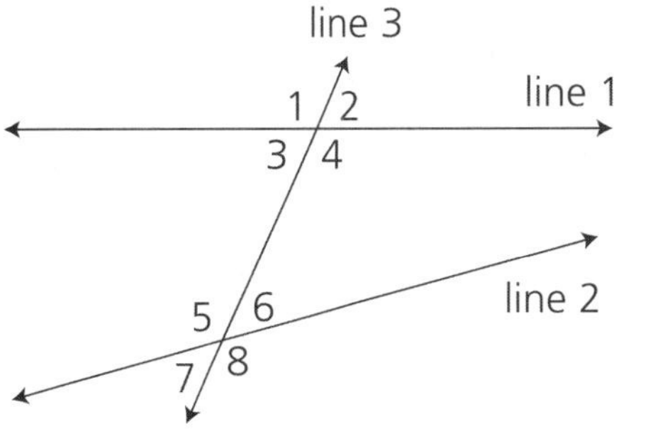

   a. Is there another pair of same-side interior angles in the figure? If so, what is it?

analysis

Angles Formed by Transversals

17. Based on what you've learned about alternate exterior angles, what two features must **same-side exterior angles** have?

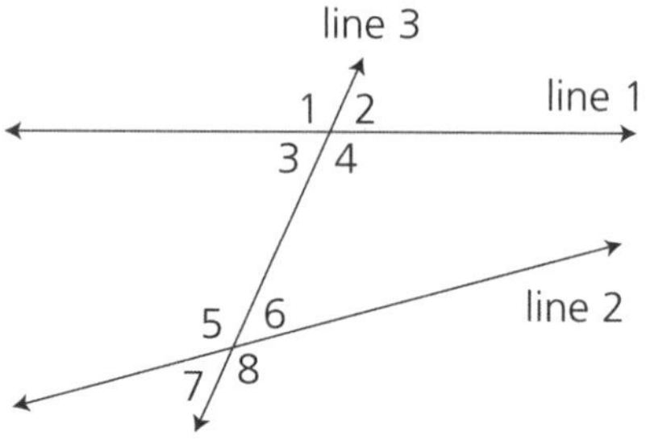

18. Name two pairs of same-side exterior angles in the diagram.

input

Angles Formed by Transversals

Corresponding angles are pairs of angles formed by the intersection of a transversal with two other lines. A pair of corresponding angles shares a similar location in relation to the transversal (line 3) and each of the other two lines. Locate any angle formed where lines 1 and 3 intersect, then look for that same location where lines 2 and 3 intersect. If you choose $\angle 1$, it is located in the upper left of the intersection of lines 1 and 3; the angle located in the upper left of the intersection of lines 2 and 3 is $\angle 5$, so $\angle 1$ and $\angle 5$ are corresponding angles.

Another example of corresponding angles is:

$\angle 4$ and $\angle 8$

Name the two other pairs of corresponding angles in the figure.

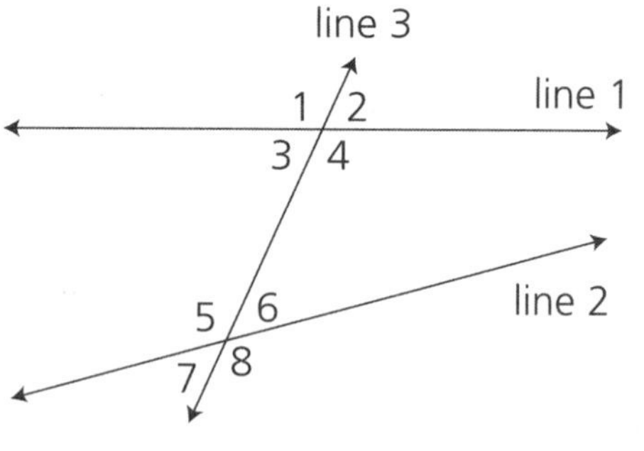

process: model

Angles Formed by Transversals

19. In the space below, create a pair of parallel lines cut by a transversal that does *not* form a 90° angle with either of the lines.

process: evaluate

Angles Formed by Transversals

20. Measure an angle where the transversal intersects each of the parallel lines. Then use these measures to calculate the measures of the remaining angles.

analysis

Angles Formed by Transversals

21. Compare your results with those of your group members. Based upon your group's results, what generalizations can you make about the relationships among angle measures when two *parallel* lines are cut by a transversal.

analysis

Angles Formed by Transversals

22. Do any of these relationships also hold true when two *non*-parallel lines are cut by a transversal? Explain.

23. Which of your group's generalizations are true only for *parallel* lines cut by a transversal?

    a. Why are those generalizations only true for parallel lines?

process: summary

Angles Formed by Transversals

24. Draw an example of a pair of vertical angles.

    a. Is it possible to draw only a single pair of vertical angles? Explain.

    b. When is a pair of vertical angles equal in measure? Explain.

process: summary

Angles Formed by Transversals

25. Draw a linear pair of angles.

    a. Is it possible to draw only a single linear pair of angles? Explain.

    b. When is a linear pair of angles equal in measure? Explain.

26. Draw a pair of supplementary angles.

    a. Is it possible to draw only a single pair of supplementary angles? Explain.

    b. When is a pair of supplementary angles equal in measure? Explain.

27. Draw a pair of alternate interior angles.

    a. Is it possible to draw only a single pair of alternate interior angles? Explain.

    b. When is a pair of alternate interior angles equal in measure? Explain.

process: summary

Angles Formed by Transversals

28. Draw a pair of alternate exterior angles.

    a. Is it possible to draw only a single pair of alternate exterior angles? Explain.

    b. When is a pair of alternate exterior angles equal in measure? Explain.

29. Draw a pair of interior angles on the same side of the transversal.

    a. Is it possible to draw only a single pair of same-side interior angles? Explain.

    b. When is a pair of same-side interior angles equal in measure? Explain.

30. Draw a pair of exterior angles on the same side of the transversal.

    a. Is it possible to draw only a single pair of same-side exterior angles? Explain.

    b. When is a pair of same-side exterior angles equal in measure? Explain.

process: summary

Angles Formed by Transversals

31. Draw a pair of corresponding angles.

    a. Is it possible to draw only a single pair of corresponding angles? Explain.

    b. When is a pair of corresponding angles equal in measure? Explain.

32. When a transversal intersects two parallel lines, which special pairs of angles must be congruent? List them.

33. When a transversal intersects two lines that are not parallel, which special pairs of angles must be congruent? List them.

process: summary

Angles Formed by Transversals

34. Harvey again misplaced his protractor after having measured only one angle in the figure. Complete the table with the missing measures of the angles and the reason why you know the measure of the angle.

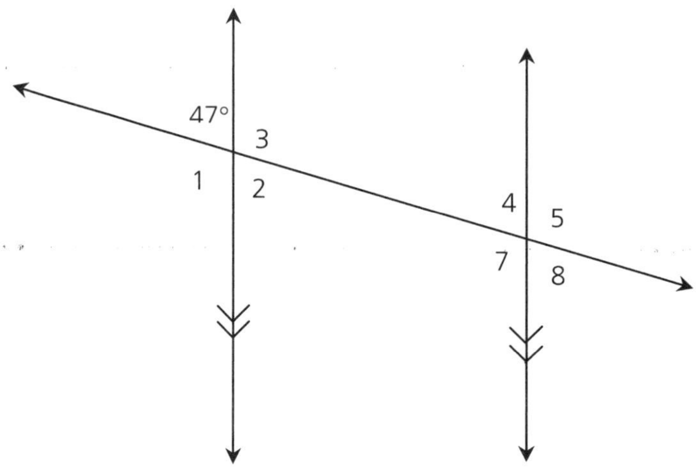

	Measure	Reason or Justification
∠1		
∠2		
∠3		
∠4		
∠5		
∠6		
∠7		

process: evaluate

Parallel Lines and Transversals

1. Measure ∠AEF. Record your measurement on the figure.

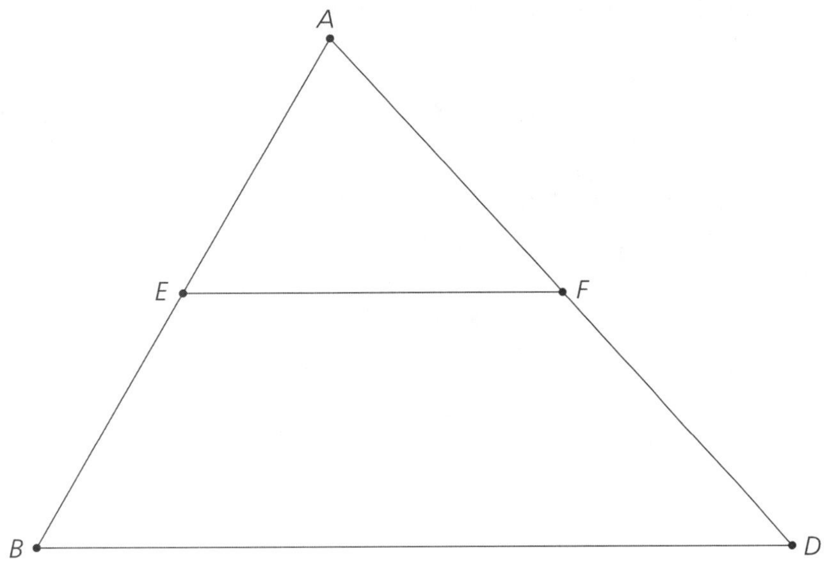

2. Given that $\overleftrightarrow{EF} \parallel \overleftrightarrow{BD}$ in the figure above, can you find any other angle measures without the use of your protractor? If so, record the measurement(s) on the figure and provide your justification here. If you cannot find any other angle measure, explain why.

3. Measure ∠ADB in the figure above. Record your measurement on the figure.

4. Use the triangles in the figure and the parallel lines to find the measures of the remaining angles in this figure. Provide a reason or explanation to describe how you found the missing angle measures.

	Measure	**Reason**
m∠ABD		
m∠AFE		
m∠A		

analysis

Parallel Lines and Transversals

5. Use a protractor to measure only ∠ABD in the following figure. Record your measurement on the figure.

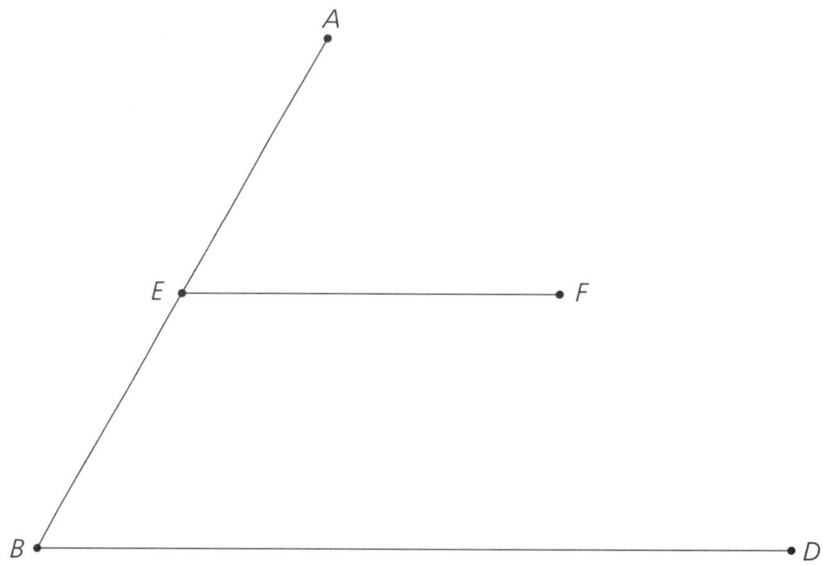

6. Is it possible to find the measure of ∠AEF (m∠AEF) in the figure above, without knowing whether $\overline{EF} \parallel \overline{BD}$? Why or why not?

7. If you are given that m∠AEF = m∠ABD in the figure on the previous page, what can you infer about $\overline{EF}$ and $\overline{BD}$? Explain your reasoning.

analysis

Parallel Lines and Transversals

8. Refer to the following figure to solve each part of this problem.

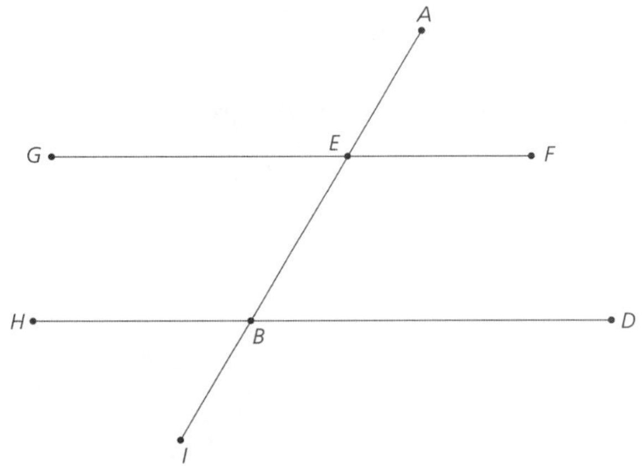

a. Knowing *only* that m∠AEG = m∠DBI, can you infer anything about the figure? If so, tell what you can infer and provide a justification. If not, tell why not.

b. Knowing *only* that m∠GEB + m∠EBH = 180°, can you infer anything about the figure? If so, tell what you can infer and provide a justification. If not, tell why not.

c. Knowing *only* that m∠FEB = m∠EBD, can you infer anything about the figure? If so, tell what you can infer and provide a justification. If not, tell why not.

d. Knowing *only* that m∠GEB = m∠EBD, can you infer anything about the figure? If so, tell what you can infer and provide a justification. If not, tell why not.

e. Knowing *only* that m∠AEF + m∠DBI = 180°, can you infer anything about the figure? If so, tell what you can infer and provide a justification. If not, why not?

Unit 4: Triangles and Quadrilaterals

Contents

preview

Properties of Kites

The figure below is called a *kite*.

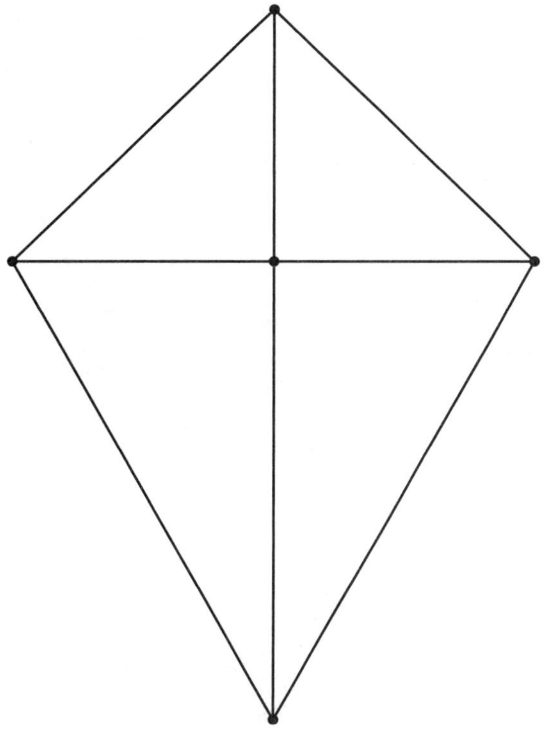

1. Label the kite in such a way that you will be able to refer to all of its parts.
2. Predict what you think may be true about the line segments, angles, and triangles formed in the kite.
3. List all of the properties you know to be true about the line segments, angles, and triangles formed in the kite. (Use the tools of your choice to complete this task.)

process: modeling

Names and Symbols for Triangles

To name a triangle or any other geometric shape, pick one vertex at which to start and name its vertices going clockwise or counterclockwise. For example, one possible name for this triangle is triangle ONE. When you name the three vertices, use capital letters, just as you do when naming any point.

1. What are two other possible names for this triangle, based upon this naming convention?

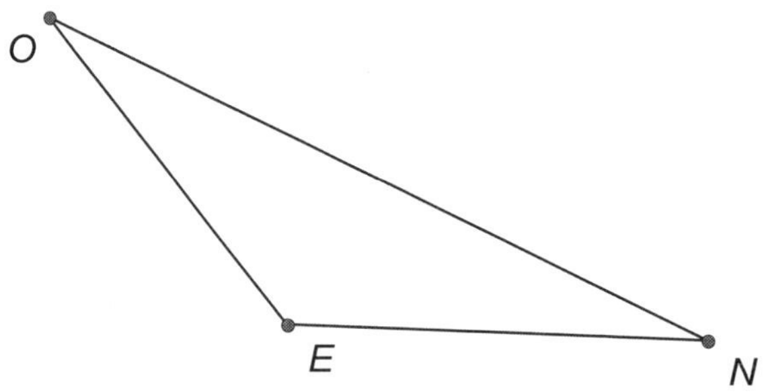

To symbolize triangle ONE, use a small triangle symbol (Δ) followed by the vertices. ΔONE means "triangle ONE."

2. Rewrite the three different names of the triangle using this notation.

process: practice

Names and Symbols for Triangles

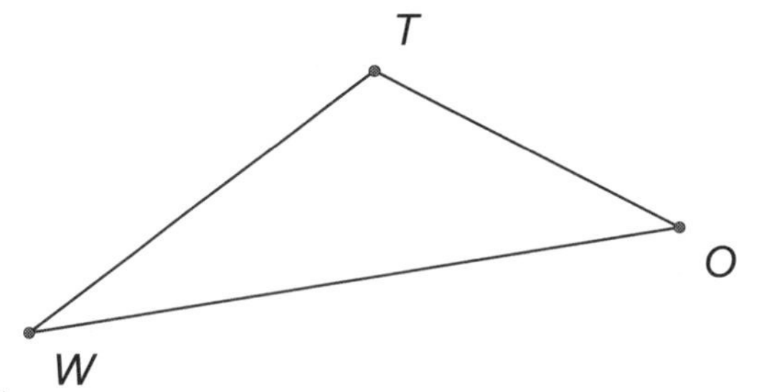

3. As in the above triangle, three points are needed to form a triangle. Name this triangle in three ways, using the shorthand notation you just learned.

Names and Symbols for Triangles

process: practice

4. Create a new triangle with vertices P, Q, and R in the space below.

   a. Name this triangle in three different ways, using the shorthand notation.

Names and Symbols for Triangles

analysis

5. Will any three points form a triangle? What must be true about the three points in order for a triangle to be formed?

Names and Symbols for Triangles

analysis

6. How many letters should we use to name a side of a triangle? Why?

   a. If you were asked to point to side T on the triangle, could you do so? Why or why not?

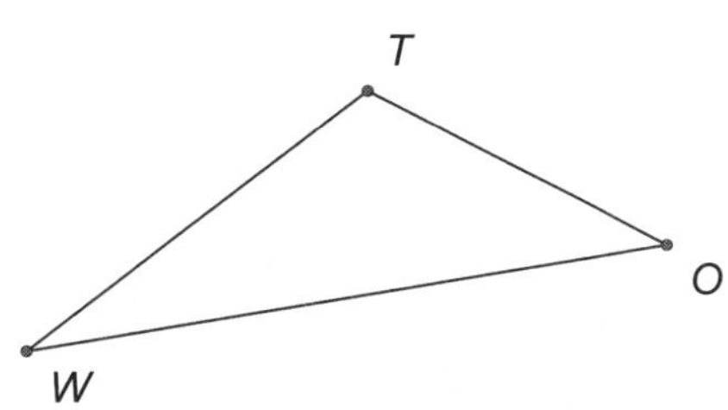

analysis

Names and Symbols for Triangles

b. If you were asked to point to side TW, could you do so? Why or why not?

c. If you were asked to point to side TWO, could you do so? Why or why not?

d. Name each side of your ΔPQR from the previous page.

7. How many letters should be used to name an angle in a triangle? Can you use one vertex? Can you use two vertices? What about three vertices? Why?

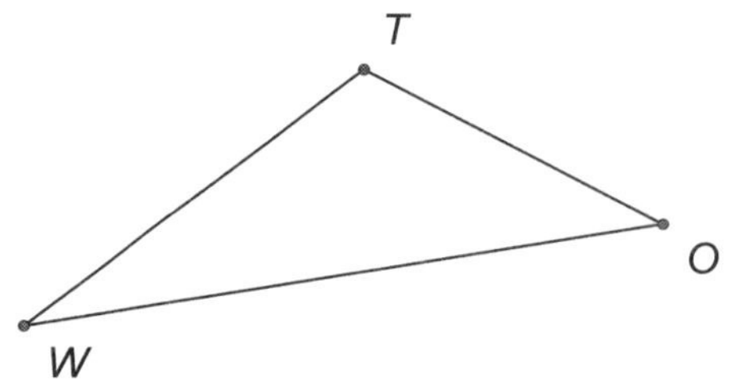

8. Name each angle in this triangle.

process: summary

Names and Symbols for Triangles

9. Name all six parts in this triangle.

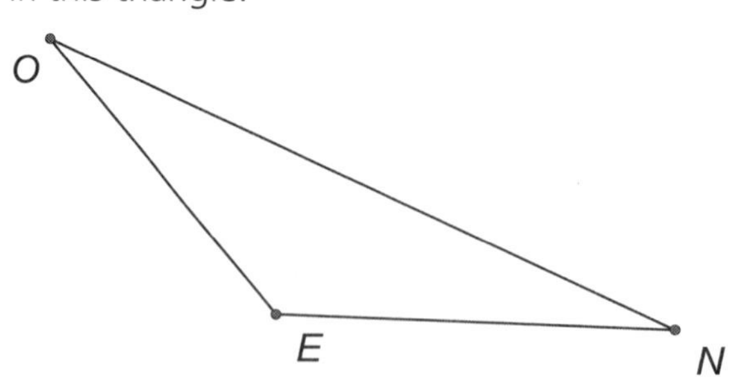

process: evaluate

Interior Angles of a Triangle

1. Measure the three angles of this triangle. Write the angle measure beside each angle. Also, write the word *acute*, *right*, *obtuse,* or *straight* beside each angle.

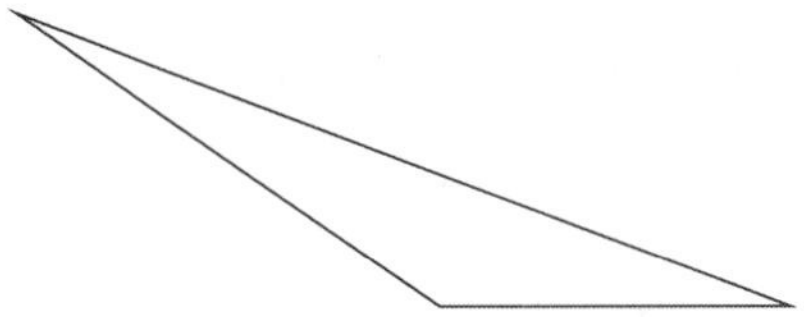

2. Measure the three interior angles of each triangle as accurately and as precisely as possible. Then find the sum of the three angle measures for each triangle.

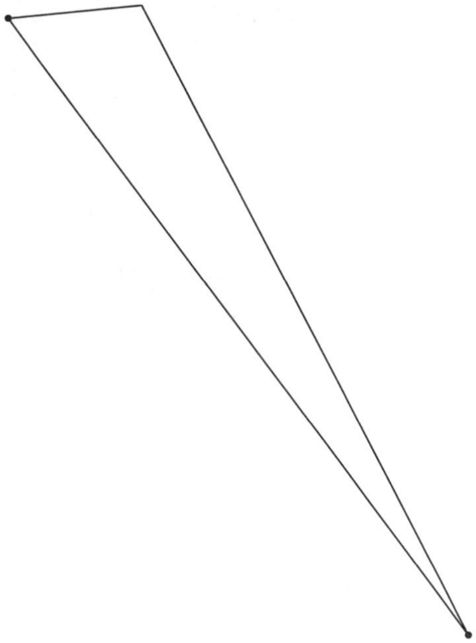

a.

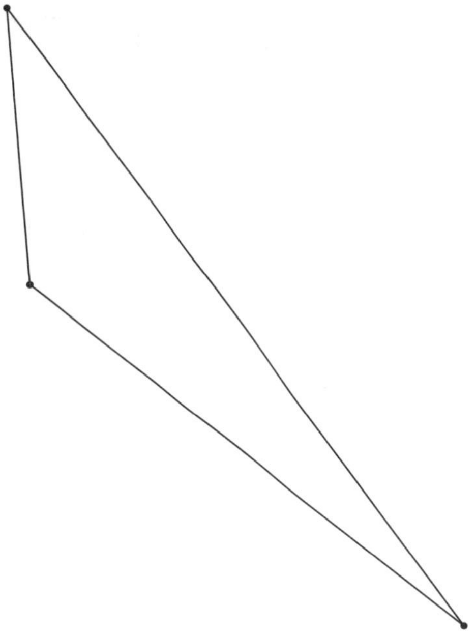

b.

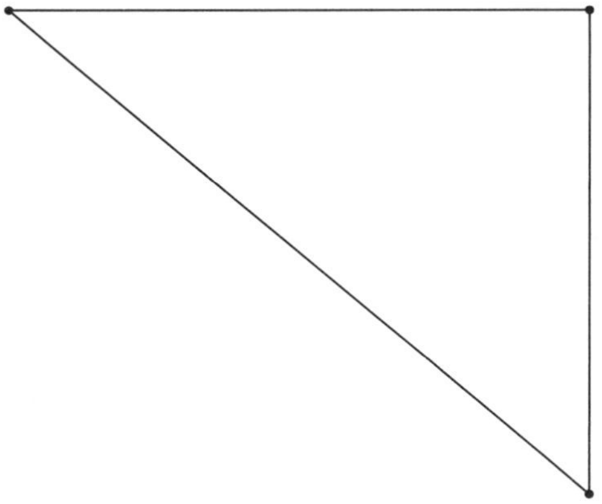

c.

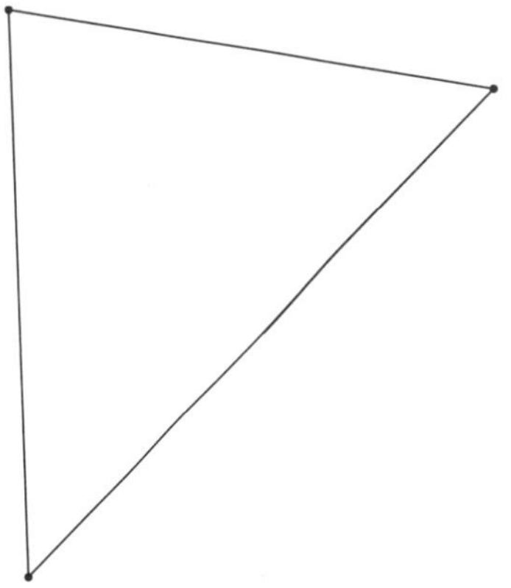

d.

process: model

Interior Angles of a Triangle

3. Summarize your findings from the four triangles in this table.

	First Angle	Second Angle	Third Angle	Sum
Triangle a				
Triangle b				
Triangle c				
Triangle d				

4. Use a straightedge to create a triangle on a separate sheet of paper.
5. Cut out the triangle.
6. Tear off the three corners of the triangle as shown in this sample.

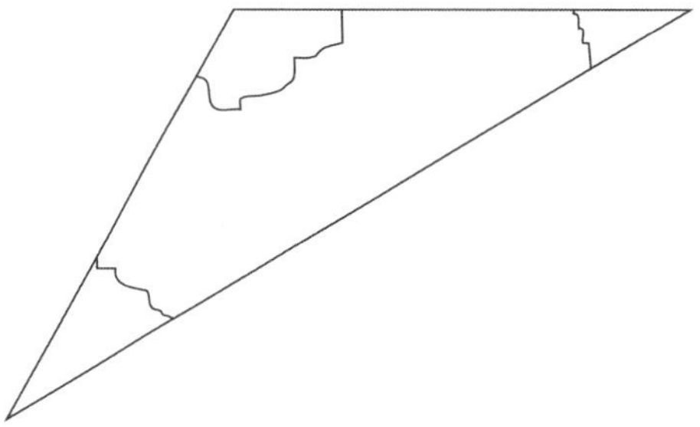

7. Now piece together the three vertices, like a puzzle. Place the angles next to each other so the right side of one angle coincides with the left side of the next angle. Tape the angles down in the space below.

8. Using what you have seen in questions 2 through 7, the sum of the measures of the three interior angles of any triangle is ____ ____ degrees.

process: model

Interior Angles of a Triangle

9. Is it possible to draw a triangle with *three* acute angles? Try it. Measure each angle and write the angle measurements on your drawing. If you find it's not possible, write complete sentences explaining why.

10. Is it possible to draw a triangle with *exactly two* acute angles? Try it. Measure each angle and write the angle measurements on your drawing. If you find it's not possible, write complete sentences explaining why.

11. Is it possible to draw a triangle with *exactly one* acute angle? Try it. Measure each angle and write the angle measurements on your drawing. If you find it's not possible, write complete sentences explaining why.

process: model

Interior Angles of a Triangle

12. Is it possible to draw a triangle with *exactly one* right angle? Try it. Measure each angle and write the angle measurements on your drawing. If you find it's not possible, write complete sentences explaining why.

13. Is it possible to draw a triangle with *exactly two* right angles? Try it. Measure each angle and write the angle measurements on your drawing. If you find it's not possible, write complete sentences explaining why.

14. Is it possible to draw a triangle with *exactly one* obtuse angle? Try it. Measure each angle and write the angle measurements on your drawing. If you find it's not possible, write complete sentences explaining why.

process: model

Interior Angles of a Triangle

15. Is it possible to draw a triangle with *exactly two* obtuse angles? Try it. Measure each angle and write the angle measurements on your drawing. If you find it's not possible, write complete sentences explaining why.

16. Is it possible to draw a triangle with *one right angle* and *one obtuse angle*? Try it. Measure each angle and write the angle measurements on your drawing. If you find it's not possible, write complete sentences explaining why.

process: summary

Interior Angles of a Triangle

17. Summarize your findings in questions 9 through 16.

Interior Angles of a Triangle

analysis

18. Explain how the sum of the three angles in a triangle could have helped you determine which triangles were and which triangles were not possible to create in questions 8 through 15.

input

Classifying Triangles by the Measures of Their Angles

As you learned in the previous section, certain angle combinations are possible in triangles. For example, creating a triangle with one obtuse angle and two acute angles is possible. Certain other angle combinations are not possible. For example, it is impossible to create a triangle with one obtuse, one right, and one acute angle.

Triangles are named according to the combination of angles within the triangle.

process: model

Classifying Triangles by the Measures of Their Angles

1. A triangle with three acute angles is called an **acute triangle.** Use the tools of your choice (protractor, ruler, compass, etc.) to draw an acute triangle. Write complete sentences explaining why it is an acute triangle.

2. A triangle with one right angle is called a **right triangle**. Use the tools of your choice to draw a right triangle. How will you validate that the triangle you create is actually a right triangle?

	Classifying Triangles by the Measures of Their Angles
analysis	3. How many non-right angles are in a right triangle, and what type of angles are these non-right angles?

	Classifying Triangles by the Measures of Their Angles
process: evaluate	4. Measure the non-right angles and record their measures on the triangle. 5. What is the sum of the measures of these two angles?

	Classifying Triangles by the Measures of Their Angles
analysis	6. Will the sum for the non-right angles in a right triangle always be the same? Explain.

process: model

Classifying Triangles by the Measures of Their Angles

7. A triangle with one obtuse angle is called an **obtuse triangle**. Use the tools of your choice to draw an obtuse triangle. How will you validate or prove that the triangle is an obtuse triangle?

analysis

Classifying Triangles by the Measures of Their Angles

8. How many non-obtuse angles are there in your triangle, and what type of angle(s) are the non-obtuse angles?

process: evaluate

Classifying Triangles by the Measures of Their Angles

9. Measure the non-obtuse angles and record their measures on the triangle.

10. What is the sum of the measures of these angles?

	Classifying Triangles by the Measures of Their Angles
analysis	11. Will the sum for the non-obtuse angles in an obtuse triangle always be the same? Explain.

	Classifying Triangles by the Measures of Their Angles
process: model	12. A triangle with three angles of equal measure is called an **equiangular triangle**. Use the tools of your choice to draw an equiangular triangle. Explain how you can verify or prove that the triangle you create is equiangular.

	Classifying Triangles by the Measures of Their Angles
process: evaluate	13. What is the sum of the measure of the angles in any triangle? 14. If the three angles are equal, what should be the measure of each angle in an equiangular triangle?

process: summary

Classifying Triangles by the Measures of Their Angles

15. Name the four ways in which triangles can be classified according to their angle measures.

    a. Sketch and label a triangle that fits each of those four classifications.

    b. For each type of triangle sketched above, provide a possible set of three angle measures.

process: summary

Classifying Triangles by the Lengths of Their Sides

In the previous section, you learned to classify triangles by their angle measures.

1. Name the four ways in which triangles can be classified according to their angles.

Triangles can also be classified according to the lengths of their sides.

process: model

Classifying Triangles by the Lengths of Their Sides

2. A triangle with three sides of equal length is called an **equilateral triangle**. Use the tools of your choice to draw an equilateral triangle. How can you verify or prove that the triangle you draw is actually an equilateral triangle?

process: evaluate

Classifying Triangles by the Lengths of Their Sides

3. Measure the three angles in your equilateral triangle. Record the measurements on the triangle.

4. Compare the measures of your angles with those of your group members.

analysis	**Classifying Triangles by the Lengths of Their Sides**
	5. What generalizations can you make about equilateral triangles?

process: model	**Classifying Triangles by the Lengths of Their Sides**
	6. A triangle with two or more sides of equal length is called an **isosceles triangle**. Use the tools of your choice to draw an isosceles triangle that is not also an equilateral triangle. How can you verify or prove that the triangle you draw is actually an isosceles triangle?

process: evaluate	**Classifying Triangles by the Lengths of Their Sides**
	7. Measure the three angles in your isosceles triangle. Record the measurements on the triangle.
	8. Compare the measures of your angles with those of your group members.

analysis

Classifying Triangles by the Lengths of Their Sides

9. What generalizations can you make about isosceles triangles?

process: model

Classifying Triangles by the Lengths of Their Sides

10. A triangle with three sides of unequal length is called a **scalene triangle**. Use the tools of your choice to draw a scalene triangle. Write complete sentences explaining why it is a scalene triangle.

analysis

Classifying Triangles by the Lengths of Their Sides

11. Measure the three angles in your scalene triangle. Record the measurements on the triangle.

12. Compare the measures of your angles with those of your group members.

analysis

Classifying Triangles by the Lengths of Their Sides

13. Based upon your group's results, can you form any generalizations about scalene triangles?

14. Using the definitions given in questions 2 and 3, are *all* equilateral triangles also isosceles triangles? Explain why or why not.

15. Using the definitions given in questions 2 and 3, are *all* isosceles triangles also equilateral triangles? Explain why or why not.

16. Is it possible to draw an *isosceles right triangle*? Try it. Explain how you know it is an isosceles right triangle, or why you cannot draw one.

Classifying Triangles by the Lengths of Their Sides

analysis

17. Is it possible to draw an *isosceles obtuse triangle*? Try it. Explain how you did it or why it cannot be done.

18. Is it possible to draw a *scalene obtuse triangle*? Try it. Explain how you did it or why it cannot be done.

19. Is it possible to draw a *scalene right triangle*? Try it. Explain how you did it or why it cannot be done.

20. Is it possible to draw a *scalene isosceles triangle*? Try it. Explain how you did it or why it cannot be done.

analysis

Classifying Triangles by the Lengths of Their Sides

21. Is it possible to draw a *scalene acute triangle*? Try it. Explain how you did it or why it cannot be done.

22. Is it possible to draw a triangle that is *equilateral* and not *equiangular*? Try it. Explain how you did it or why it cannot be done.

process: summary

Classifying Triangles by the Lengths of Their Sides

23. Summarize your findings in questions 16 through 22.

process: evaluate

The Triangle Inequality

To begin this activity, you need a piece of strand pasta (like linguine). Break the pasta at two random points to make three pieces from the one strand.

Try to form a triangle from these three pieces with the pieces touching end to end.

Collect the class' results for attempting to form a triangle from the three broken pieces.

1. How many students are in your class?

2. How many students' pasta pieces formed a triangle when the three pieces were connected end to end?

3. How many students' pasta pieces did not form a triangle when the three pieces were connected end to end?

4. From the results you just gathered, show your work and calculate the estimated probability that a piece of pasta, when broken into three pieces, will form a triangle.

process: model

The Triangle Inequality

5. Measure each of your pieces of pasta and record those measurements below.

process: model

The Triangle Inequality

6. Gather the measurements of the others in your class along with whether their pieces formed a triangle.

Piece 1 in cm	Piece 2 in cm	Piece 3 in cm	Triangle?

analysis

The Triangle Inequality

7. Examine the measurements of the pieces that did form a triangle compared to the measurements of the pasta pieces that did not form a triangle. What observations can you make?

   a. Under what conditions can a triangle always be formed?

   b. Under what conditions can a triangle never be formed?

8. Based upon your observations, would the following measurements create a triangle? Explain your response to each.

   a. 2cm, 5.1 cm, 2.4 cm

   b. 9.2 cm, 7 cm, 1.9 cm

process: model

The Triangle Inequality

Next, we will simulate the activity just performed using graphics calculators, so that you can gather more data in a short amount of time to estimate the probability of forming a triangle from pasta pieces. Gathering more data will add strength to your argument about when a triangle can or cannot be formed.

For this activity we will assume that the length of a piece of pasta is 100 cm. Using your calculator, generate two random numbers between 0 and 100. Use each of these random numbers to represent the measurements of two of the broken pieces of the 100-cm -long piece of pasta.

9. How will you determine the length of the third piece of pasta?

10. Record your measurements in the following chart. Determine if your pasta will form a triangle.

Trial number	Piece 1	Piece 2	Piece 3	Triangle?
1.				
2.				
3.				
4.				
5.				
6.				
7.				
8.				
9.				
10.				
11.				
12.				

process: model

The Triangle Inequality

13.				
14.				
15.				
16.				
17.				
18.				
19.				
20.				

11. Use the results from your chart to estimate the probability of creating a triangle. Show your work and calculate the estimated probability that a triangle will be formed.

12. Gather the class results for triangles that could be formed. How many students are in your class? How many trials would that be altogether?

13. How many triangles could be formed?

14. Show your work and calculate the estimated probability of forming a triangle based upon the class' data?

The Triangle Inequality

analysis

15. Does the class' result from the random numbers in the calculator differ from the class' result when you physically broke the pasta? How might you account for any differences you have identified?

16. Does the class' result for the random numbers differ from your individual result for the probability using the random numbers?

17. How might you account for any differences you see? Are the differences you noted in question 16 similar to, greater than, or less than the differences you saw when comparing to the class data for actually breaking the pasta (question 15)?

18. Re-examine your conjecture about when a triangle could or could not be formed. Do your results here agree with or contradict that conjecture?

19. If your results agree, explain why your conjecture works. If your results disagree, reformulate your conjecture below.

process: evaluate

The Triangle Inequality

The probability that the pieces of pasta will form a triangle can also be found by using a combination of some algebra and geometry. Think about the 100-cm-long piece of pasta from the last part of this activity. Say that the first piece of pasta is x cm long. The second piece of pasta is y cm long.

20. How long is the third piece of pasta? If you are having trouble answering this, think about some specific examples (like 36 cm and 42 cm for the lengths of the first two pieces) before you try to generalize with x and y.

analysis

The Triangle Inequality

21. Why does x+y have to be less than 100 in our situation?

22. Because of this fact, what possible values can x take on?

23. What possible values can y take on? (Make sure to account for the x-cm-long piece that already exists.)

24. Now explain why each of the following must be true according to your conjecture about how two of the lengths must compare to the third length in order to form a triangle.

    a. $y > 100 - (x + y)$

    b. $[100 - (x + y)] > y$

    c. $[100 - (x + y)] > x$

The Triangle Inequality

process: evaluate

25. Simplify each of these three inequalities.

    a. $x + y > 100 - (x + y)$

    b. $x + [100 - (x + y)] > y$

    c. $y + [100 - (x + y)] > x$

26. Graph each of the simplified inequalities on the grid below. (Pay attention to your scaling and label each axis appropriately.)

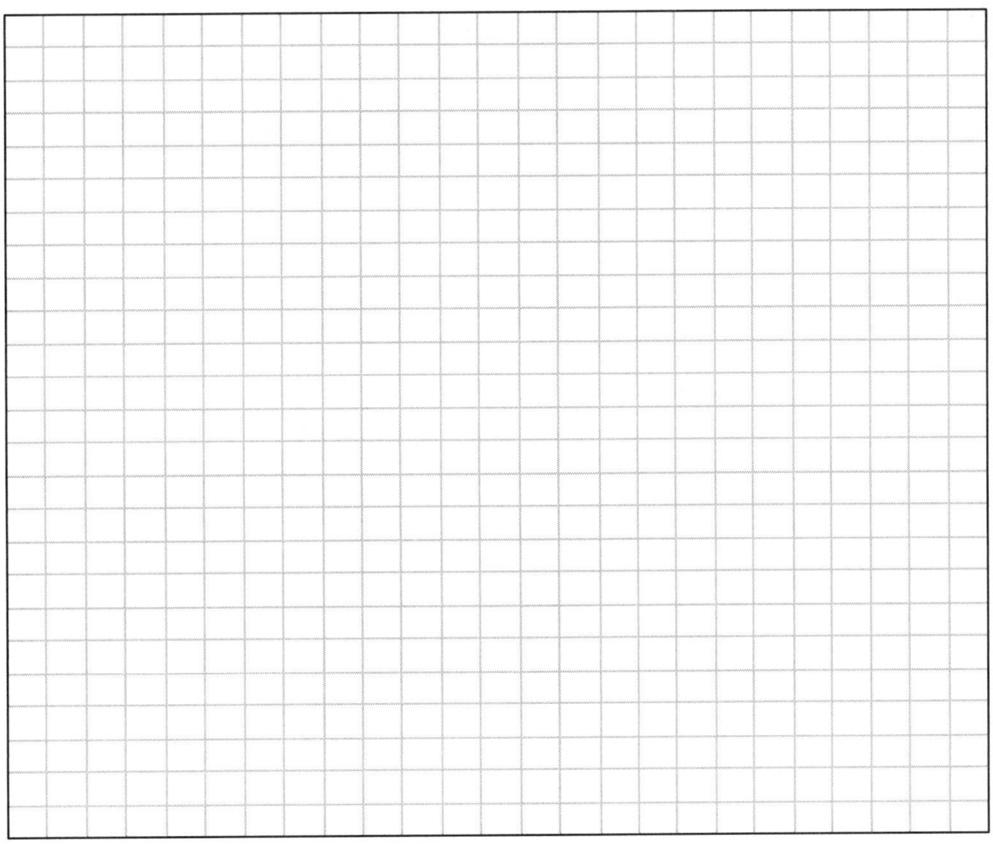

process: evaluate

The Triangle Inequality

27. Shade the area of the graph satisfying the system of three inequalities.

28. Calculate the area of the shaded region, which satisfies the system of inequalities.

29. Use this calculation to find the probability of forming a triangle.

$$P(\text{forming a triangle}) = \frac{\text{area of the shaded region}}{\text{area of the region for the entire domain}}$$

input

Exterior Angles

Triangles and other polygons have interior angles and exterior angles associated with them. As you may suspect, **interior angles** in geometric figures are enclosed within the figure like so:

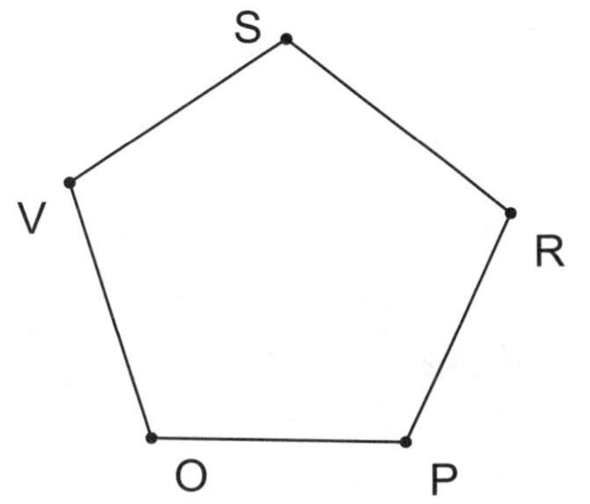

Each of the angles within this pentagon is an interior angle.

Exterior angles are the angles that appear on the outside of the polygon when you extend the side of the polygon in one direction. Three exterior angles of this triangle are marked *E*.

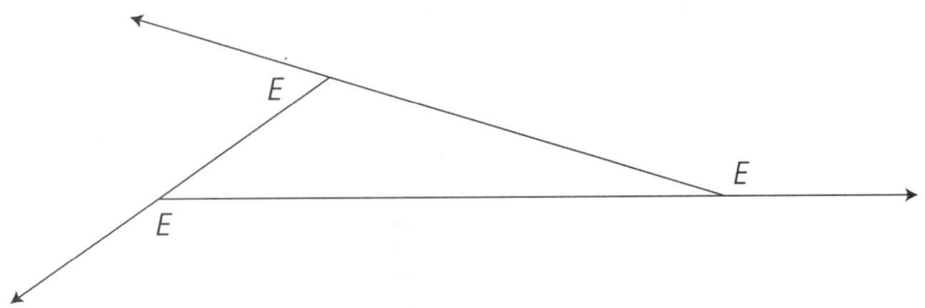

process: practice

Exterior Angles

1. Shade the exterior angle or angles shown in each figure.

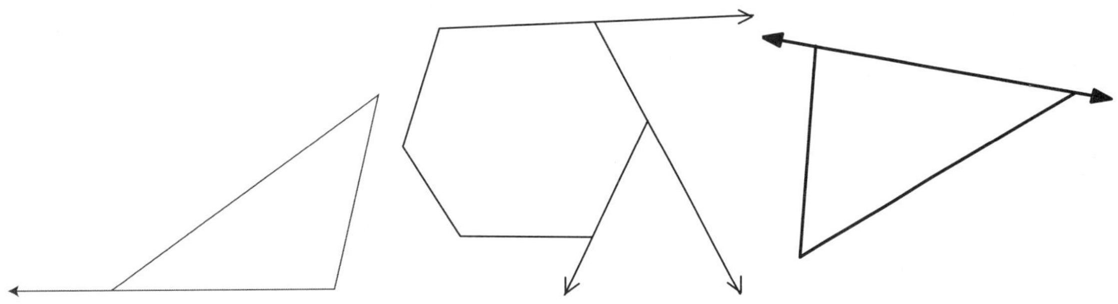

Exterior Angles

analysis

2. One of your group members is absent today. When he returns, it will be your job to explain to him how to locate exterior angles in a polygon. Write your explanation in complete sentences.

3. If you join an exterior angle with the interior angle that is next to it, what type of angle pair do you have?

4. What is true about the sum of the measures of an exterior angle and its adjacent interior angle?

    a. Use the figure below as a guide to write a mathematical statement expressing what you just wrote in words.

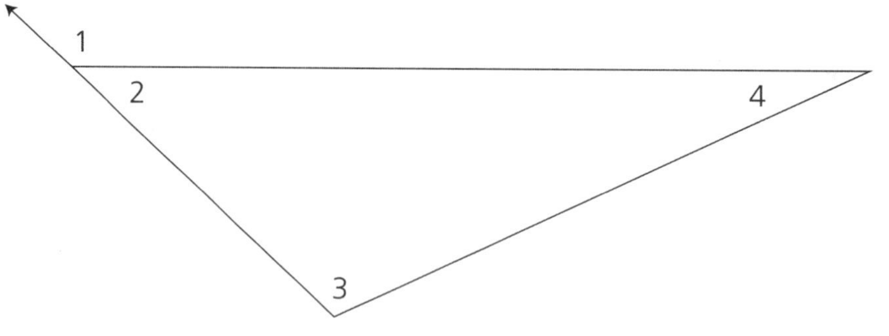

Exterior Angles

analysis

5. What is true about the sum of the measure of the three interior angles of a triangle?

   a. Using the figure on the previous page as a guide, write a mathematical statement to express your answer to question 5.

6. What do you notice about the sum of $\angle 1$ and $\angle 2$ as compared to the sum of $\angle 3$, $\angle 2$, and $\angle 4$?

   a. Write a mathematical equation to express this fact.

7. Which angle is common to both sides of your equation in question 6a?

   a. Subtract this common angle from both sides of your equation to form a new equation. Write this new equation below.

8. When you draw an exterior angle of a triangle, the two interior angles that do not share a vertex with the exterior angle are called **remote interior angles**. Name the two remote interior angles in the diagram.

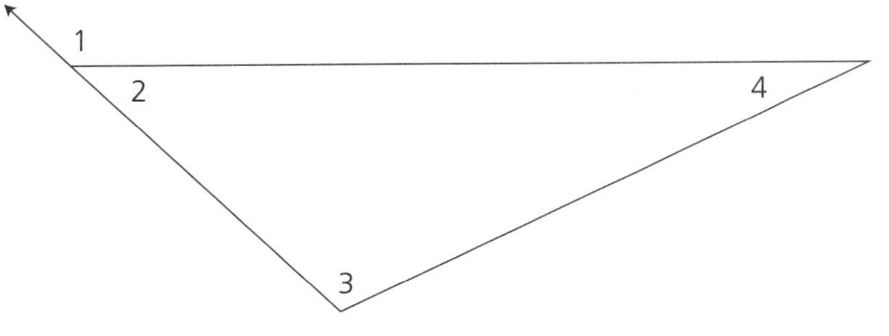

analysis

Exterior Angles

9. How is the sum of the measure of these two remote interior angles related to the measure of the exterior angle?

process: practice

Exterior Angles

10. Examine the figure below.

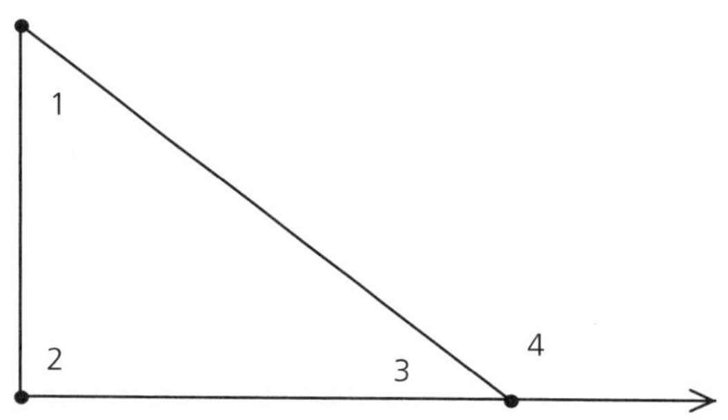

a. Name the exterior angle in this figure.

b. Name the remote interior angles.

c. Will the sum of the measures of the remote interior angles equal the measure of the exterior angle in this triangle? Explain.

process: practice

Exterior Angles

11. Create your own triangle with one exterior angle shown. Will the sum of the measures of the two remote interior angles equal the measure of the exterior angle in your triangle?

analysis

Exterior Angles

12. Will the sum of the measure of the remote interior angles always equal the measure of the exterior angle in a triangle? Explain.

process: practice

Exterior Angles

13. Use a protractor to find the sum of the measures of the exterior angles of each of these triangles. Write each angle measure by the angle.

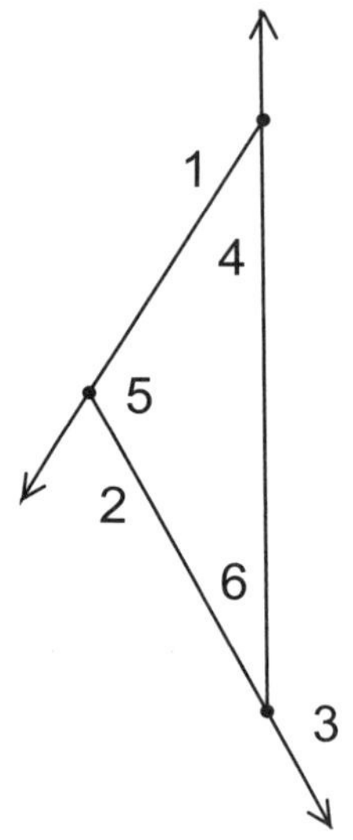

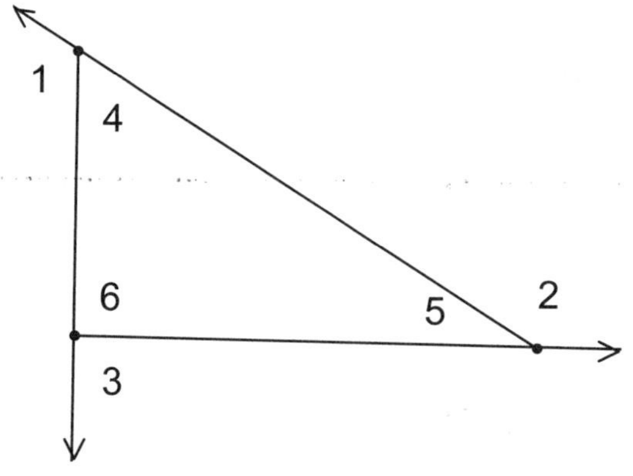

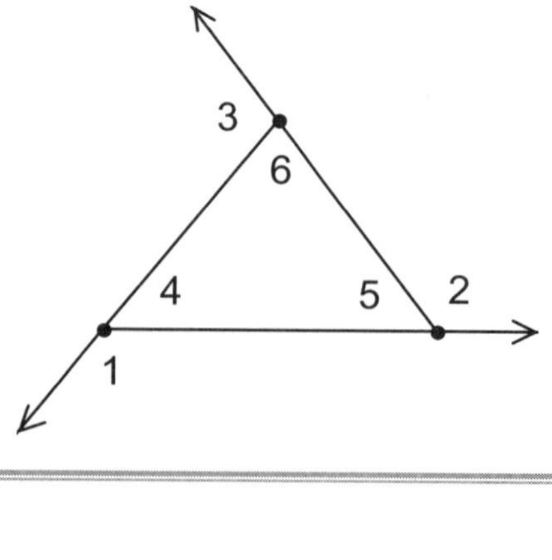

analysis

Exterior Angles

14. In each of the three preceding triangles, what is:

    $m\angle 1 + m\angle 4$

    $m\angle 2 + m\angle 5$

    $m\angle 3 + m\angle 6$

    How do you know?

15. In each of the three preceding triangles, what is:

    $(m\angle 1 + m\angle 4) + (m\angle 2 + m\angle 5) + (m\angle 3 + m\angle 6)$?

    How do you know?

16. In each of the three preceding triangles, what is:

    $m\angle 4 + m\angle 5 + m\angle 6$?

17. Based upon your answers to question 15 and 16 above, what is: $m\angle 1 + m\angle 2 + m\angle 3$? Explain your reasoning.

process: summary

Exterior Angles

18. Summarize your understanding of exterior angles.

input

Bisectors

In the word *bisect,* both the prefix "bi" and the root "sect" are derived from Latin. In Latin, the prefix "bi" means *two*, while the root "sect" means *to cut*.

analysis

Bisectors

1. Using these Latin meanings, create a definition for the English term, **bisect.**

input

Bisectors

To **bisect an angle** means to divide or cut the angle into two angles of equal measure.

An **angle bisector** is a line, ray, or segment that bisects an angle.

process: evaluate

Bisectors

2. If you bisect a 90° angle, what is the measure of each of the two resulting angles?

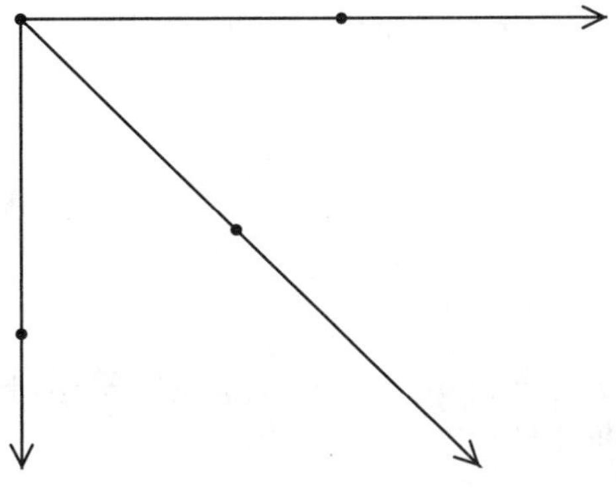

3. If you bisect a 45° angle, what is the measure of each of the two resulting angles?

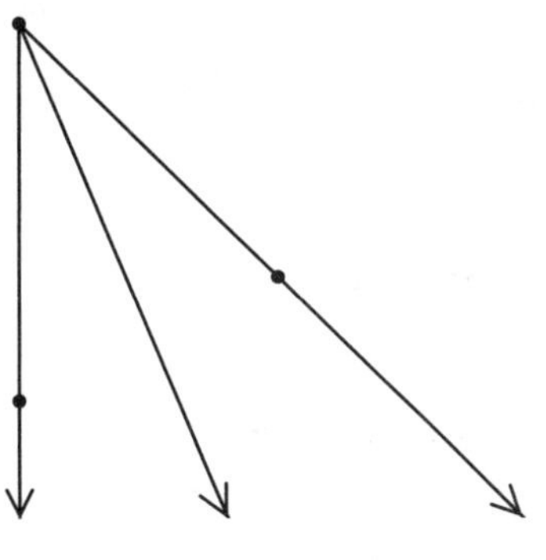

process: evaluate

Bisectors

4. If you bisect a 22.5° angle, what is the result? Using your protractor, create an accurate diagram to illustrate this.

5. If you bisect a straight angle, what is the result? Using your protractor, create an accurate diagram to illustrate this.

process: model

Bisectors

6. Use your protractor to bisect the three interior angles of this triangle.

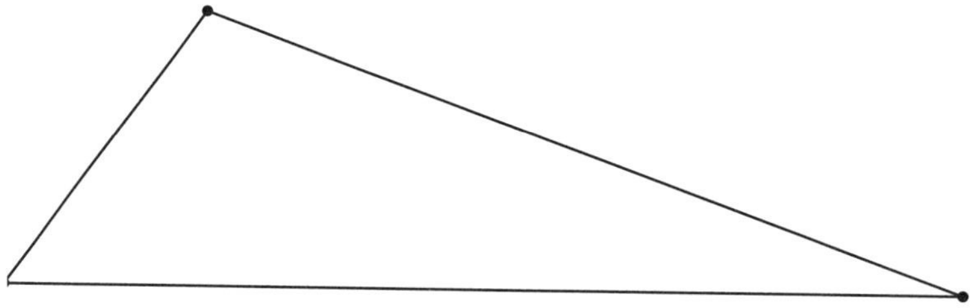

7. Draw a different triangle and use your protractor to bisect the three interior angles.

analysis

Bisectors

8. Can you think of an angle that cannot be bisected? Can all angles be bisected? Why or why not?

9. Based upon what you know about angle bisectors, what does it mean to **bisect a segment**?

process: evaluate

Bisectors

10. If you bisect a 6-cm line segment, what are the lengths of the smaller segments?

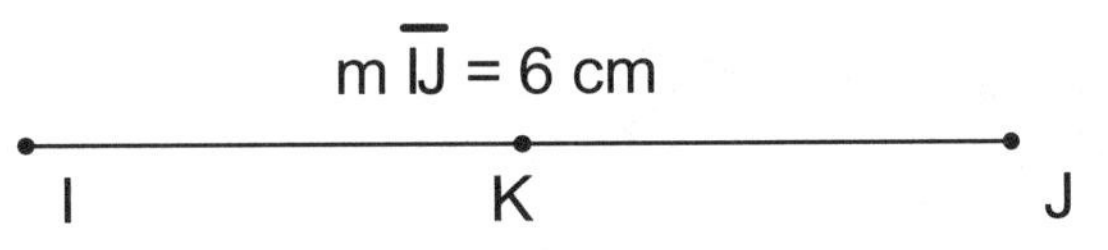

In other words, if IJ = 6 cm, and this segment is bisected at K, then IK = _ _,and KJ = _ _.

11. If you bisect a 9-inch line segment, what are the lengths of the smaller segments?

12. If you bisect an 11-cm line segment, what are the lengths of the smaller segments?

13. If you bisect an 8.5-inch line segment, what are the lengths of the smaller segments?

Bisectors

analysis

14. If you bisect a line segment of length x units, what is the length of each of the smaller segments?

15. Can you think of a line segment that cannot be bisected? Can all line segments be bisected? Explain why or why not.

16. Is it possible to bisect a line? Why or why not?

17. Is it possible to bisect a ray? Why or why not?

Bisectors

process: summary

18. Refer to this figure to solve each part of this problem.

   a. Draw a bisector of segment *ES*. Use *M* to label the point where the segment has been bisected.

   b. Is the measure of $\overline{EM}$ equal to the measure of $\overline{SM}$? Explain why or why not.

   c. Is the measure of $\overline{EM}$ plus the measure of $\overline{SM}$ equal to the measure of $\overline{ES}$? Explain why or why not.

input

Bisectors

If $EM = SM$ and $EM + SM = ES$, then point M is called the **midpoint** of $\overline{ES}$.

In other words, the **midpoint** of a segment is a point on the segment, which is the same distance from each of the endpoints of the segment.

extension

Bisectors

19. Why isn't it enough to know that EM + SM = ES to call M the midpoint of segment ES?

20. Why isn't it enough to know that EM = SM to call M the midpoint of segment ES?

In triangle PAB, M is the midpoint of $\overline{AB}$. That means $\overline{AB}$ was bisected. When the midpoint of a side of a triangle is connected to the opposite vertex, that segment is called a **median** of the triangle. $\overline{MP}$ is a median of ΔPAB.

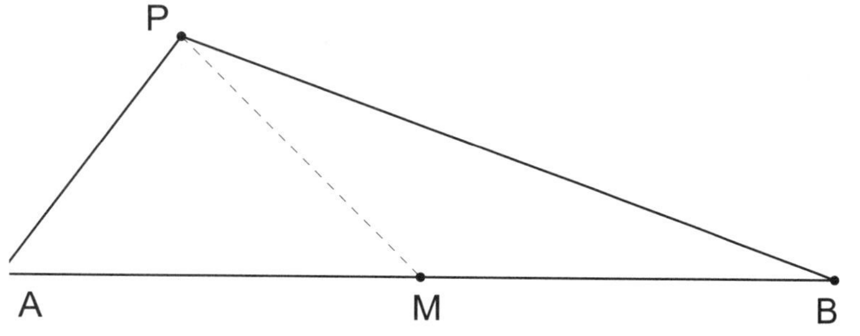

21. How many different medians can you draw in triangle PAB? Draw and label them. Use markers on segments that have the same length.

process: model

Parts of a Quadrilateral

1. Using a straightedge, draw a quadrilateral.

   a. How many parts does it have?

   b. Label the quadrilateral and name each part.

2. Use complete sentences to answer questions about the quadrilateral below.

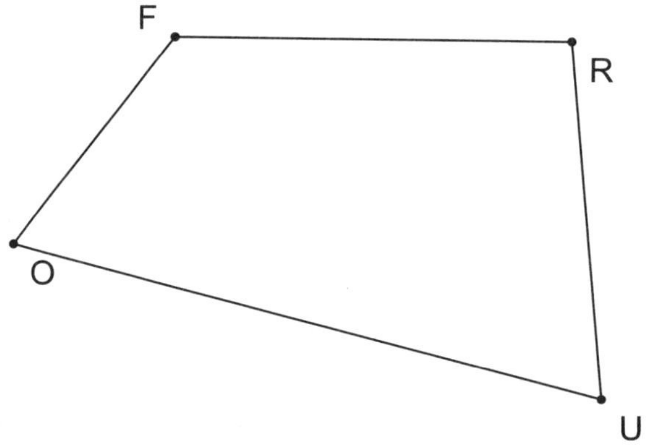

   a. Name the quadrilateral and explain how you chose the name.

   b. Name all the sides of the quadrilateral.

   c. Name all the angles of the quadrilateral.

analysis

Parts of a Quadrilateral

3. Use complete sentences to describe the differences and similarities between naming the sides of a quadrilateral and naming the sides of a triangle.

4. Describe the difference and the similarity between naming the angles of a quadrilateral and naming the angles of a triangle.

5. Describe the difference and the similarity between naming a quadrilateral and naming a triangle.

process: summary

Special Quadrilaterals

1. Many quadrilaterals have special names implying special properties. List any special quadrilaterals with which you are familiar.

analysis

Special Quadrilaterals

2. Circle the most specific special name that applies to this figure.

   a. Square
   b. Rectangle
   c. Quadrilateral
   d. Parallelogram
   e. Rhombus
   f. Trapezoid

3. Circle the most specific special name that applies to this figure.

   a. Square
   b. Rectangle
   c. Quadrilateral
   d. Parallelogram
   e. Rhombus
   f. Trapezoid

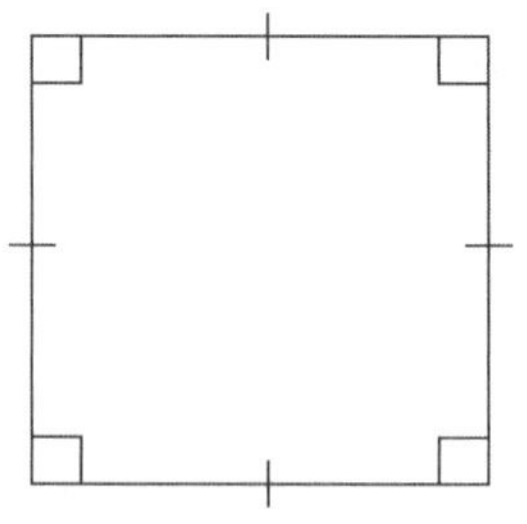

analysis

Special Quadrilaterals

4. Circle the most appropriate special name that applies to this figure.

    a. Square
    b. Rectangle
    c. Quadrilateral
    d. Parallelogram
    e. Rhombus
    f. Trapezoid

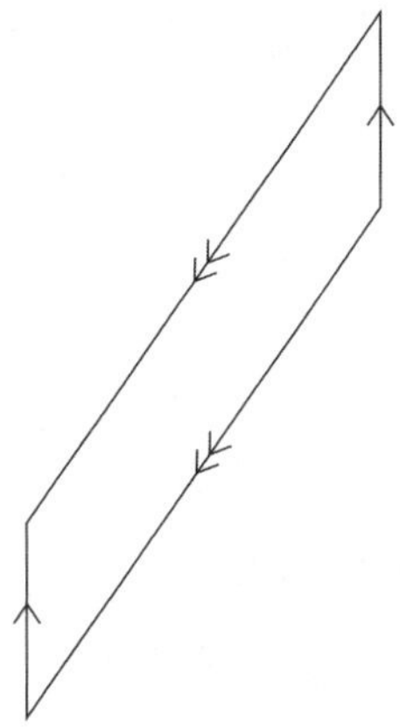

5. Circle all the possible names that apply to this figure. (There may be more than one answer.)

    a. Square
    b. Rectangle
    c. Quadrilateral
    d. Parallelogram
    e. Rhombus
    f. Trapezoid

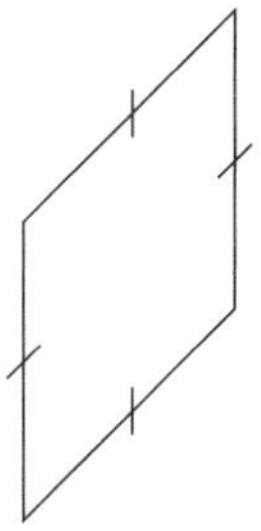

6. Circle all the possible names that apply to this figure. (There may be more than one answer.)

    a. Square
    b. Rectangle
    c. Quadrilateral
    d. Parallelogram
    e. Rhombus
    f. Trapezoid

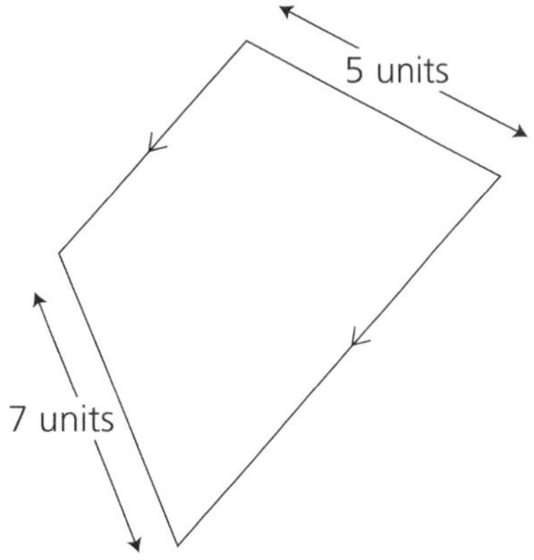

process: model

Special Quadrilaterals

7. Based on the work you did in questions 2 through 6, write definitions for the geometric figures below. Make a small sketch beside each definition.

   a. Quadrilateral

   b. Parallelogram

   c. Trapezoid

   d. Rhombus

   e. Rectangle

   f. Square

analysis

Special Quadrilaterals

8. Is *KING* a quadrilateral? Why or why not?

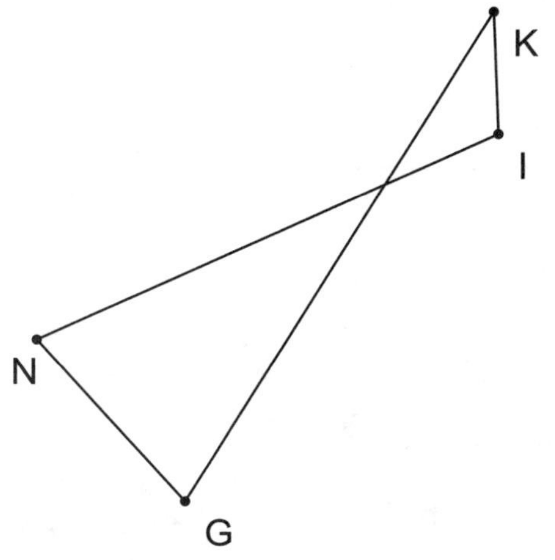

9. Is polygon *WARP* a quadrilateral? Why or why not?

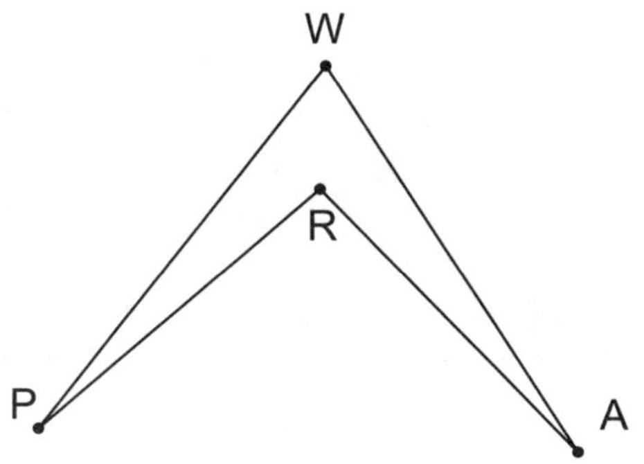

input

Parallelograms

A quadrilateral with two pairs of parallel sides is called a **parallelogram**. Refer to this parallelogram to answer the questions below.

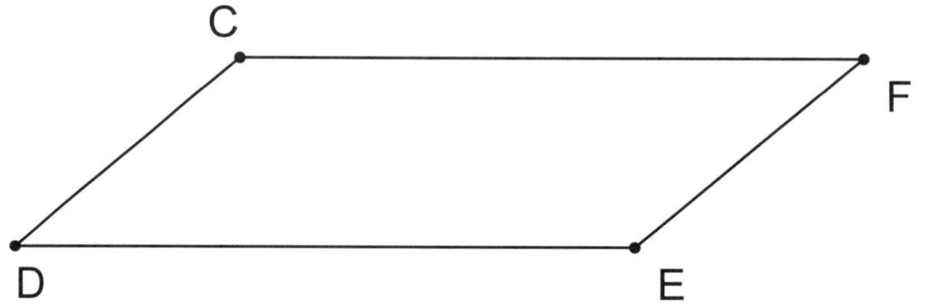

process: model

Parallelograms

1. Two sides of a parallelogram that do not intersect are called a pair of **opposite sides**. Name two pairs of opposite sides.

2. Two sides of a parallelogram that do intersect are called a pair of **consecutive sides**. Name four pairs of consecutive sides.

3. Two angles of a parallelogram that share a common side are called a pair of **consecutive angles**. Name four pairs of consecutive angles.

process: model

Parallelograms

4. Two angles of a parallelogram that do not share a common side are called a pair of **opposite angles**. Name two pairs of opposite angles.

analysis

Parallelograms

5. Predict the measures of and relationships between the pairs of sides and pairs of angles mentioned in questions 1 through 4. Which pairs are congruent? Which pairs are supplementary? Which pairs are parallel?

process: evaluate

Parallelograms

6. Measure each side and each interior angle of the parallelogram. Record your measurements on the figure.

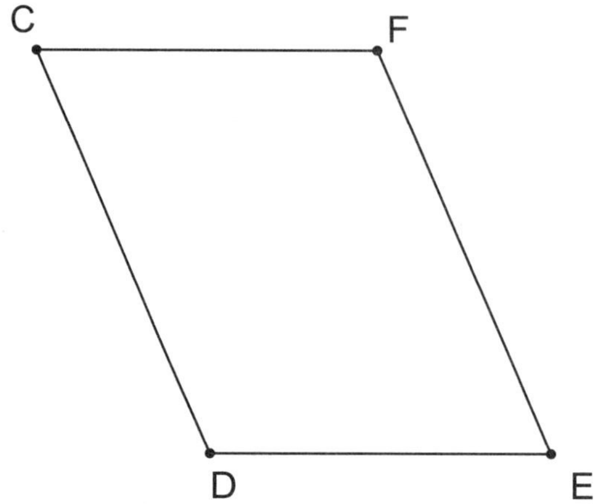

analysis

Parallelograms

7. Which of your predictions from question 5 are supported by your measurements?

8. Which of your predictions from question 5 are not supported by your measurements?

process: evaluate

Parallelograms

Remember that a diagonal is a line segment that connects a pair of opposite vertices in a quadrilateral. The parallelogram below contains one diagonal.

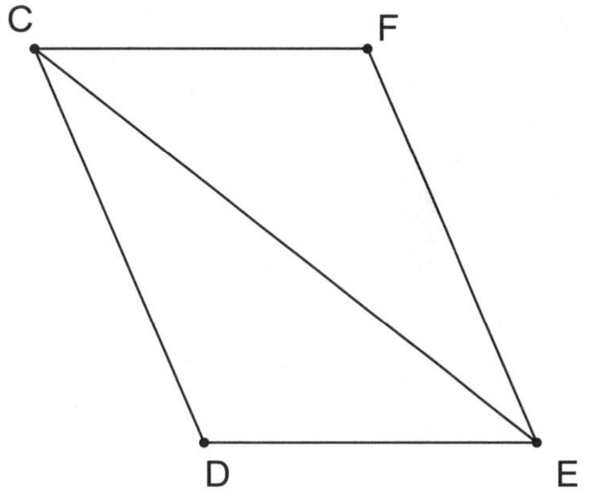

9. Measure every angle and every line segment in the parallelogram. Record your measurements on the figure.

analysis

Parallelograms

10. Are the two triangles formed by this diagonal congruent? Explain why or why not.

process: evaluate

Parallelograms

11. Draw the other diagonal so that your parallelogram has both diagonals. Measure any new segments and angles and record their measurements.

	Parallelograms
analysis	12. Compare the lengths related to the diagonals. What do you notice?

	Parallelograms
process: summary	13. Based on your investigations of parallelograms, make a complete list of all the properties of parallelograms.

Rhombi

input

A parallelogram with four congruent sides is called a **rhombus**. The plural of *rhombus* is *rhombi*.

Rhombi

process: model

1. Draw a rhombus that does not have four congruent angles. Label your rhombus.

Rhombi

analysis

2. Rhombi share all the properties of parallelograms. What additional properties do you think apply to rhombi?

process: evaluate

Rhombi

3. Measure each side and each interior angle in this rhombus. Record your measurements on the figure.

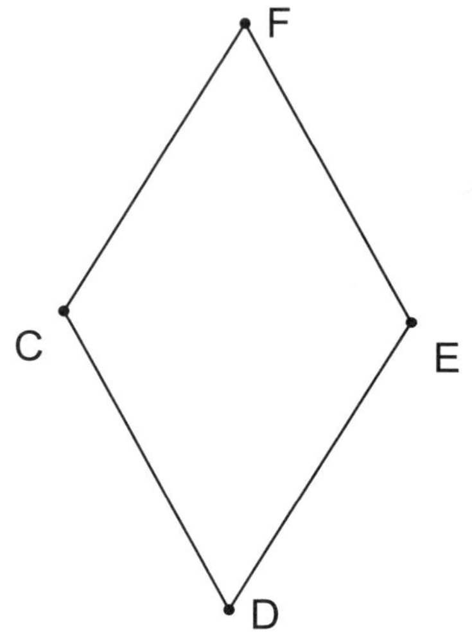

analysis

Rhombi

4. Which of your predicted additional properties about rhombi are supported by your measurements?

5. Which of your predicted additional properties are not supported by your measurements?

	Rhombi
process: evaluate	6. The rhombus below includes one diagonal. Measure every angle and every line segment in the rhombus. Record your measurements on the figure.

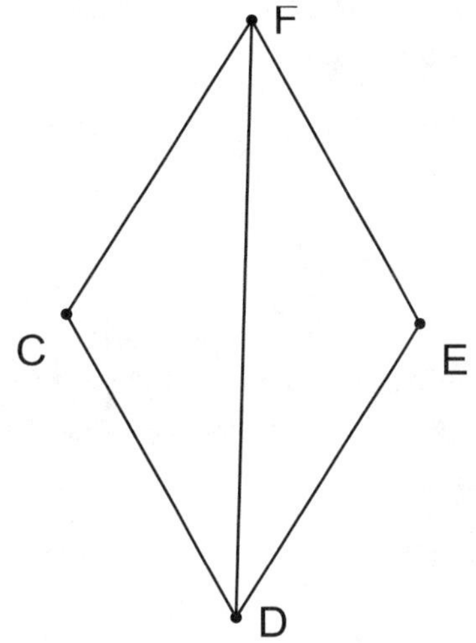

	Rhombi
analysis	7. Are the triangles formed by this diagonal congruent? Explain why or why not.

	Rhombi
process: evaluate	8. Now draw the second diagonal so that your rhombus has both diagonals. Measure any new segments and angles. Record these measurements on the figure.

analysis

Rhombi

9. Label the point at which the two diagonals intersect. Considering only the measurements of the segments of each diagonal, what relationship can be noted about the diagonals?

10. Notice the angles that are formed where the diagonals intersect. Considering the angles formed by the diagonals' intersection, what else can be said about the relationship between the two diagonals?

11. Consider the angles that are formed where the diagonals intersect the vertices of the rhombus. In a complete sentence, describe what the diagonals do to the angles in the rhombus.

process: summary

Rhombi

12. Summarize your findings about the properties of rhombi.

input

Rectangles

A parallelogram with four congruent angles is called a **rectangle**.

process: model

Rectangles

1. Draw a rectangle that does not have four congruent sides. Label your rectangle.

analysis

Rectangles

2. All properties of parallelograms hold true for rectangles. What additional properties do you think hold true for rectangles?

process: evaluate

Rectangles

3. Measure each side and each interior angle in this rectangle. Record your measurements on the figure.

analysis

Rectangles

4. Which of your predicted additional properties about rectangles are supported by your measurements?

5. Which of your predicted additional properties are not supported by your measurements?

process: evaluate

Rectangles

6. This rectangle includes one diagonal. Measure every angle and every line segment in the rectangle. Record your measurements on the figure.

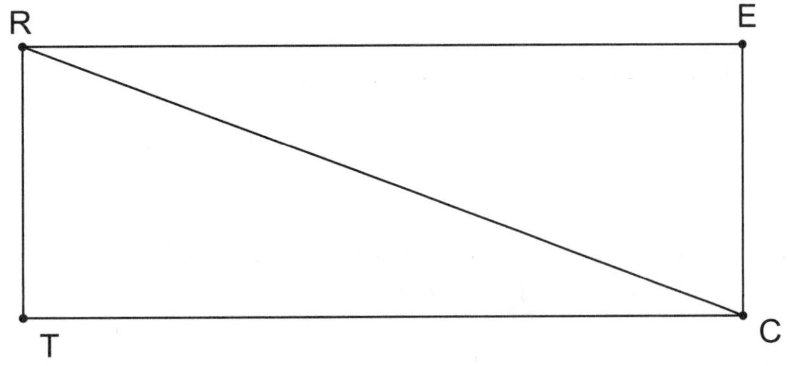

7. Draw the second diagonal, so that your rectangle has both diagonals. Measure any new segments and angles, and record their measurements on the figure.

analysis

Rectangles

8. Label the point at which the two diagonals cross. What relationships can be noted about these diagonals?

9. What relationships can be noted about the parts of the diagonals?

10. What relationships can be noted about the parts of the angles formed where the diagonals intersect the vertices?

process: summary

Rectangles

11. Summarize your findings about the properties of rectangles.

input

Squares

A parallelogram with four congruent angles and four congruent sides is called a **square**.

process: model

Squares

1. Draw a square and label it.

analysis

Squares

2. All properties of parallelograms hold true for squares. What additional properties do you think hold true for all squares?

3. A square can also be defined as a rectangle with four congruent sides. Why?

process: evaluate

Squares

4. Measure each side and each interior angle in this square. Record your measurements on the figure.

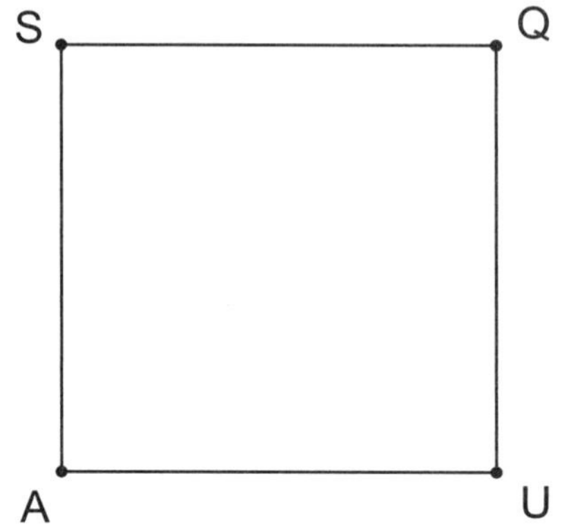

analysis

Squares

5. Which of your predicted additional properties about squares are supported by your measurements?

6. Which of your predicted additional properties are not supported by your measurements?

process: evaluate

Squares

7. This square includes one diagonal. Measure every angle and every line segment in the square. Record your measurements on the figure.

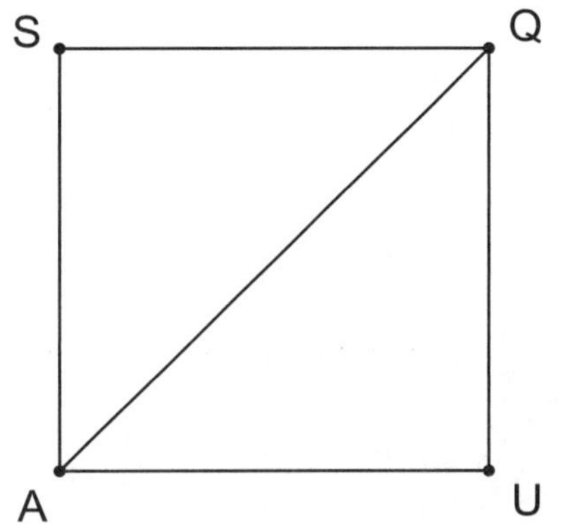

8. Draw the second diagonal, so your square has both diagonals. Measure any new segments and angles, and record their measurements on the figure.

analysis

Squares

9. Label the point at which the two diagonals intersect. Describe, in complete sentences, any relationships you observe.

10. Is there anything special about the angles formed by the diagonals at the vertices of the square? If so, describe it in a complete sentence.

11. Do you see anything special about the lengths of the two diagonals? If so, describe your observations in complete sentences.

	Squares
process: summary	12. Summarize your findings about the properties of squares.

analysis

Quadrilateral Relationships

1. Decide whether each of the following statements is true or false. If the statement is true, explain why. If the statement is false, correct the statement to make it true.

   a. All rectangles are squares.

   b. All squares are rectangles.

   c. All trapezoids are parallelograms.

   d. All rectangles are parallelograms.

   e. All squares are rhombi.

Quadrilateral Relationships

analysis

f. All parallelograms are quadrilaterals.

g. All quadrilaterals are parallelograms.

h. All rhombi are rectangles.

i. A square is an equilateral rectangle.

j. A square is an equiangular rhombus.

k. A rectangle is an equiangular parallelogram.

l. A rectangle is an equilateral parallelogram.

analysis

Quadrilateral Relationships

m. A rhombus is an equilateral parallelogram.

n. A rhombus is an equiangular parallelogram.

o. A parallelogram is an equilateral quadrilateral.

p. A parallelogram is an equiangular quadrilateral.

process: summary

Quadrilateral Relationships

2. Based on the concepts developed in this unit, draw a diagram illustrating the relationship between the following types of figures: parallelogram, quadrilateral, rectangle, rhombus, square, and trapezoid.

input

Sum of the Interior Angle Measures in a Polygon

You can use what you know about the sum of the interior angle measures of a triangle to calculate the sum of the interior angle measures of a quadrilateral, as well as those of other polygons.

A line segment that connects two opposite vertices of a quadrilateral is called a **diagonal** of the quadrilateral.

process: model

Sum of the Interior Angle Measures in a Polygon

1. Draw a quadrilateral with one diagonal included.

process: evaluate

Sum of the Interior Angle Measures in a Polygon

2. What is the sum of the measures of the interior angles of any triangle?

analysis

Sum of the Interior Angle Measures in a Polygon

3. Notice that the diagonal divides the quadrilateral into two triangles. If you know the sum of the measures of the interior angles in each triangle, what is the sum of the measures of the four interior angles of the quadrilateral? Explain your response.

process: model

Sum of the Interior Angle Measures in a Polygon

This definition of a diagonal applies to any polygon: A line segment that connects non-consecutive vertices of a polygon is called a **diagonal** of the polygon.

4. Draw a pentagon. Select one of the vertices and draw all the diagonals that connect to that vertex.

process: evaluate

Sum of the Interior Angle Measures in a Polygon

5. How many triangles have been formed by the diagonals drawn in the pentagon?

6. What is the sum of the measures of the three interior angles of each triangle?

	Sum of the Interior Angle Measures in a Polygon
process: evaluate	7. What is the sum of the measures of the interior angles in all the triangles? 8. What is the sum of the measures of the five interior angles in the pentagon?

	Sum of the Interior Angle Measures in a Polygon
process: model	9. Draw a hexagon. Select one of the vertices and draw all the diagonals that connect to that vertex.

	Sum of the Interior Angle Measures in a Polygon
process: evaluate	10. How many triangles have been formed by the diagonals drawn in the hexagon? 11. What is the sum of the measures of the three interior angles in each triangle?

process: evaluate	**Sum of the Interior Angle Measures in a Polygon**
	12. What is the sum of the measures of the interior angles in all the triangles? 13. What is the sum of the measures of the six interior angles in the hexagon?

process: model	**Sum of the Interior Angle Measures in a Polygon**
	14. Draw a heptagon. Select one of the vertices and draw all the diagonals that connect to that vertex.

process: evaluate	**Sum of the Interior Angle Measures in a Polygon**
	15. How many triangles have been formed by the diagonals drawn in the heptagon? 16. What is the sum of the measures of the three interior angles in each triangle?

process: evaluate	**Sum of the Interior Angle Measures in a Polygon**
	17. What is the sum of the measures of the interior angles in all the triangles? 18. What is the sum of the measures of the seven interior angles in the heptagon?

process: model	**Sum of the Interior Angle Measures in a Polygon**
	19. Draw an octagon. Select one of the vertices and draw all the diagonals that connect to that vertex.

process: evaluate	**Sum of the Interior Angle Measures in a Polygon**
	20. How many triangles have been formed by the diagonals drawn in the octagon? 21. What is the sum of the measures of the three interior angles in each triangle?

process: evaluate

Sum of the Interior Angle Measures in a Polygon

22. What is the sum of the measures of the interior angles in all the triangles?

23. What is the sum of the measures of the eight interior angles in the octagon?

process: model

Sum of the Interior Angle Measures in a Polygon

24. Complete the following table based on your answers to questions 1 – 23.

Name of polygon	Number of sides	Number of diagonals from one vertex	Number of triangles formed	Sum of interior angles in a triangle	Sum of interior angles in this polygon
triangle	3	0	1		
quadrilateral					
pentagon					
hexagon					
heptagon					
octagon					
n-gon	n				

25. Describe any patterns you see in the table.

26. Complete the last row in the table based on the patterns you found in the table.

process: practice

Sum of the Interior Angle Measures in a Polygon

27. Use the method you found for calculating the sum of the measures of the interior angles in a polygon to find the sum of the measures of the interior angles in a decagon.

28. Use the patterns you found to determine the sum of the measures of the interior angles in a polygon with 100 sides. Explain this process.

process: summary

Sum of the Interior Angle Measures in a Polygon

29. In a paragraph, explain how to calculate the sum of the interior angles in any polygon.

input

Sum of the Exterior Angle Measures in a Polygon

To calculate the sum of the exterior angle measures of a quadrilateral or other polygons, you can use what you know about the sum of the exterior angle measures of a triangle.

Remember that an exterior angle can be formed by extending one side at each vertex of a polygon. (You drew exterior angles of triangles in the activity starting on page 3-67.)

process: model

Sum of the Exterior Angle Measures in a Polygon

1. Use a straightedge to create a triangle with one exterior angle at each vertex.

process: evaluate

Sum of the Exterior Angle Measures in a Polygon

2. What is the sum of the measures of the exterior angles of any triangle?

	Sum of the Exterior Angle Measures in a Polygon
process: model	3. Draw a quadrilateral with one exterior angle located at each vertex.

	Sum of the Exterior Angle Measures in a Polygon
analysis	4. The sum of the measures of the four *interior* angles of a quadrilateral is 360°. How can you use this property to calculate the sum of the measures of the four *exterior* angles? (Hint: Think about linear pairs of angles.)

	Sum of the Exterior Angle Measures in a Polygon
process: evaluate	5. What is the sum of the measures of the four exterior angles of any quadrilateral?

	Sum of the Exterior Angle Measures in a Polygon
process: model	6. Draw a pentagon with one exterior angle located at each vertex.

Sum of the Exterior Angle Measures in a Polygon

analysis

7. The sum of the measures of the five interior angles of a pentagon is 540°. How can you use this property to calculate the sum of the measures of the five exterior angles? (Hint: Think about linear pairs of angles.)

Sum of the Exterior Angle Measures in a Polygon

process: evaluate

8. What is the sum of the measures of the five exterior angles of any pentagon?

process: model

Sum of the Exterior Angle Measures in a Polygon

9. Draw a hexagon with one exterior angle located at each vertex.

analysis

Sum of the Exterior Angle Measures in a Polygon

10. The sum of the measures of the six interior angles of a hexagon is 720°. How can you use this property to calculate the sum of the measures of the six exterior angles? (Hint: Think about linear pairs of angles.)

process: evaluate

Sum of the Exterior Angle Measures in a Polygon

11. What is the sum of the measures of the six exterior angles of any hexagon?

analysis

Sum of the Exterior Angle Measures in a Polygon

12. Based on your answers to questions 1 – 11, predict the sum of the measures of the exterior angles of a decagon. Explain your reasoning.

process: summary

Sum of the Exterior Angle Measures in a Polygon

13. Describe how to find the sum of the measures of the exterior angles of any polygon.

Unit 5: Transformations and Symmetry

Contents

preview

Urban Sprawl

Many cities in the United States are experiencing "urban sprawl." Urban sprawl is a consequence of population growth in these cities. As the population rapidly increases, not only do the number of houses and apartment buildings increase, the demand for services and amenities will rise, so that the number of shops, businesses, and offices also rises.

One problem associated with "urban sprawl" is ensuring that electricity can be delivered to all the new homes, businesses, and shops in the growing communities. This requirement is where some of the mathematics of community planning comes into play.

Below is a scaled-down version of some town planning documents. Line PW represents the current power line. Point S represents a new shopping district that is developing, while point R represents a new residential plan of single-family homes that is in the process of being built. Currently, no electricity is being supplied to either of the new areas. Your job is to determine where to place a new electrical substation so that the length of new power line is minimized.

Where on line PW should the electrical substation be placed so that the distance from R to the substation to S is minimized?

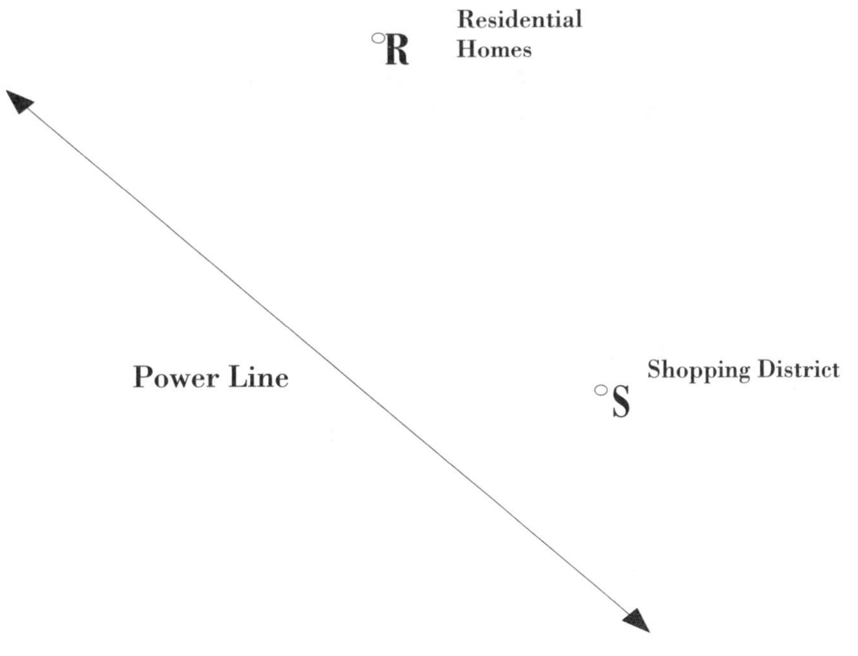

preview

Urban Sprawl

1. Explain how you determined the placement of the substation along the power line.

2. Measure the total distance (in centimeters) from the residential area to the substation and then to the shopping district. How far is it on your scaled-down drawing?

3. If one centimeter on this plan is equivalent to 3 miles, how much electric line must be strung to connect both the shopping district and the new neighborhood to the transformer? Make sure to use full sentences in your response.

Introduction to Transformations

input

A geometric figure is said to be *transformed* if the figure has been changed or moved such that all the points of the figure are changed in the same way. These transformations are sometimes referred to as flips, slides, and turns. So that you may communicate mathematically, let's examine some terms that are more technical then these informal terms.

Reflection: Flip all of the points over a given line.

Translation: Shift or slide all of the points horizontally, vertically, or both.

Rotation: Turn all of the points around one given point.

Dilation: Stretch or shrink the figure.

Introduction to Transformations

process: practice

1. Pair the following examples with the descriptions above:

a. b. c. d.

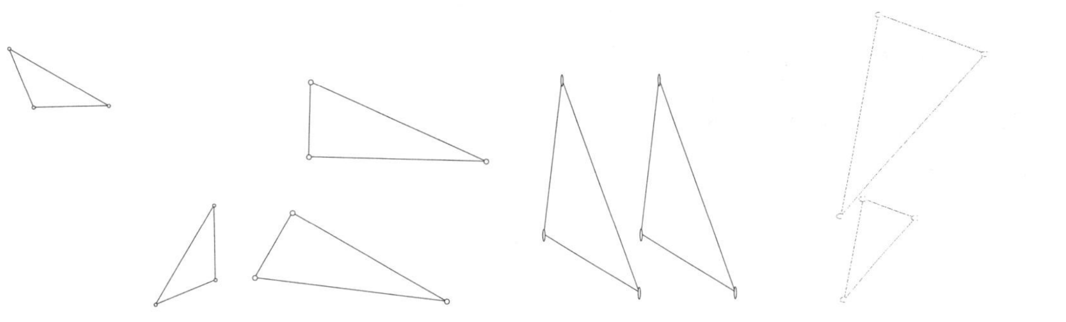

Introduction to Transformations

analysis

Translations, reflections, and rotations are all grouped together and called **isometries**.

2. Using the explanations and examples above, can you think of a reason why these three transformations are grouped together with a common name?

analysis

Introduction to Transformations

3. What makes a dilation different from the other three types of transformations mentioned above?

4. *Transformational Geometry*, which you will be studying in this unit, is often called "motion geometry." Why do you think this is so?

process: model

Introduction to Transformations

5. Examine the pictures below. Each of these pictures shows one or more type(s) of transformation. Identify the type of transformation(s) being shown. Note that more than one type of transformation may be appropriate.

a. 

b. 

c. 

d. 

e. 

process: model

Introduction to Transformations

6. Use a crayon, pencil, pen, or marker to color in the block letter **P** below. *Hint: This activity will work better if you make your coloring very dark.* After you have colored the letter, fold the paper along the vertical line and rub the other side of the paper over the area where the colored letter **P** is located.

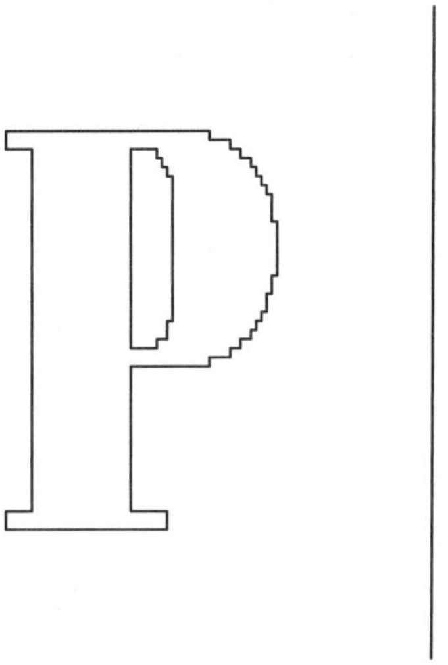

analysis

Introduction to Transformations

7. Which of the transformations did you just perform in question 6?

8. Is this transformation an isometry? How do you know? As usual, use complete sentences in your response.

THIS PAGE HAS BEEN INTENTIONALLY LEFT BLANK SO YOU CAN SEE YOUR LETTER.

input

Introduction to Transformations

In mathematical terms, the original letter **P** above is called the **pre-image** or the **parent** figure. The letter **P** you created is called the **image**. To designate that this new letter **P** is the image, use the notation **P′** ("P prime") to show this is different from the original **P**.

You are currently working for a new startup technology company called Acme. The company is designing its corporate stationery. What you see pictured below is the current draft of the stationery, showing the company logo in the upper left-hand corner and a border design along the right side of the front. On the back there is a watermark of the company's name.

Front Back

The new administrative assistant is having a difficult time determining how to put the stationery into the laser printer in order to get a letter to print correctly. You decide to lend a hand.

process: model

Introduction to Transformations

9. First you try flipping the stationery over to the back. Illustrate what you see as you place the paper in the printer, and tell which transformation is being used.

   a. If the **F** symbol is at the top left before you flip the paper to the back (in the pre-image), where is it after you flip the paper (in the image)?

process: model

Introduction to Transformations

b. If the border was on the right before you flip the paper to the back (in the pre-image), where is it after you flip the paper (in the image)?

c. If the watermark was face down before you flip the paper (in the pre-image), where is it afterwards (in the image)?

10. When the letter in question 9 was printed, it appeared upside down on the stationery, so you use a fresh piece of stationery. Instead of flipping it over to the back, you keep it face up but turn it around. Illustrate what you see as you place the paper in the printer. Tell which transformation is being used.

a. If the **F** symbol is at the top left before you turn the paper (in the pre-image), where is it after you turn the paper (in the image)?

b. If the border was on the right in the pre-image, where is it in the image?

c. If the watermark was face down in the pre-image, where is it in the image?

analysis

Introduction to Transformations

11. This time when the letter was printed, the words appeared on the watermarked side instead of the front. Using the information from the previous two printing attempts, what transformation(s) should you perform to the piece of stationery to make the letter print correctly? Illustrate what you will see as you place the paper in the printer. Explain how you knew this.

process: model

Introduction to Transformations

Fold an 8 ½ by 11 piece of paper in order to make the largest square possible.

12. Explain how you know that what you created is a square.

Cut away the part of the paper that is not included in your square. Fold your square to create two identical triangles and cut along the fold.

13. Use your two triangles to create one larger triangle. Sketch your resulting figure below, showing each of the smaller individual triangles.

    a. If you started with only one of the smaller triangles, what transformation could have been performed to create the larger triangle?

    b. Could a different transformation have been performed to get the same result? If so, which transformation?

process: model

Introduction to Transformations

14. Use your triangles to create a parallelogram that is not a rectangle. Sketch the resulting figure, showing the two triangles you used.

    a. If you started with only one of the smaller triangles, what transformation could have been performed to create the parallelogram?

process: model

Introduction to Transformations

b. Could a different transformation have been performed to get the same result? If so, which transformation?

15. Sketch your original square below, showing the two triangles.

a. If you started with only one of the smaller triangles, what transformation could have been performed to create the square?

b. Could a different transformation have been performed to get the same result? If so, which transformation?

process: practice

Introduction to Transformations

16. Tell which isometry can be used to map one quadrilateral directly onto another in the picture below.

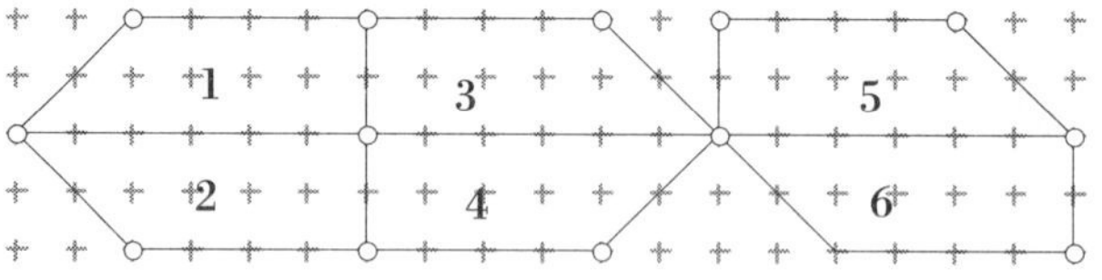

a. Quadrilateral 1 onto quadrilateral 2

b. Quadrilateral 1 onto quadrilateral 4

c. Quadrilateral 2 onto quadrilateral 6

process: practice

Introduction to Transformations

d. Quadrilateral 1 onto 5

e. Quadrilateral 3 onto 6

f. If quadrilateral 2 is moved to another position on the grid using a translation, what are the possible positions?

g. If quadrilateral 4 is moved via a rotation, what other positions could it occupy?

h. If quadrilateral 1 is moved via a reflection, what other positions can it occupy?

i. Can you use dilations to transform a quadrilateral in this figure into another quadrilateral in this figure? Explain your response.

process: summary

Introduction to Transformations

17. Describe how an image is related to its pre-image when a reflection has taken place? (What stays the same and what changes?) Use complete sentences.

18. Describe how an image is related to its pre-image when a translation has taken place?

19. Describe how an image is related to its pre-image when a rotation has taken place?

20. Describe how an image is related to its pre-image when a dilation has taken place?

Reflections

input

Have you ever looked in the mirror, seen some dirt on your face, or a piece of food in your teeth, and reached to remove it only to find you had reached for the wrong side of your face? Whether you are trying to fix your hair or follow your instructor's image at a karate or aerobics class, mirror images can be tricky. By the conclusion of this lesson, you will discover properties of reflections, learn some vocabulary about reflections, and perform a few reflections of your own.

Let's start with a variation of a simple activity that you may have done as a child.

Creating the pre-image and the image

Write your name in large block letters on the top half of a piece of regular-sized paper. Use a dark pen or marker to do this. (It makes no difference whether you write your name across or down.)

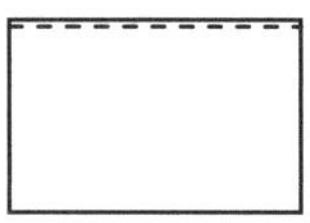

After you have completed your name, fold the paper in half so that your name is now inside the fold and hidden from view.

Trace your name on the outside of the fold, being very precise. (You should still be reading your name from left to right.)

Now open the paper to its full size and trace your name again on the blank half of the paper by looking through the paper at your first tracing. Again, be very precise.

Reflections

analysis

1. Which of the isometries relates your original name or pre-image to the second tracing or image? (This image is the one that appears on the same side of the paper as your original name.)

2. Cite some similarities and differences between your pre-image and your image.

process: practice

Reflections

3. Pick a specific point on the pre-image and its corresponding point on the image. Label the original point on the pre-image point A. What is the correct mathematical label for the corresponding point on the image? Label the image.

4. Translate this mathematical notation into words?

analysis

Reflections

Using a ruler, connect your pre-image point to its image point.

5. What is the label for the line segment you have just created?

6. Why is it logical that your line segment intersects the fold of your paper?

process: evaluate

Reflections

Distances

7. Name the point where your line segment intersects your fold as point Z. Measure the distance from your pre-image point to point Z. Record the distance on your paper and here.

8. Measure the distance from your image point to point Z. Record the distance on your paper and record it here.

analysis

Reflections

9. What do you notice about the distances you measured? Does this observation make sense? If so, explain why. If not, explain why not, making sure to use complete sentences in your explanation.

10. What is the mathematical name for point Z on your line segment connecting the pre-image point to the image point?

11. Pick another point on your fold and call it point Y. Connect this point to your pre-image point and to the image point. Measure these distances and record them.

12. Do these measurements confirm what you wrote about point Z?

Angles: Part 1

13. Measure the angles formed by segment AA′ and the fold in your paper. Record the measurements here and record them on your paper.

14. What do you notice? Do you think your observation will always hold true? Explain.

15. Using the fact that Z is the midpoint of AA′ and the fact that the angle formed between AA′ and the fold is 90°, give a more mathematical name to the fold in your paper. *Hint: What does the fold do to your connecting line segment?*

process: evaluate

Reflections

Angles: Part 2

16. Measure the angle created when you connect point Y to your pre-image point A and then to point Z. Record the measurement on your paper and here.

17. Measure the corresponding angle on your reflection after connecting point Y to your first image point A′ and then to point Z. Record your measurement.

analysis

Reflections

Other Properties

18. Create two more points, B and C, on your pre-image so that A, B, and C are collinear (all in the same line). Will the corresponding three image points be collinear? Why or why not? Test your conjecture by reflecting points B and C.

19. Using points A, B, and C, tell which point is between the other two. Will the image of this point lie between the images of the other two points? Explain. Test your conjecture.

20. So far, we have noted many properties that are the same from a pre-image to an image. What *differences* do you notice from the pre-image to the image?

	Reflections
analysis	21. Trace the outline of one of the letters in your pre-image name. As you trace, do you move in a clockwise or counterclockwise direction? 22. Trace the corresponding letter in your image name. Make sure to start at the point corresponding to your start point in the last question and move along the corresponding path. As you trace, do you move in a clockwise or counterclockwise direction? 23. The direction in which you move as you trace the outline of a letter or a polygon is called its **orientation**. Is the orientation or direction the same or different from your pre-image to your image?

	Reflections
process: summary	24. Use complete sentences to summarize your findings from this activity.

process: practice

Reflections

25. In the last activity, you needed to be precise. In the next activity, you will continue to explore reflections, but you will be sketching the reflections, instead of trying to be very precise. Starting with the three circles below, use a line segment to show where a mirror should be placed so that when you take the circles and the reflection together, you see the number of circles requested. Then sketch the reflection image plus the pre-image.

a. Six circles

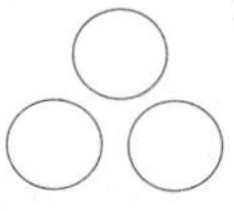

b. Five circles

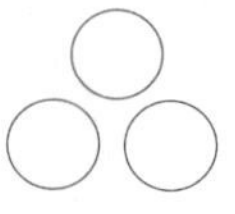

c. Four circles

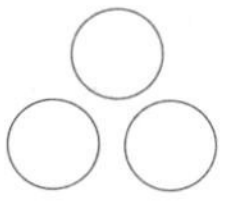

d. Three circles

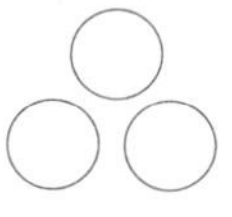

e. Two circles

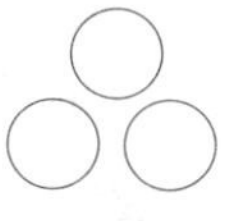

f. One circle

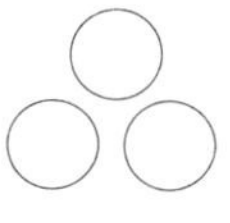

Reflections Using a Ruler and Protractor

input

By now, you have used paper folding, sketches, and perhaps mirrors or Mira® tools to create reflections. Next, you will use a protractor and a ruler to make your reflections more precise than you can by sight.

In this activity, we will reflect a point, a segment, and then a triangle using a new method.

Earlier in this section, you determined that the mirror or reflection line was actually the perpendicular bisector of the line connecting the pre-image to the image. Our new method will use this important fact.

If the reflection line is a perpendicular bisector of the segment that connects A to A′, what sort of angle must be formed between the line of reflection and segment AA′?

Align the edge of your protractor with point A and align the 90° marks with line BC.

Use the edge of your protractor to create a segment that is perpendicular to line BC and extend this segment. Label the point where this segment intersects line BC as point M.

Use your ruler to measure the distance from point A to line BC. Now place a point on the extended segment that will be the reflection of A, labeling it appropriately.

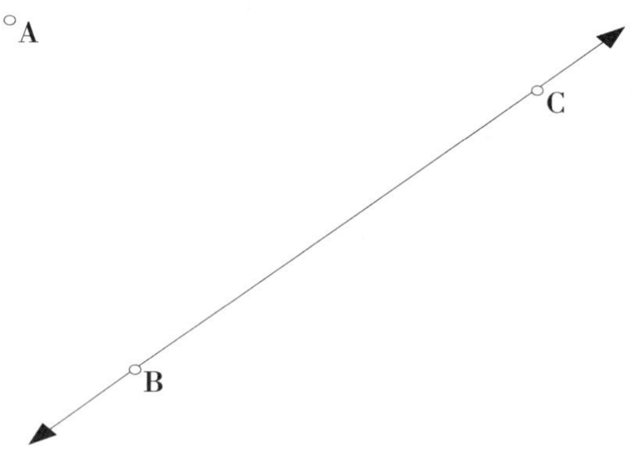

analysis

Reflections Using a Ruler and Protractor

1. How can you make sure that your point is actually the reflection of A over $\overline{BC}$?

2. Does this method of using a protractor and ruler to create a reflection make sense? Why or why not?

3. Complete the following statement:

   A′ is the reflection of A over a mirror line if and only if the mirror line is the ________ of segment AA′.

input

Reflections Using a Ruler and Protractor

Mathematicians use some shorthand notation to write about reflections. The symbol **r** is used to signify a reflection. For the example you just completed, the notation used would be $\mathbf{r}_{BC}(A) = A'$. Reading this in words, you would say, "The reflection of point A over line BC is point A prime."

	Reflections Using a Ruler and Protractor
process: evaluate	4. Complete the following mathematical statements based on the figure on 5-20: a. $\mathbf{r}_{BC}(B) =$ b. $\mathbf{r}_{BC}(A') =$ c. $\mathbf{r}_{BC}(C) =$

	Reflections Using a Ruler and Protractor
analysis	5. Make a generalization about points *on* the line of reflection, based on the answers you gave above.

process: practice

Reflections Using a Ruler and Protractor

Practice the method you used above to create the reflection of $\overline{MN}$ over line CD.

6. Which points will you need to reflect to make sure the whole line segment gets reflected correctly?

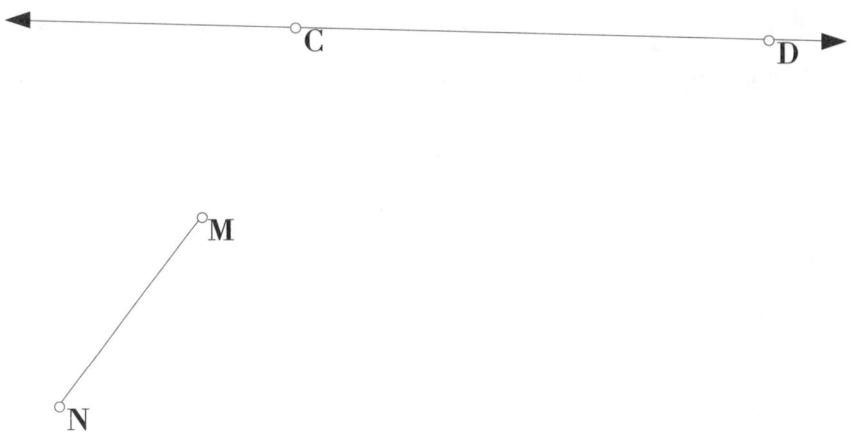

7. Use the reflection notation you just learned to write mathematically about the reflection of point M and then about the reflection of point N.

8. Complete the following:
   a. $\mathbf{r}_{DC}$(MN) =
   b. $\mathbf{r}_{DC}$(NM) =
   c. $\mathbf{r}_{DC}$(DC) =
   d. $\mathbf{r}_{DC}$(M′) =
   e. $\mathbf{r}_{DC}$(N′) =
   f. $\mathbf{r}_{DC}$(M′N′) =

analysis

Reflections Using a Ruler and Protractor

9. Make a generalization about points and their reflections based on the reflections you have done.

process: practice

Reflections Using a Ruler and Protractor

Use the ruler and protractor method to reflect ΔXYZ over line DE.

10. Which points will you need to reflect in order to create the image triangle?

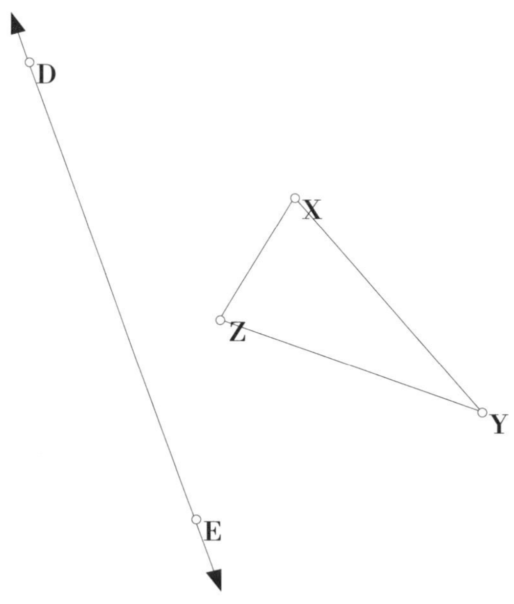

11. Use the notation for reflections to represent the reflection of the figure above.

analysis

Reflections Using a Ruler and Protractor

12. What is the same about the pre-image and the image triangles?

13. What is different about the two triangles?

process: practice

Reflections Using a Ruler and Protractor

14. Complete the following mathematical statements:

    a. $\mathbf{r}_{DE}(Z) =$

    b. $\mathbf{r}_{DE}(YX) =$

    c. $\mathbf{r}_{DE}(\Delta X'Y'Z') =$

    d. $\mathbf{r}_{DE}(\angle YZX) =$

    e. rDE(Z′X′) =

15. Complete this statement:

    If $\mathbf{r}(\Delta ABC) = \Delta A'B'C'$, then $\mathbf{r}(A) =$ __ ___, $\mathbf{r}($ $) = B'$, and $\mathbf{r}($ $) = ($ $)$

process: summary

Reflections Using a Ruler and Protractor

Your best friend was absent from class today. Use the instant messenger on your computer to give your friend instructions about how to perform a reflection without using paper folding or a mirror. You can't show your friend any pictures. You can only use words to describe this process.

Instant Messenger:

Reflections in the Coordinate Plane: Part 1

process: model

1. Give the coordinates for the vertices of the following triangle. Note them on the figure.

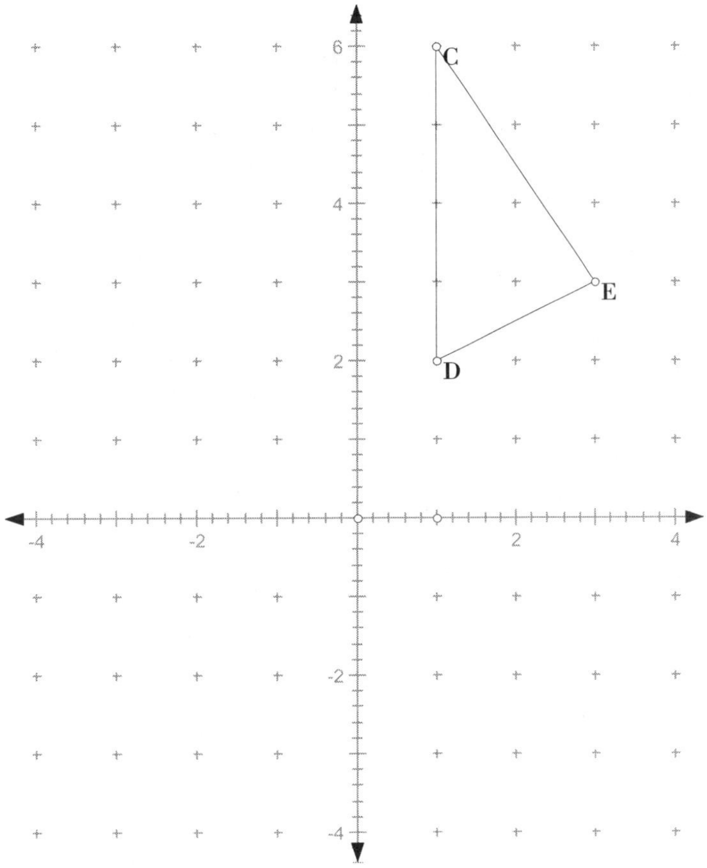

Reflections in the Coordinate Plane: Part 1

process: model

2. Reflect each of the points over the y-axis.
3. Label the image points using the correct notation for transformed points.
4. Connect the image points to form the image triangle.
5. Describe this reflection in the space below by using the shorthand mathematical notation for reflections.

   Δ Δ

6. Give the coordinates of your image points on the image triangle.

input

Reflections in the Coordinate Plane: Part 1

Another notation that is used with transformations helps to show how points "map" onto one another (move from one location to another). This notation uses an arrow between the labels or coordinates of the two points.

For example, if you want to show that the pre-image point A corresponds to (or maps onto) point A', you would write: **A→A'**. If you want to show that (1,3) maps onto (-1,3), you would write: **(1,3)→(-1,3).**

7. Use this notation to write about your three pre-image to image pairs in the triangle in this task.

analysis

Reflections in the Coordinate Plane: Part 1

8. Describe your observations. Is a pattern evident when you compare the coordinates of the image points with the coordinates of the pre-image points?

process: model

Reflections in the Coordinate Plane: Part 1

9. Create a new triangle on the coordinate plane on the previous page. Give it the coordinates: X(-2,1), Y(-2,-4), and Z(-3,0).
10. Reflect each of the points over the y-axis and label these points.
11. Connect the image points to form the image triangle.
12. Give the coordinates of your image points on the image triangle.
13. Use the mapping or arrow notation to write about the pre-image to image pairs in the space below.

analysis

Reflections in the Coordinate Plane: Part 1

14. Describe your observations in examining the two triangles and their images. Is a pattern evident when the coordinates of image and pre-image are compared?

process: practice

Reflections in the Coordinate Plane: Part 1

15. Use the patterns you found to reflect ΔKLM over the y-axis.

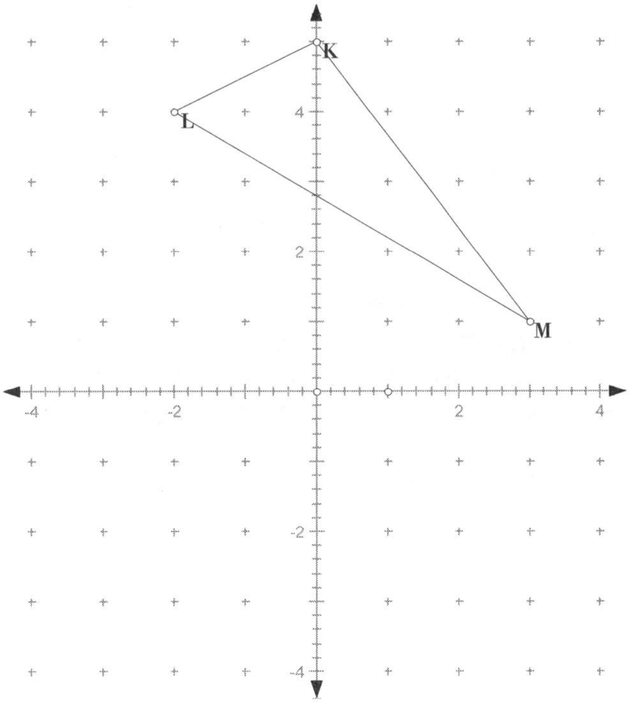

Reflections in the Coordinate Plane: Part 1

analysis

16. If a point had the coordinates (x,y), what would be the coordinates of its reflection over the y-axis? Use the mapping notation to write your response.

17. If a point had the coordinates (-x,y) what would be the coordinates of its reflection over the y-axis? Use the mapping notation to answer.

18. What if a point had the coordinates (x,-y)? Use the mapping notation.

19. What about a point with coordinates (-x,-y)?

Reflections in the Coordinate Plane: Part 1

analysis

20. Does the pattern you noted earlier for reflections over the y-axis apply to all points? In other words, could you use your pattern to answer the last four questions? If so, explain why this pattern occurs. If not, use this space to revise your generalization so that it applies to all points being reflected over the y-axis.

Reflections in the Coordinate Plane: Part 2

process: model

1. Give the coordinates of the vertices of this quadrilateral.

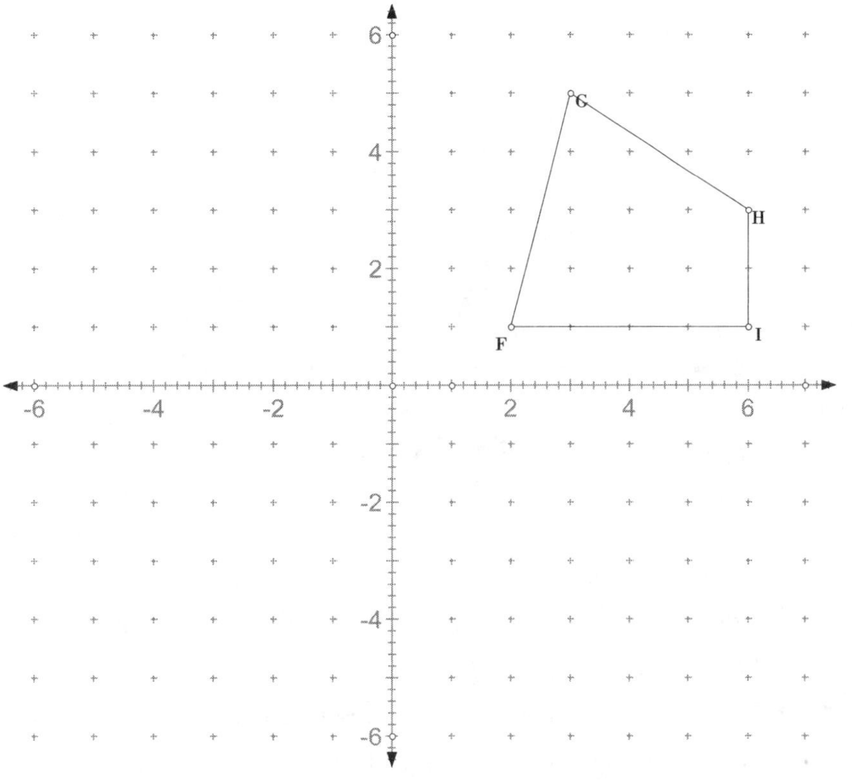

Reflections in the Coordinate Plane: Part 2

process: model

2. Reflect each of the points over the **x-axis** and label the image points correctly.
3. Connect the image points to form the image quadrilateral.
4. Describe this reflection using the shorthand mathematical notation for reflections (the **r** notation).
5. Place the coordinates of your image points on the image quadrilateral.
6. Use the mapping (arrow) notation to describe your pre-image to image pairings for the four vertices of the quadrilateral.

analysis	**Reflections in the Coordinate Plane: Part 2**
	7. What patterns do you notice in these coordinate pairings?

process: model	**Reflections in the Coordinate Plane: Part 2**
	8. Create a new quadrilateral on the coordinate plane with the coordinates: N(-6,4), O(0,0), P(-3,-4), and Q(-6,0). 9. Reflect each of the points over the **x-axis**. 10. Label and connect the image points to form the image quadrilateral. 11. Place the coordinates of your image points on the quadrilateral. 12. Use the mapping notation with these four pairs of points.

analysis	**Reflections in the Coordinate Plane: Part 2**
	13. Do you notice the same patterns as you did with quadrilateral FGHI? If so, explain. If not, how might you revise your initial observations so they can apply to both pairs of quadrilaterals?

Reflections in the Coordinate Plane: Part 2

analysis

14. If a point had the coordinates (x,y), what would be the coordinates of its reflection over the x-axis? Use the mapping notation to answer this question.

15. If a point had the coordinates (-x,y), what would be the coordinates of its reflection over the x-axis? Use the mapping notation to answer this question.

16. What about a point with coordinates (x,-y)?

17. What about a point with coordinates (-x,-y)?

Reflections in the Coordinate Plane: Part 2

analysis

18. Does the pattern you noted earlier for reflections over the x-axis apply to all points? In other words, could you use your pattern to answer the last four questions? If so, explain why this pattern occurs. If not, use this space to revise your generalization so that it applies to all points being reflected over the x-axis.

Reflections in the Coordinate Plane: Part 3

process: model

1. Give the coordinates of the following polygon. Note them on the coordinate grid.
2. What type of polygon is this? (Be as specific as possible.)

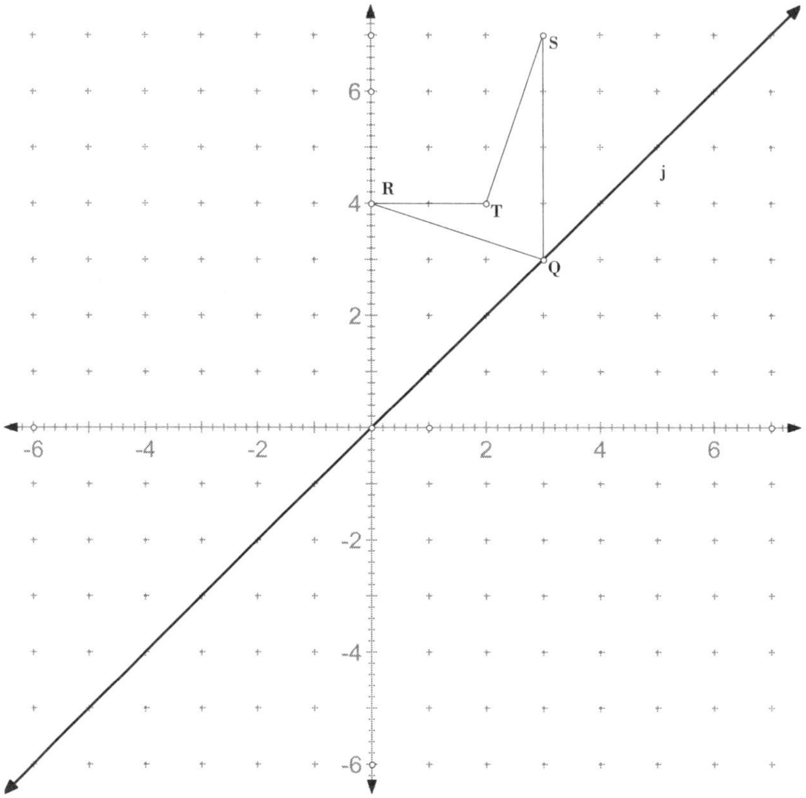

Reflections in the Coordinate Plane: Part 3

process: model

3. What is the equation of line j, which is graphed above?
4. How do you know?
5. Reflect each of the points over the y = x line.
6. Label the image points using the correct notation for transformed points.
7. Connect the image points to form the image polygon.

process: model

Reflections in the Coordinate Plane: Part 3

8. Describe this reflection using the shorthand mathematical notation for reflections (the **r** notation).

9. Provide the coordinates of your image points on the image polygon.

10. Use the mapping notation to write about your four pre-image to image pairs from the polygon above.

analysis

Reflections in the Coordinate Plane: Part 3

11. Describe your observations. Is a pattern evident when you compare the coordinates of the image points with the coordinates of the pre-image points?

process: model

Reflections in the Coordinate Plane: Part 3

12. Create a new polygon on the same coordinate plane. Give it the coordinates: A(-3,4), B(-3,-5), and C(-6,-5).
13. Reflect each of the points over the y = x line.
14. Label and connect the image points to form the image triangle.
15. Place the coordinates of your image points on the image triangle.
16. Use the mapping (arrow) notation to write about the pre-image to image pairs.

Reflections in the Coordinate Plane: Part 3

analysis

17. If a point had the coordinates (x,y), what would be the coordinates of its reflection over the line y = x? Use the mapping notation to answer this question.

18. What about a point with coordinates (-x,y)? Use the mapping notation.

19. What about a point with coordinates (x,-y)?

20. What about a point with coordinates (-x,-y)?

Reflections in the Coordinate Plane: Part 3

analysis

21. Does the pattern you noted earlier for reflections over the y = x line apply to all points? Could you use your pattern to answer the last four questions? If so, explain why this pattern occurs. If not, revise your generalization.

Reflections in the Coordinate Plane: Part 3

process: summary

It's the night before the exam. Your friend calls you in desperation. She says, "I have something in my notes that doesn't make sense. I can't figure it out." You ask what she has in her notes, and she says, "When you reflect over an axis, the coordinate for that axis changes to negative, and when you reflect over the line y = x, both coordinates change." Explain what mistake she made in her notes. Tell her what she should have written and explain what it means. Use examples to help with your explanation.

input

Translations

You recently started working at a new company as the webmaster. One of your first responsibilities is to redesign the introductory screen for the company's website. Currently, there is a gif with the company's pentagonal logo in the upper left corner. Your idea is to animate this gif so that it moves to the right and down. In order to make this happen, you have to know the starting position of the gif, the ending position, and each intermediate step. Below is a model of how you want this logo to move across the screen.

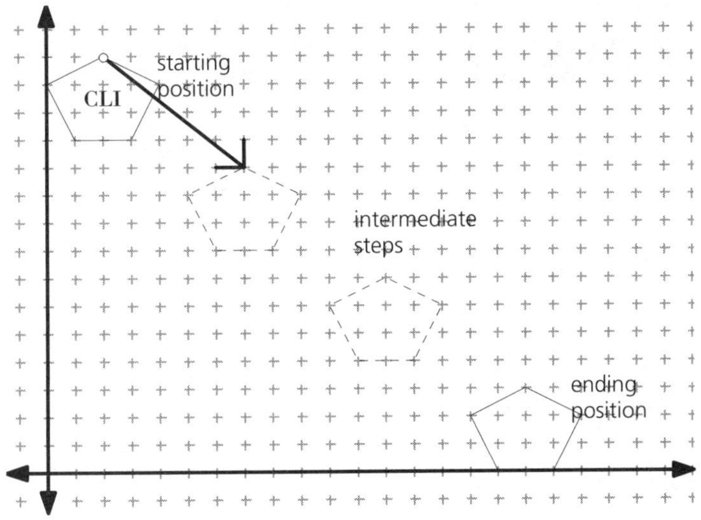

1. Describe to your technical assistant how you want the logo to move across the screen. Use full sentences to write your description.

   a. Connect one of the starting vertices to its corresponding vertex on the first intermediate step. This is called the "slide arrow." A more technical term for this arrow is a **vector**.

analysis

Translations

2. What two things does this vector tell you about the translation?

process: model

Translations

3. Create axes on the grid and give the coordinates of the vertices on your original logo. Record them below. Record the vertices for the 1st, 2nd, and final image.

Pre-image	First image	Second image	Final image

4. Use the mapping notation you learned in the ***Reflections in the Coordinate Plane*** section to show how the points in the pre-image map to the points in the *first image*.

analysis

Translations

5. If a point in the original gif had the coordinates (x,y), what would be the coordinates of its first image? Use the mapping notation to answer this question.

analysis

Translations

6. Examine the coordinates in the pre-image compared to the coordinates of the points in the *second image*. Describe any patterns you see.

7. If a point in the original gif had the coordinates (x,y), what would be the coordinates of its second image? Use the mapping notation to answer this question.

8. Examine the coordinates in the pre-image compared to the coordinates of the points in the *final image*. Describe any patterns you see.

9. If a point in the original gif had the coordinates (x,y), what would be the coordinates of its final image? Use the mapping notation to answer this question.

input

Translations

Just like with reflections, there is a shorthand mathematical notation used with translations. Instead of using an "**r**", use a "**T**". If point A is translated to point A*, you will write **T**(A) = A*.

process: model

Translations

10. Use this notation to write about the translation of the initial gif to the final position.

11. Sometimes the mapping notation is even combined with the shorthand notation. Interpret the following, using full sentences to respond.

$$\mathbf{T}(x,y) = (x - 3.5, y + 9)$$

12. Use the combined notation to write about the translation of the initial pre-image to its *first image*.

input

Translations

You performed so well at the CLI company, you decide to do some freelance work. In each of the jobs you get, you are moving a figure across the computer screen. However, each company gives you the information in a different form.

In the end, you must still analyze the situation to have all the information about the figure's movement at your immediate disposal.

process: model

Translations

The first company for which you freelance is a manufacturer of kites. The company wants its kite figure to move up the screen as it moves from left to right. The criteria for the kite's appearance on the screen says that it should be **4 centimeters to the right** and **3 centimeters further up the screen** every time it reappears.

13. The initial kite is pictured below. Model its movement on the screen.

14. Show a vector on the coordinate grid that would represent each successive movement of the kite.

15. Describe the movement of the kite using the shorthand notation for translations combined with the mapping notation. (**T**(x,y) =)

process: model

Translations

The second side contract you secure is for a company that makes honey. The logo for this company is a part of a honeycomb. You decide to use the simplest part of the honeycomb, a hexagon, to model the movement across the computer screen. This company provides you with only the starting positions of the vertices: (0,0); (5,0); (8, -4); (5, -8); 0, -8); and (-3, -4), and the mathematical notation for the movement:

$$\mathbf{T}(x,y) = (x-8, y+10)$$

16. Use the coordinate grid to model the hexagon's movement.

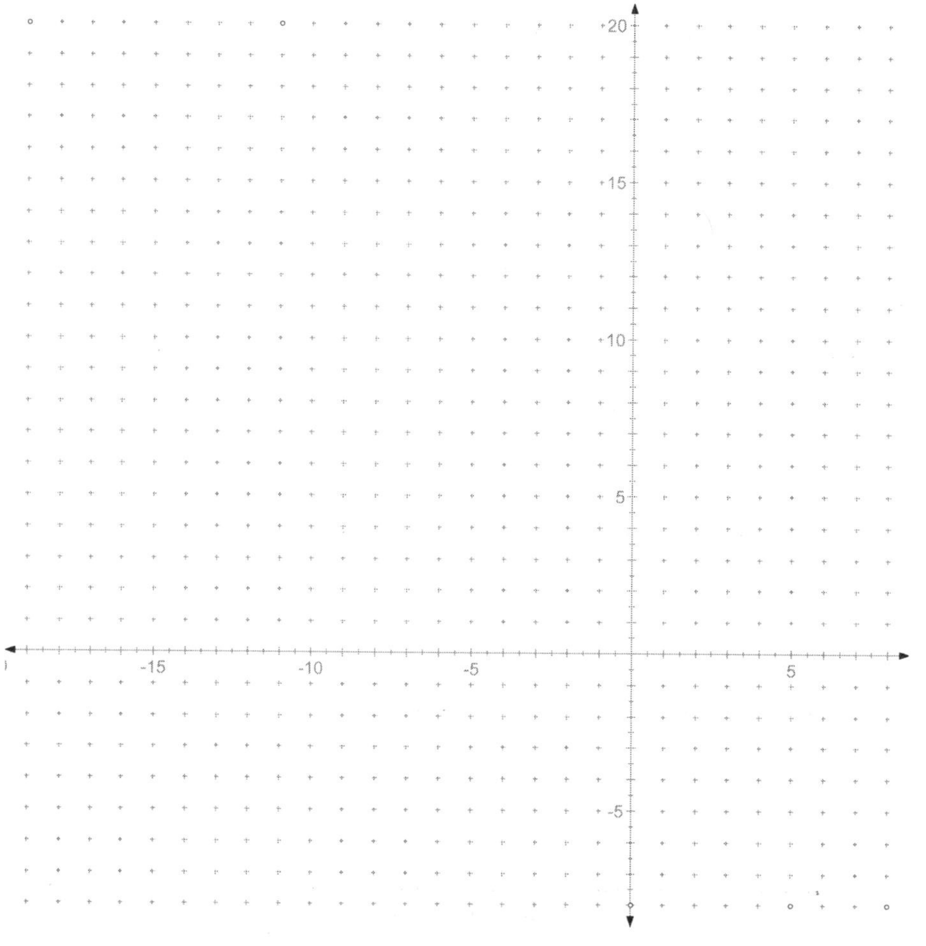

17. Show a vector on the coordinate grid that would represent each successive movement of the hexagon.

18. Describe, in full sentences, how the hexagon will move across the screen.

process: model

Translations

The third and final side job you get is for a school district. The district's Chief of Technology provides you with the coordinates of the vertices of the triangular emblem of the school. They are: (7,7); (15, 7); and (11, 15). You are also provided with the vector showing how the school's emblem should move across the screen. The vector is pictured below. (Note that the vector does *not* begin at a vertex, so it is showing only the **distance** and **direction** of the movement, *not* the original location of the triangle.)

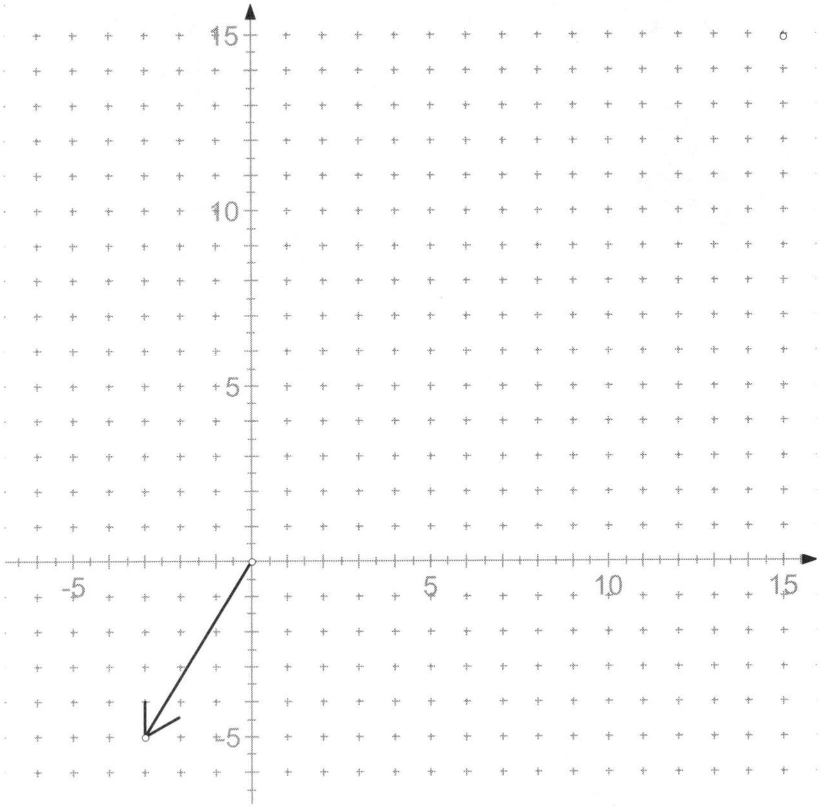

19. Use the coordinate grid to model the triangle's movement across the screen.

20. Describe, in full sentences, how the triangle moves across the screen.

21. Use the **T**(x,y) notation to describe the movement of the triangle.

process: summary

Translations

22. What are two different ways to perform a translation?

23. Imagine that a point (x,y) is translated horizontally by "a" units and translated vertically by "b" units. Use complete sentences to explain how you can go about finding the new position of this point. Then give the new coordinates of the point and use the notation you just learned to describe the translation.

Rotations

input

Our ability to live on the Earth's surface is based on the fact that the Earth rotates. The Earth's *rotation around the sun* places us at a distance from this energy source so that we don't "burn up," but permits us to derive enough heat to survive. The Earth's *rotation on its own axis* creates a force that keeps the atmosphere in place so that we have the right mix of oxygen and nitrogen (and other elements) to stay alive.

In this section, you will perform some rotations around points on the figures, similar to the Earth rotating on its axis, as well as around points outside figures, similar to how the Earth rotates around the sun.

Rotations

analysis

1. The physical representation of a rotation you are familiar with is that of turning. Cite at least two examples of everyday objects that involve rotations. Describe the rotation that is taking place in each of the examples you used.

2. Perhaps the simplest example to use is the model of the hands of a clock. Around what are the hands of the clock rotating?

   a. Illustrate which direction is clockwise and which direction is counterclockwise.

Rotations

input

Rotations are typically measured in degrees. Rotations are positive when they occur in a **counterclockwise** direction, and rotations are negative when they move in a **clockwise** direction.

process: model

Rotations

3. If the hand of a clock moves from 9 to 12, what is the degree of this rotation? Is this rotation positive or negative?

4. Describe and show an example of the hands of a clock moving -60°.

process: model

Rotations

In this task, you will rotate objects around another object, which is usually a point.

5. Place a piece of tracing paper over the picture below. Trace the triangle.
6. Place your pencil point on the center of rotation.

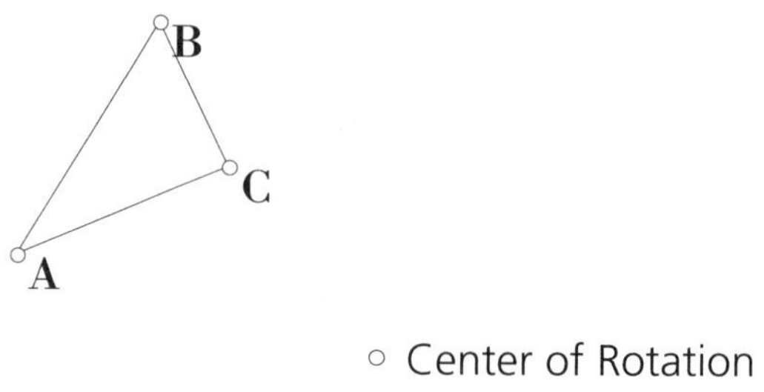

a. Turn the tracing paper by 90°. Return to the original starting position.

b. Turn the tracing paper by 180°. Return to the original starting position.

c. Turn the tracing paper by 270°. Return to the original starting position.

process: model

Rotations

7. Change the center of rotation so that it is point A. Place your pencil point on A.
   a. Use your tracing paper to rotate the triangle by approximately 60°. Return to the starting position.
   b. Rotate the triangle by about 45°.
   c. Rotate the triangle by -45°.

8. Continue rotating the triangle using other points on the triangle or outside the triangle as the center of rotation. Make sure to rotate in both positive and negative directions. List the centers of rotation and the angles of rotation that you used.

analysis

Rotations

9. What properties of the triangle are preserved under rotations?

10. What properties of the triangle change when rotations are performed?

process: model

Rotations

In the last activity, you experimented with rotating a triangle, but because you used tracing paper, the rotated triangles do not appear on your paper. In this activity, you will rotate a few geometric figures using your compass and protractor.

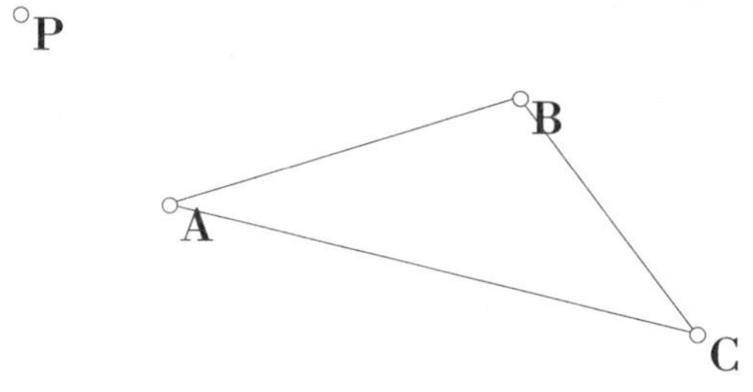

11. Point P is your center of rotation, so use your compass to create a circle that is centered at P and passes through A. Create segment AP, which is the radius of the circle.
12. You will rotate the triangle 50°, so you will use your protractor to measure 50° along the circle.
    a. Center the protractor at P
    b. Align the edge of the protractor with $\overline{AP}$. Make sure to move in the positive direction.
    c. Mark 50° along the edge of the protractor.
    d. Connect P with your 50° mark, using the straightedge of the protractor.

    You've now created a 50° angle, so the point where this segment intersects the circle is A′.
13. Starting with the creation of circles, follow the same directions in order to rotate point B and point C by 50°.

With these three circles drawn, you have completed half of any other rotation you might perform around point P.

process: practice

Rotations

14. Apply the procedure you just learned for rotating an object. Use the following quadrilateral. Choose one of the four vertices of the quadrilateral as your center of rotation. Choose a number of degrees and a direction for your rotation. (Use your compass and protractor.)

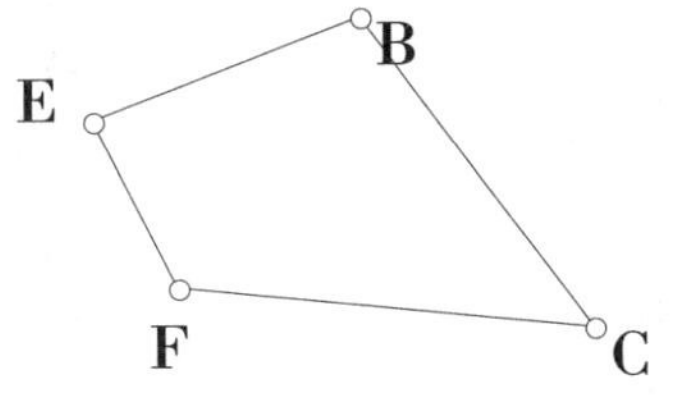

process: model

Rotations in the Coordinate Plane

As you saw with reflections and with translations, rotations can also take place in the coordinate plane.

1. Label the vertices of the triangle and give the coordinates of each of the vertices on the coordinate grid.
2. Rotate the triangle by -90° around the origin. If you use only positive numbers, what is this measure?

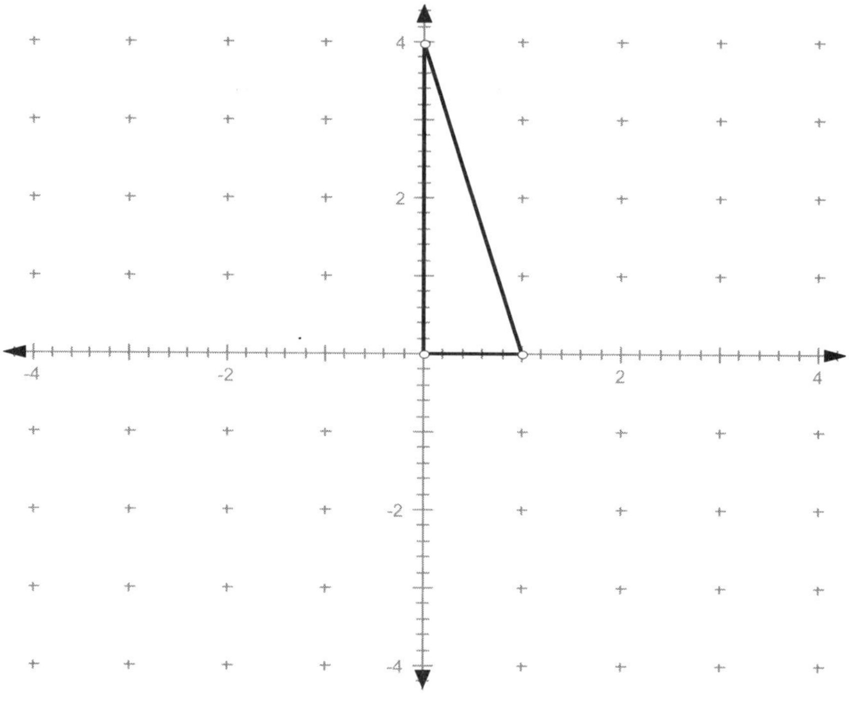

3. Label the vertices of the rotated triangle and give the coordinates of the vertices on the coordinate grid.
4. Use the mapping notation [(x,y)→] to describe where each of the original points goes under the rotation.
5. Use the mathematical notation for rotations to describe this rotation.

process: model

Rotations in the Coordinate Plane

6. Give the coordinates of the vertices of the quadrilateral on the coordinate grid.
7. Rotate the quadrilateral by 90° around the origin. (You may want to use tracing paper for help on this one.) If you use negative numbers in place of 90°, what is the measure of this rotation?

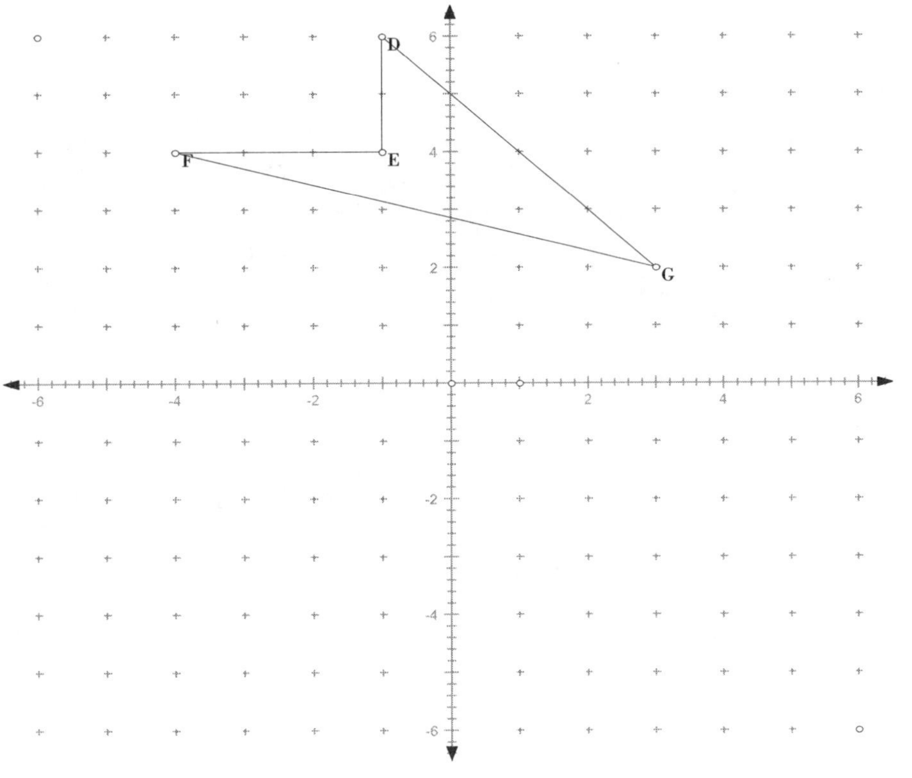

8. Label the vertices and give the coordinates of the rotated quadrilateral on the coordinate grid.
9. Use the mapping notation [(x,y)→] to describe where each of the original points goes under the rotation.

10. Use the shorthand notation for rotations (**R**) to describe this rotation.

process: model

Rotations in the Coordinate Plane

11. Give the coordinates of the vertices of the pentagon on the coordinate grid.
12. Rotate the pentagon by 180° around the origin. If you use negative numbers in place of 180°, what is the measure of this rotation?

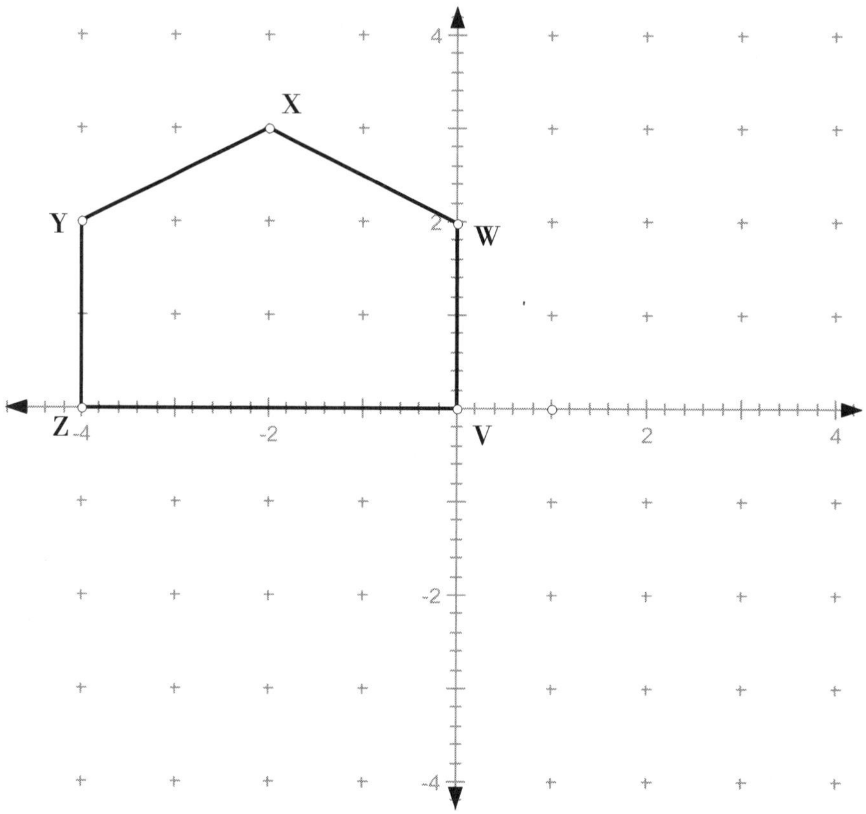

13. Give the coordinates of the transformed pentagon on the coordinate grid.
14. Use the mapping notation [(x,y)→] to describe where each of the original points goes under the rotation.

15. Use the shorthand notation for rotations to describe this rotation.

analysis

Rotations in the Coordinate Plane

16. Looking back over the three rotations you have just performed, what patterns do you see when you compare the new coordinates to the old coordinates? Rotate *each* of the three figures by 90°, 180°, and 270° in order to confirm the patterns you think you see. Describe any and all of the patterns you see using full sentences.

17. Use the mapping notation with (x,y), instead of with specific points, to describe the patterns you see.

process: evaluate

Rotations in the Coordinate Plane

As with translations, this notation can also be combined with the mapping notation. So, you might write:

$$\mathbf{R}_{180°}(x,y) = (-x, -y).$$

18. Complete these mathematical statements.

a. $\mathbf{R}_{90}(x,y) =$ b. $\mathbf{R}_{180}(x,y) =$ c. $\mathbf{R}_{270}(x,y) =$

19. Explain why these patterns exist.

process: summary

Rotations in the Coordinate Plane

20. Summarize your findings about isometries thus far. Include the relationships you discovered between reflections, translations, and rotations. Also make sure to note any patterns you discovered with these isometries and the coordinate plane.

21. Complete the following chart by filling in the boxes to indicate whether a given property stays the same or changes under a given isometry.

		ISOMETRY		
		Reflections	*Translations*	*Rotations*
PROPERTY	*Length*			
	Angle measure			
	Position			
	Orientation			
	Shape			
	Size			

preview

Dilations

Thus far in the unit on transformations and symmetry, you have discussed reflections, rotations, and translations in detail. Each of these transformations is an isometry.

1. What is common to all of the transformations you have discussed?

2. In this section, we will discuss a new transformation that is not an isometry. Thinking about the definition of an isometry, what might be different about this transformation?

input

Dilations

Have you ever used a flashlight to create shadow puppets on the wall? The light makes the image of your hands, which are formed into a make-believe rabbit or alligator, appear larger. The image of your "puppet" is a dilation of your hands.

A model or scaled-down version of a new building, mall, or park that is about to be built is a dilation, representing a smaller version of the actual object.

3. In your own words, describe a dilation and give some other examples of dilations in everyday situations.

In this section, you will be dilating some geometric figures in order to discover relationships between dilated figures.

process: model

Dilations

4. Given ΔABC on the grid below, place a point Y to the left of it. Imagine that Y is the flashlight and ΔABC is your hand when you create the shadow puppet.

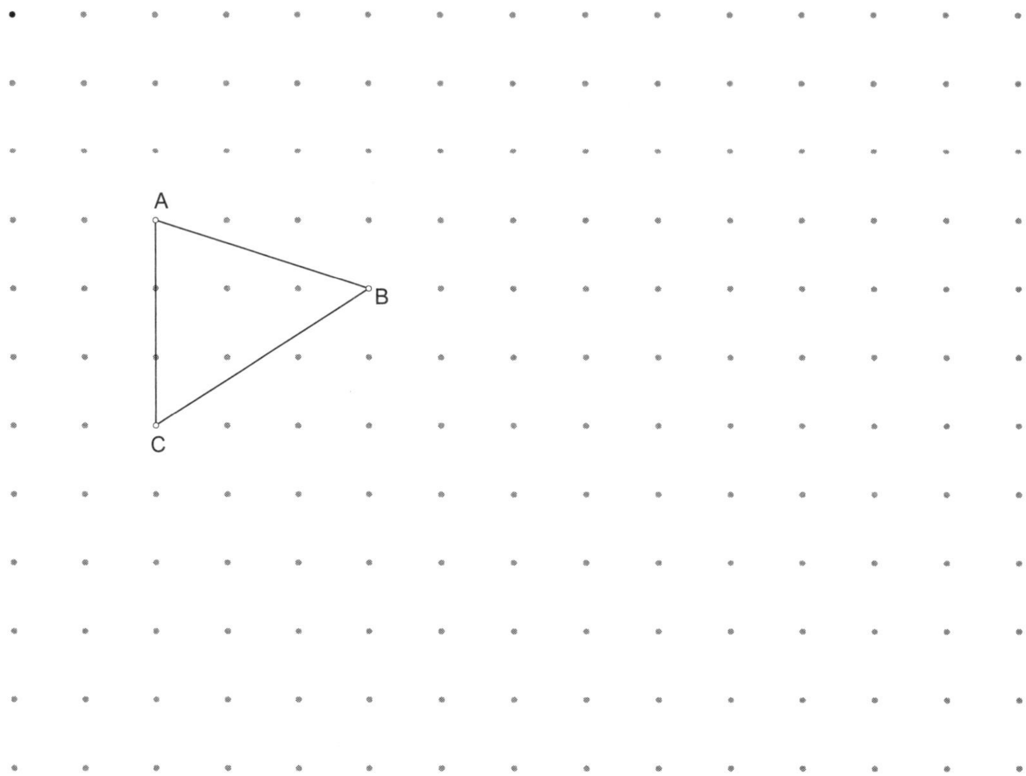

5. Create three line segments, $\overline{YA}$, $\overline{YB}$, and $\overline{YC}$. Measure the distances YA, YB, and YC. Record them on your picture.
6. Create three rays, $\overrightarrow{YA}$, $\overrightarrow{YB}$, and $\overrightarrow{YC}$, representing the rays of light from the flashlight. Extend the rays as far as you can on the page.
7. Triple the distances you found in step 2. Use these tripled distances to create new segments along your rays. YA′ should be 3*YA, YB′ should be 3*YB, and YC′ should be 3*YC.
8. What is the ratio of YA to YA′?

9. What is the ratio of YB′ to YB?

10. Connect your new points A′, B′, and C′ to create a new triangle. This new triangle is like the image of your shadow puppet on the wall.

	Dilations
process: evaluate	You should now have $\Delta A'B'C'$, which is the image of a dilation of ΔABC with the center of dilation Y. 11. Measure the three angles of ΔABC and record them on the figure. 12. Measure the three angles of $\Delta A'B'C'$ and record them on the figure. 13. Carefully measure the three sides of ΔABC. Record the measurements. 14. Carefully measure the three sides of $\Delta A'B'C'$ and record these measurements. 15. What is the ratio of AB to A'B'? 16. What is the ratio of B'C' to BC?

	Dilations
analysis	17. How do the measures of the angles compare? 18. Do you think the same will be true in every dilated figure? Why or why not? 19. How do the measures of the three sides compare? 20. Do you think the same will be true in every dilated figure? Explain.

analysis

Dilations

21. The ratio you found on the previous page is called the **scale factor** or dilation factor. Why do you think the ratio of the lengths is called the scale factor?

input

Dilations

Dilations have a shorthand notation associated with them. The notation for recording a dilation is the symbol **S**, (not the symbol D, as you may have thought). So, with regard to the dilation you just completed, you might write

S(A) = A′

or

S(BC) = B′C′

22. Write the shorthand notation for three other dilations that occurred in your ΔABC.

process: model

Dilations in the Coordinate Plane

Like other transformations, dilations can occur in the coordinate plane.

In the last activity, the center of dilation was outside the triangle you dilated. In this activity, the center of dilation will be a point *on* the figure. It will be point C on the quadrilateral.

1. Use a straightedge to create rays from point C, passing through points D, E, and F respectively.
2. Measure the distances CD, CE, and CF. Record these distances on the figure.
3. Double the distances. Use these doubled distances to create segments CD', CE', and CF' along the rays you drew.
4. Create a dilated quadrilateral, CD'E'F'.

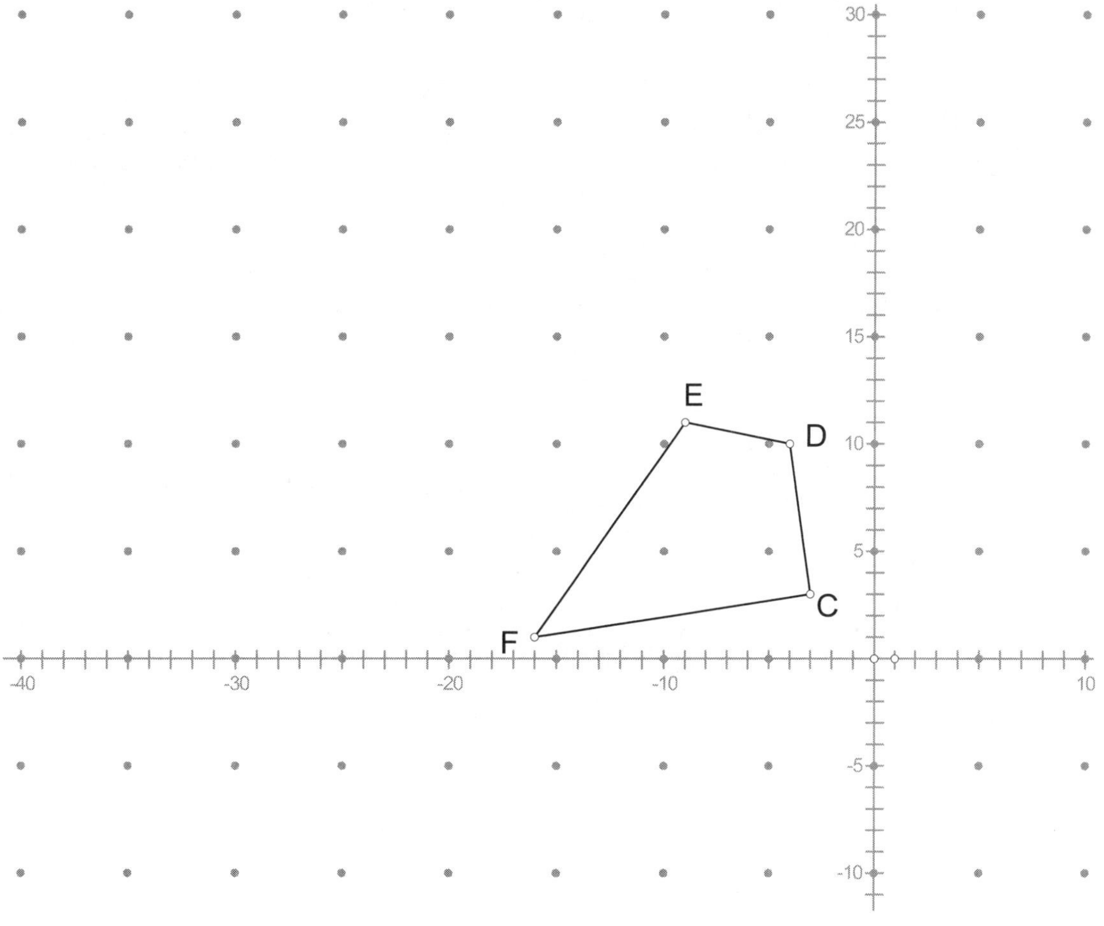

process: evaluate

Dilations in the Coordinate Plane

5. Measure the four angles of quadrilateral CDEF and record them on the figure.
6. Measure the four angles of quadrilateral C′D′E′F′ and record them on the figure.
7. Measure the four sides of quadrilateral CDEF. (Two of these should already have been measured in step 2.) Record the measurements on the figure.
8. Measure the four sides of quadrilateral C′D′E′F′ and record these measurements.
9. What is the ratio of DE to D′E′?

10. What is the ratio of E′F′ to EF?

analysis

Dilations in the Coordinate Plane

11. How do the measures of the angles compare?

12. Do these angle measures confirm or dispute your findings from the first dilation you performed? Why?

13. How do the measures of the sides compare?

14. Do these measurements and ratios confirm or dispute your conjecture from the first dilation? Explain. Revise your conjecture if necessary.

15. Use your findings from these three dilations to create a more formal definition of dilation.

input

Symmetry

When you see objects like flowers, ice crystals, butterflies, and starfish, what do you observe about their form? The balance that you observe in these objects and many others is known as **symmetry**. You will see symmetry in some works of art. You will notice it in the way some people decorate their homes. You may even notice symmetry being used in community planning.

Certain isometries are linked to a type of symmetry. Reflections are linked to **mirror symmetry** or **bilateral symmetry**. The objects pictured below contain bilateral symmetry.

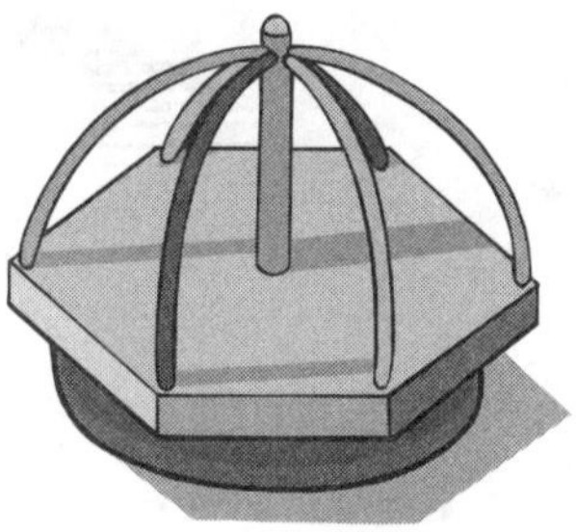

The objects pictured below do NOT contain bilateral symmetry.

analysis

Symmetry

1. Based on the examples and non-examples you viewed above, attempt to create your own definition for **mirror** or **bilateral symmetry**.

input

Symmetry

Another type of symmetry is **turning** or **rotational symmetry**, which is linked to rotations. Examples of objects containing rotational symmetry:

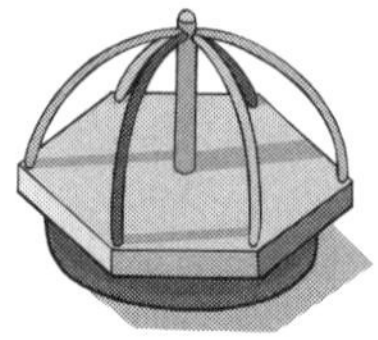

Non-Examples of rotational symmetry

analysis

Symmetry

2. Based on the examples and non-examples you viewed above, attempt to create your own definition for **turning** or **rotational symmetry**.

input

Symmetry

A figure displays **reflectional symmetry** when a line can be drawn to split the figure into two identical halves, so that each half is a mirror reflection of the other. The line that splits the figure in half is called the **line of symmetry** or the **axis of symmetry**. Some figures that display reflectional symmetry have only one line of symmetry, while others may have several lines of symmetry.

analysis

Symmetry

3. How many lines of symmetry are there in an isosceles triangle? Sketch any and all lines of symmetry in the following isosceles triangle.

4. How many lines of symmetry are there in a square? Sketch any and all lines of symmetry in the following square.

process: practice

Symmetry

5. In each figure below, sketch any and all lines of symmetry.

a. 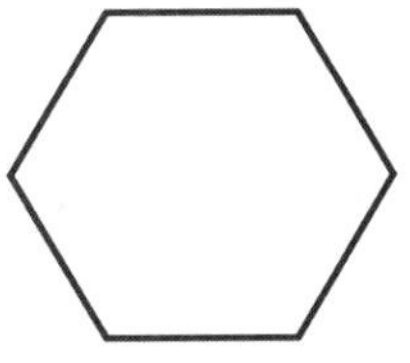

b. 

c. 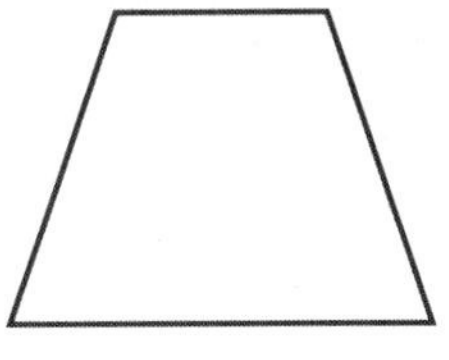

d. 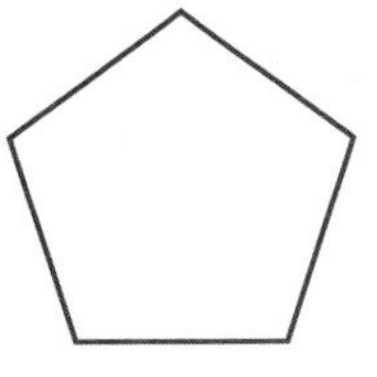

e. 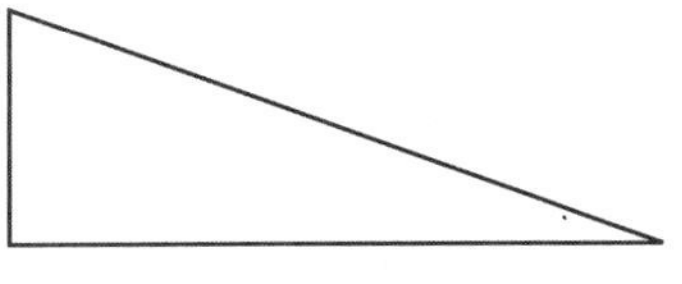

process: practice

Symmetry

6. On each coordinate system below, you are given part of a figure and its line of symmetry. Sketch the rest of each figure, based on the given line of symmetry.

   a. The line of symmetry is the y-axis.

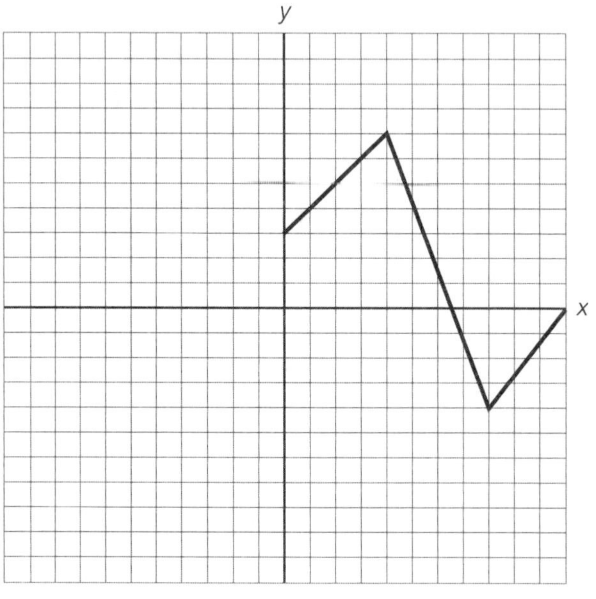

   b. The line of symmetry is the vertical line $x = 3$.

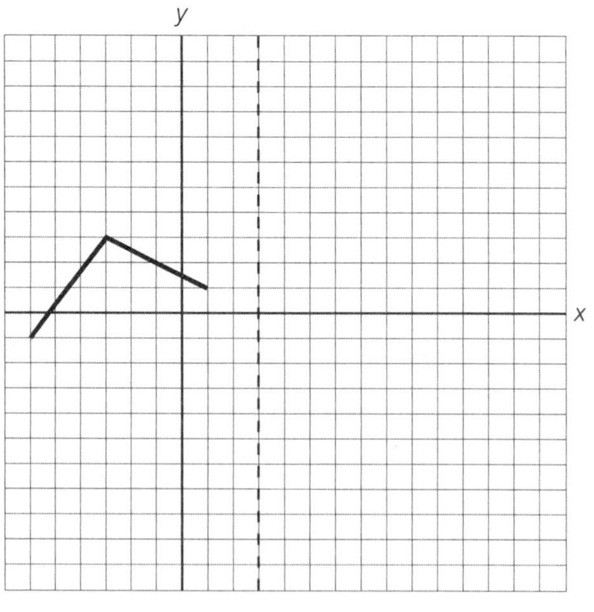

input

Symmetry

A figure displays rotational symmetry when it can be rotated a certain number of degrees about a point so that after the rotation, the figure is in the same position as it started. Some figures that display rotational symmetry have one rotational symmetry, while others may have several rotational symmetries. When a figure can be rotated 180° about a point and the resulting figure coincides with the original, the figure is said to have **point symmetry**.

Examples:

Ellipse

An ellipse has one rotational symmetry: 180° about its center.

initial position　　　　180° counterclockwise rotation

Equilateral Triangle

An equilateral triangle has two rotational symmetries: 120° and 240° about its center point.

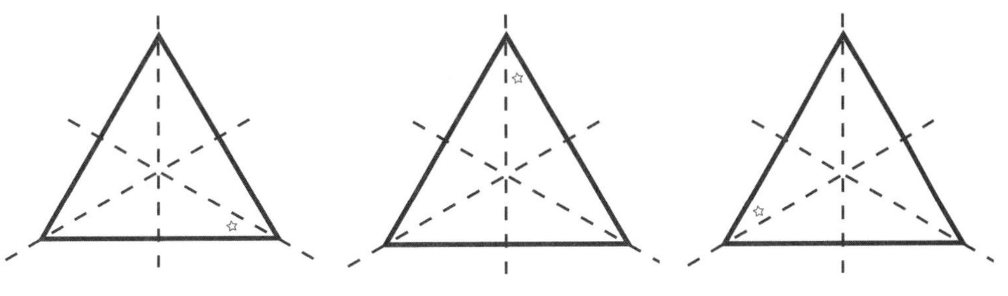

initial position　　　　120° rotation　　　　240° rotation

Symmetry

analysis

7. For each figure below, tell how many degrees the figure must be turned to show rotational symmetry. Keep in mind that some of these figures have more than one rotational symmetry. Stop when you have rotated 360°.

   a. Square

   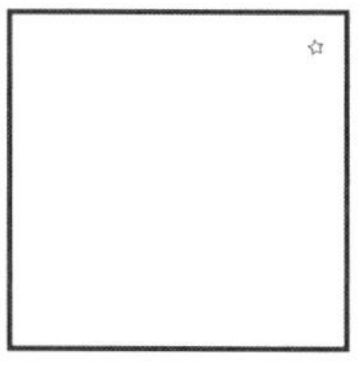

   Rotational Symmetries:

   b. Pentagon

   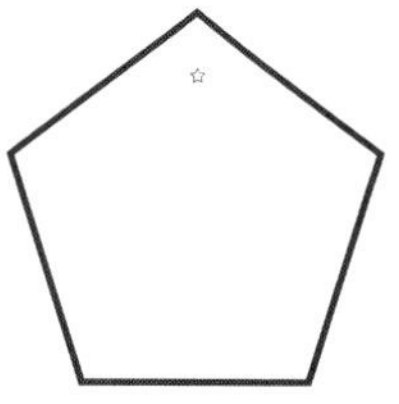

   Rotational Symmetries:

Symmetry

process: practice

8. On each coordinate system below, you are given part of a figure and information on the rotational symmetry of the full figure. Sketch the rest of the figure, based on the information about its rotational symmetry.

   a. Three rotational symmetries: 90°, 180°, and 270° about the origin

   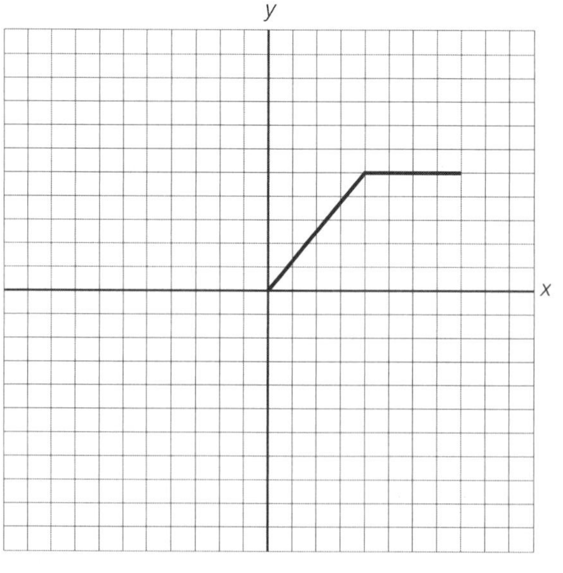

input

Revisiting the Urban Sprawl Problem

Now that you have studied transformations and symmetry, it is time to revisit our problem from the first day of this unit to assess what you have learned.

Below is a scaled-down version of some town planning documents. Line PW represents the current power line. Point S represents a new shopping district that is developing, while point R represents a new residential plan of single-family homes that is in the process of being built. Currently, no electricity is being supplied to either of the new areas. Your job is to determine where to place a new electrical substation so that the length of new power line is minimized.

analysis

Revisiting the Urban Sprawl Problem

Where on line PW should the electrical substation be placed so that the distance from R to the substation to S is minimized?

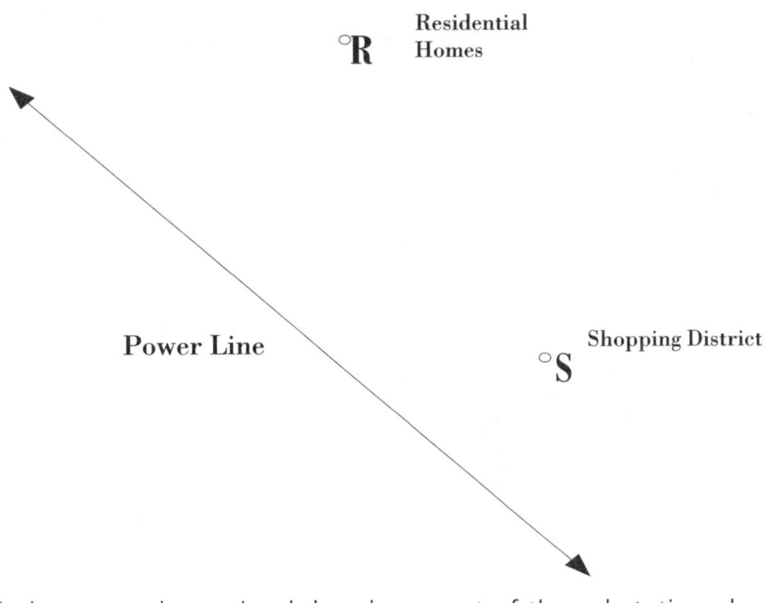

1. Explain how you determined the placement of the substation along the power line.

Unit 6: Similarity

Contents

preview

Washington Monument Problem

Is it possible to determine the height of the Washington Monument using only a tape measure?

Imagine you are standing with your back to the Lincoln Memorial gazing into the reflecting pool at the reflection of the Washington Monument. To measure the height of the monument:

- You assume your eyes are 6 feet above the ground.
- You see the very top of the monument in the reflecting pool that is situated between the Lincoln Memorial and the Washington Monument.
- You measure the distance from the spot where you are standing to the spot where you see the top of the monument in the reflecting pool. That distance is 12 feet.
- You measure the distance from the location in the reflecting pool where you see the top of the monument to the base of the monument. That distance is 1,110 feet.

With this information, try to find the height of the Washington Monument. If you have trouble getting started, reread the problem carefully. Draw a diagram of the situation as you read. Show your work as you attempt to solve this problem.

process: summary

Similar and Congruent Figures

In the previous unit on Transformations and Symmetry, you studied isometries as well as dilations.

1. Define the term "isometry" as you understand it.

   a. Of the transformations you have studied, which of them are isometries?

   b. Sketch at least two examples of isometries.

2. Define the term "dilation" as you understand it.

   a. Sketch at least two different examples of dilations.

Similar and Congruent Figures

analysis

3. How is an isometry different from a dilation?

4. How is an isometry the same as a dilation?

In this unit, you will continue to learn about isometries and dilations, but you will do so from some different perspectives.

5. The following two figures are similar figures. Do these figures illustrate an isometry or dilation or a combination of isometries and dilations? Explain your response.

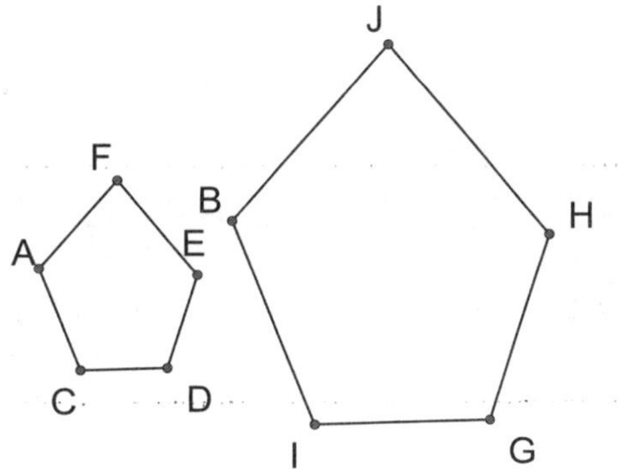

6. The following figures are congruent figures. Do these figures illustrate an isometry or dilation or a combination of isometries and dilations? Explain your response.

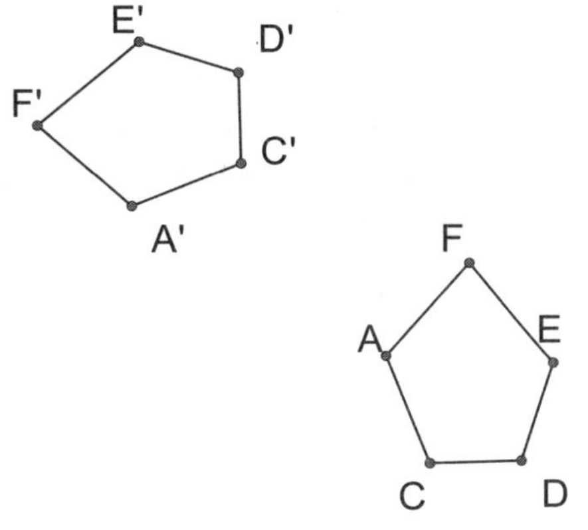

analysis

Similar and Congruent Figures

7. Based on your experiences with the dilation of figures or objects, what are two properties of similar figures that you already know?

8. Based on your experiences with the isometries, what are two properties of congruent figures that you already know?

input

Similar and Congruent Figures

There are many examples of the uses of similar figures and congruent figures outside your math class. For example, a realistic model train is similar to an actual locomotive. A photograph that is enlarged for framing may be similar to the original. A set of baskets of different sizes that nest inside one another are similar objects.

9. List at least two more examples of similar objects that you have encountered.

Congruent objects are identical to one another. For example, many of the desks in your classroom are congruent to one another. Plastic children's toys that are made by using a mold are also examples of congruent objects.

10. List at least two other examples of congruent objects.

process: evaluate

Similar and Congruent Figures

Let's use similar and congruent figures to find some missing measurements.

Artists and architects often use scaled drawings or models before working on the life-sized objects.

11. The sculptor of the statue of Crazy Horse that is being constructed in South Dakota used a scaled-down model of the sculpture before executing the construction of the real thing. In the completed sculpture, the face will be 87 feet 6 inches high, while the total height of the sculpture will be 563 feet. What is the height of the model used, if the length of Crazy Horse's face in the model is 20 inches (1.75 feet)? Create a sketch of the model and the sculpture, then show your work as you solve for the unknown height.

Similar and Congruent Figures

input

You have already learned about special mathematical notation for segments and rays, for parallel lines and perpendicular lines, and for various transformations. There is also a special mathematical notation used to denote similar figures and congruent figures.

The symbol that means similar is a small wavy figure that looks like ~. It is sometimes called a tilde.

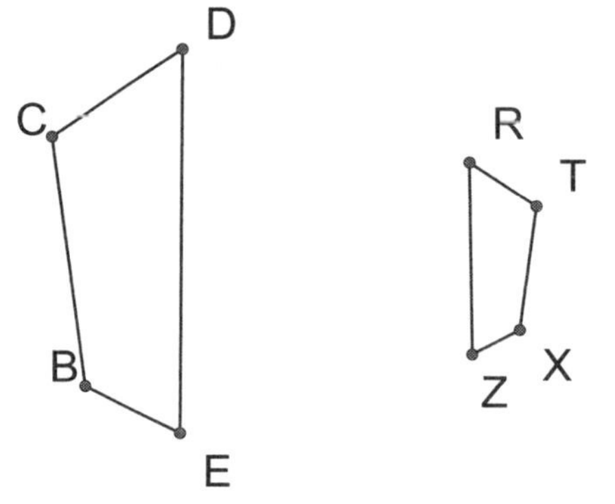

To describe the relationship between the two figures above, mathematicians would write: Quadrilateral BCDE ~ Quadrilateral XTRZ. This notation is read as, "Quadrilateral BCDE is similar to quadrilateral XTRZ."

In the similarity statement, the vertices are named in corresponding order. Because vertex B pairs with vertex X when the vertices are "matched," they are both named first. Vertex C matches with vertex T, so they are each named next. Notice that the vertices of both quadrilaterals are named in opposite directions (clockwise or counterclockwise). Since quad. BCDE was dilated and then reflected to get to quad. XTRZ, the orientations are different, making one orientation clockwise and one counterclockwise.

Similar and Congruent Figures

analysis

12. Are the quadrilaterals above illustrating an isometry, dilation, or a combination of isometries and dilations?

Similar and Congruent Figures

process: model

13. Write a similarity statement to describe the relationship between the following similar triangles.

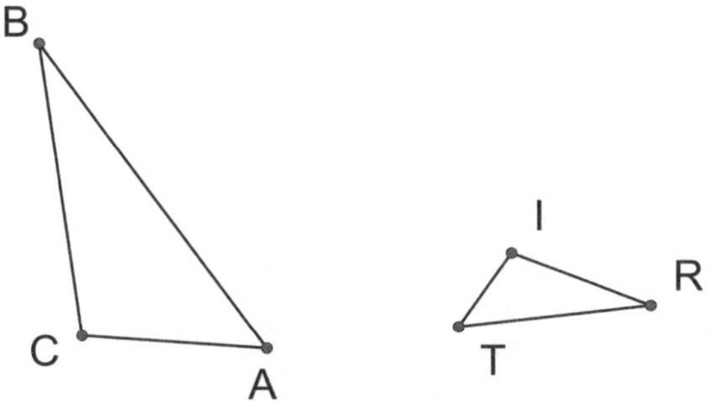

14. Sketch two figures to illustrate the following statement: $\Delta MOL \sim \Delta XYZ$.

The symbol for congruency is the same symbol as the one used for similarity except an equals sign is written below it. That symbol looks like this: $\cong$

15. Write the words that correspond to the following shorthand description of these two congruent triangles. $\Delta ABC \cong \Delta WEQ$.

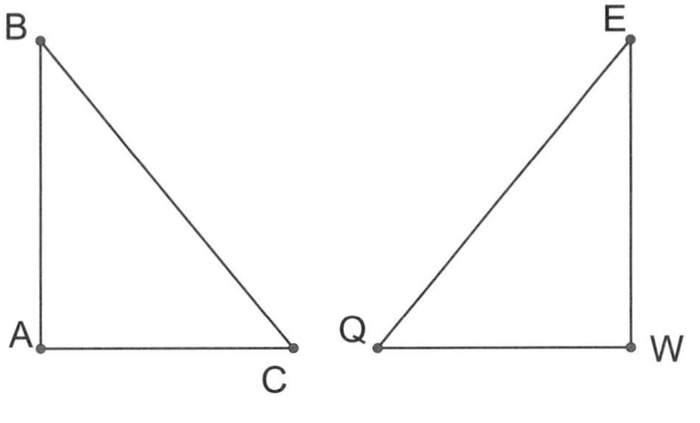

Similar and Congruent Figures

analysis

16. Are the two quadrilaterals pictured below similar? Use your ruler and protractor to help you determine the answer. Explain your reasoning. If the figures are similar, write a similarity statement to describe them.

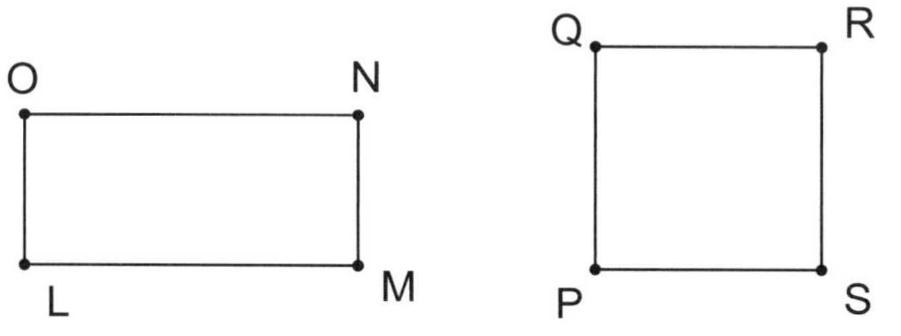

a. Are these quadrilaterals congruent? Explain your reasoning. If the figures are congruent, write a congruency statement to describe them.

Similar and Congruent Figures

analysis

17. Are the two quadrilaterals pictured below similar? Use your ruler and protractor to help you determine the answer. Explain your reasoning. If the figures are similar, write a similarity statement to describe them.

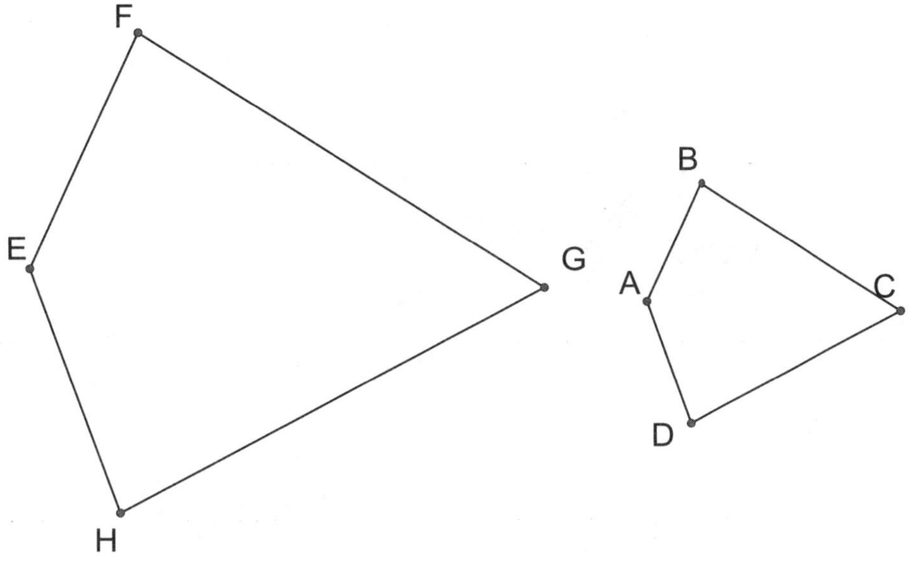

a. Are these two quadrilaterals congruent to one another? Explain your answer. If the quadrilaterals are congruent, write a congruency statement to describe this relationship.

analysis

Similar and Congruent Figures

18. Are the two hexagons pictured below similar? Use your ruler and protractor to help you determine the answer. Explain your reasoning. If the figures are similar, write a similarity statement to describe them. .

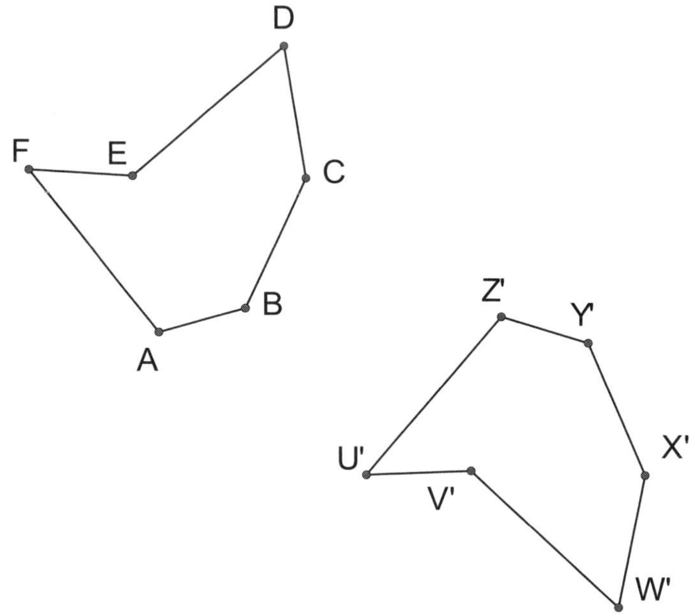

a. Are these two hexagons congruent to one another? Explain your answer. If they are congruent, write a congruency statement describing this relationship.

analysis

Similar and Congruent Figures

19. Describe the relationship between the corresponding angles of two similar figures.

20. Describe the relationship between the corresponding sides of two similar figures.

21. Describe the relationship between the corresponding angles of two congruent figures.

22. Describe the relationship between the corresponding sides of two congruent figures.

23. *If two triangles are similar, then they are congruent.* Is this statement true or false? Explain.

24. *If two triangles are congruent, then they are similar.* Is this statement true or false? Explain.

Similar and Congruent Figures

analysis

25. Is it possible to draw two *rectangles* that are not similar? If so, give an example, label the figures, and explain why they are not similar. If all rectangles are similar, explain why.

26. Is it possible to draw two *squares* that are not similar? If so, give an example, label the figures, and explain why they are not similar. If all squares are similar, explain why.

Similar and Congruent Figures

process: summary

27. Can two similar figures be used to illustrate an isometry?

28. Can two similar figures be used to illustrate a dilation?

29. Can two similar figures illustrate a combination of isometries and dilations?

30. Can two congruent figures illustrate an isometry?

31. Can two congruent figures illustrate a dilation?

32. Can two congruent figures illustrate a combination of isometries and dilations?

process: summary

Similar and Congruent Figures

33. Your little brother is watching you do your geometry homework on similar and congruent figures. He is curious about the meaning of different terms. You want to satisfy his curiosity, but since he is only six years old, you cannot use formal definitions. Instead, you need to define the terms using everyday language. Write an explanation of similar and congruent figures so that your little brother will understand. Make sure to include pictures and everyday examples.

process: practice

Finding Unknown Parts of Similar and Congruent Polygons

The two kites below are similar to one another.

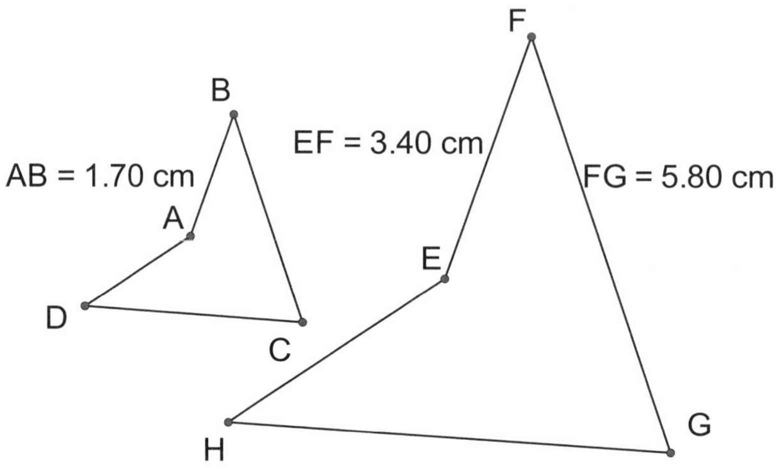

1. Write a similarity statement describing their relationship.

2. Using the given measurements, determine the scale factor between the two figures. Explain what you did to determine the scale factor.

3. Using the scale factor you just found, determine the length of segment BC. Explain what you did to determine $\overline{BC}$.

process: practice

Finding Unknown Parts of Similar and Congruent Polygons

4. The following triangles are similar to one another with a scale factor of 0.25.

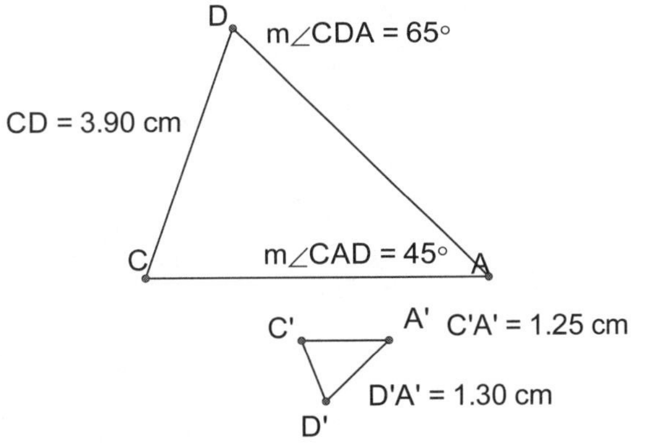

   a. Write a similarity statement about these triangles.

   b. Find the missing sides and the missing angles using what you know about similar figures.

5. The following triangles are congruent to one another.

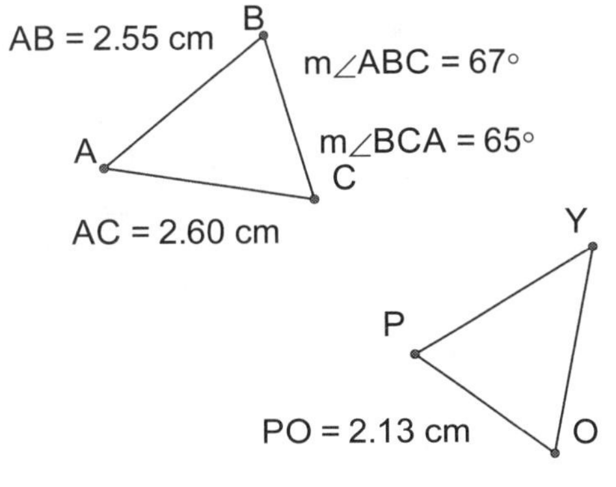

   a. Write a congruency statement for these triangles.

   b. Find the missing measures of the sides and angles.

analysis

Finding Unknown Parts of Similar and Congruent Polygons

6. Based upon the information given in the diagram, are the following triangles congruent? Why or why not?

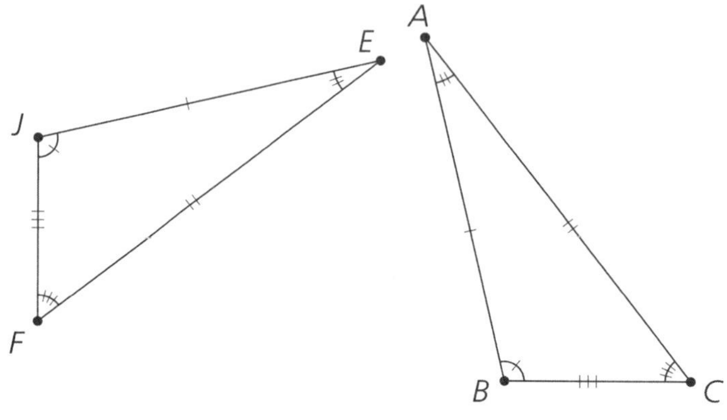

7. Based upon the given information, are the following quadrilaterals similar? Why or why not?

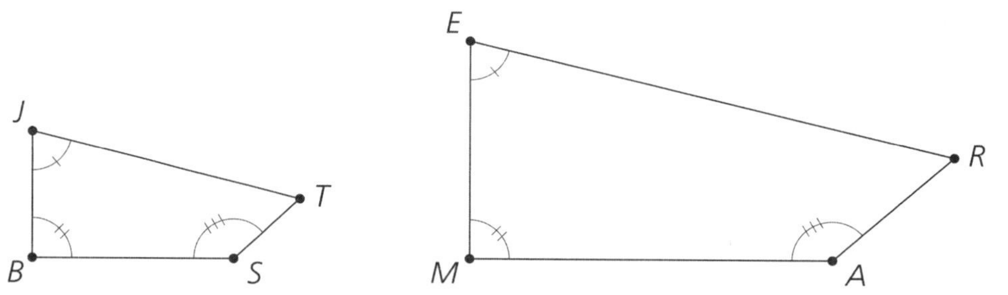

$$\frac{BS}{MA} = \frac{ST}{AR} = \frac{TJ}{RE} = \frac{JB}{EM}$$

process: evaluate

Finding Unknown Parts of Similar and Congruent Polygons

8. One day you were wandering around your town. You noticed a large pine tree that you remember helping to plant with your elementary school class. You were curious about how tall it had grown. You made some measurements that you think you can use in conjunction with your knowledge of similar triangles to find the tree's current height.

   To start the problem, you made a sketch, and in the sketch ΔMEP ~ ΔTRP.

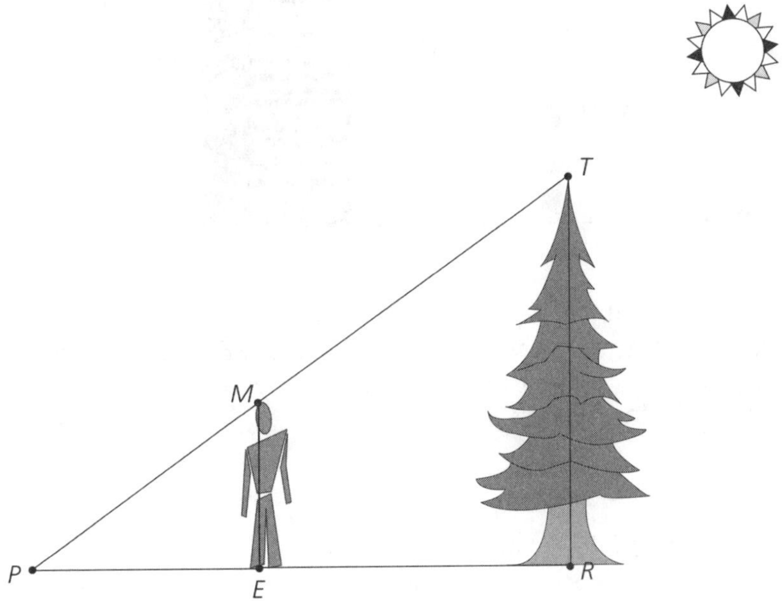

Line segment *ME* represents you at your height of 5 feet. Line segment *TR* is the tree and line segment *RP* is the tree's shadow, which is 42 feet in length. Line segment *PE* is your shadow measuring 7 feet. You and the tree are both perpendicular to the ground.

Without climbing the tree, calculate its height. Show your work.

process: evaluate

Finding Unknown Parts of Similar and Congruent Polygons

9. Years after finding the height of the tree in the previous problem, you were interviewing for a job in the USX tower in Pittsburgh, PA. You wondered about the height of this skyscraper. Use the following sketch and the given measurements to find the height of the building.

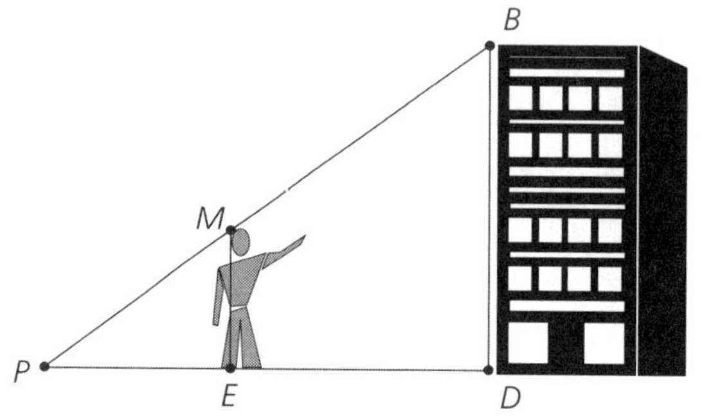

$\Delta MEP \sim \Delta BDP$

Line segment *ME* represents you, and your height is now 6 feet. Line segment *BD* represents the USX tower. Line segment *PE* is your shadow, which is 4′6″ in length. Line segment *PD* is the shadow of the building, which is 630′8″ long. Find the height of the tower. Show your work.

process: summary

Finding Unknown Parts of Similar and Congruent Polygons

10. Write a summary of the general process you used in solving the previous two problems. How did you make use of the similar triangles and given measurements in order to find the requested information?

input

Congruent Triangles

When you studied isometries in the last unit, and as you studied congruent figures in the first part of this unit, you saw that if every pair of sides and every pair of angles were equal when two figures were compared, then the figures were congruent.

In this activity, your goal is to find alternative methods for creating congruent triangles, so all six pairs of parts (three angles and three sides) do not need to be shown equal before concluding that the triangles are congruent.

process: model

Congruent Triangles

1. As carefully and completely as you can, define **congruent triangles**. Use any illustrations that may be helpful.

process: evaluate

Congruent Triangles

2. On a separate sheet of paper, use your straight edge to draw a triangle. Label the triangle ABC. After drawing the triangle, measure each angle and each side. Record the information on this page, in the space below. *Do not record this information on the actual triangle.* (You do not want your group members to know these measurements.) Measure as accurately as you can. (If you measure accurately, what should be true about the sum of the angles?)

   m ∠A = _____ AB = _______

   m ∠B = _____ CB = _______

   m ∠C = _____ AC = _______

   Exchange triangles with a member of your group.

process: model

Congruent Triangles

3. Examine the triangle given to you. On a clean sheet of paper, try to create an exact copy of the triangle without making a tracing and without asking your partner for the measurements of the triangle. However, you can certainly choose to make some of your own measurements. Do not use reflections to complete this activity.

4. When you are done copying your partner's triangle, label it A'B'C'.

5. Record the specific steps that you took to copy your partner's triangle in the space below. Be very specific, as if you are writing a recipe or an instruction manual for someone to follow.

Congruent Triangles

process: evaluate

6. Record the measurements of your new triangle (ΔA'B'C') below.

   m ∠A' = ________ A'B' = ______

   m ∠B' = ________ B'C' = ______

   m ∠C' = ________ C'A' = ______

7. Compare these measurements with the original measurements recorded by your partner before you traded triangles. Verify that the original triangle created by your partner and the copied triangle created by you are indeed congruent. How do they compare? If the triangles are congruent, what conditions must be met?

process: model

Congruent Triangles

8. Repeat the last exercise with your partner. Draw a triangle on a new piece of paper. This time, label it ΔXYZ, and make it a different triangle than your first triangle.

process: evaluate

Congruent Triangles

9. Record the measurements here.

m ∠X = _____ XY = _______

m ∠Y = _____ YZ = _______

m ∠Z = _____ ZX = _______

process: model

Congruent Triangles

10. Switch papers. This time when you copy your partner's triangle on a separate sheet of paper, try to copy it with the minimum number of steps. Your goal is to create congruent triangles using the fewest number of steps and without having to use all three pairs of corresponding sides and all three pairs of corresponding angles.

    Record the set of procedures (or "recipe" / "instruction manual") for creating congruent triangles in the space below. Label the new triangle X'Y'Z'. Your directions should be very specific and tell the reader exactly what to do and in what order.

    Hint: At some point in the process of creating these congruent triangles, you can stop measuring angles and sides in order to make the triangles congruent. At that point, when you draw the rest of the parts, they automatically fall into place or match up, without you purposefully making them match.

11. If you believe you have found one way to create congruent triangles, do not stop. Try to find another way. In fact, it may be possible to find 3 or 4 ways to make any triangle congruent to a given triangle.

process: model

Congruent Triangles

12. Create another triangle. Read your instructions to the other students within your group (or in your class) so they can create a copy of their triangle using your instructions. Be aware of which parts of the triangle you are copying before everything else falls into place.

13. Describe the methods you and your classmates developed by naming the parts that *had* to be copied before the rest of the triangle "fell into place." (Name the parts in the order that you encounter them as you work your way around the triangle, not necessarily in the order the parts were copied.)

process: summary

Congruent Triangles

14. If all the corresponding angles of two triangles are congruent and all the corresponding sides are congruent, then the triangles are congruent. In this activity, you found that you can determine whether two triangles are congruent using less information. How much information is sufficient for creating congruent triangles?

15. Summarize your class findings about creating copies of triangles (congruent triangles) without necessarily copying all six pairs of parts.

input

Proving Triangles Congruent

In the previous section of this unit, you developed methods for showing that triangles are congruent. One of these methods uses two sides of the triangle and the angle *included between* the two sides to show that two triangles are congruent. Once these three parts of the triangle are copied, the rest of the parts of the triangle automatically end up being equal, so the new triangle is automatically congruent to the original triangle. This method for showing that triangles are congruent is called the Side – Angle – Side method. (Note that the angle is *between* the two sides.) This method is abbreviated as SAS.

analysis

Proving Triangles Congruent

1. Look at the other methods you used to show that two triangles are congruent. Try to name these methods in a similar way to SAS.

2. The method outlined above uses three parts of the original triangle to create the new triangle. What is the minimum number of parts that need to be copied in order to assure that the original triangle and the new triangle are congruent?

3. Could you have copied only one part of the original triangle to create a congruent triangle? Why or why not?

analysis

Proving Triangles Congruent

4. Were you able to copy two parts of the original triangle and have the other parts "fall into place" to create a triangle congruent to the original? If so, explain which parts you used and explain the method. If not, tell why two parts were not enough to copy in order to create a congruent copy.

5. Even if you were unable to use one or two parts to create congruent triangles, you know at least one method (from the beginning of this section) that uses three parts to create congruent triangles. Are there other methods that use three parts to create congruent triangles? If so, what are those method(s)? If not, tell why SAS is the only method to use three parts.

process: practice

Proving Triangles Congruent

6. Use the methods you and your classmates created for showing that triangles are congruent to explain why the following triangles are congruent. Make sure to mark any parts that are given as equal before you start your explanation. Also, make sure to use full sentences in a thorough and complete explanation.

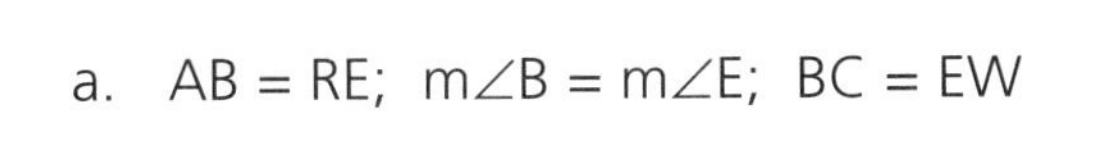

a. AB = RE; m∠B = m∠E; BC = EW

process: practice

Proving Triangles Congruent

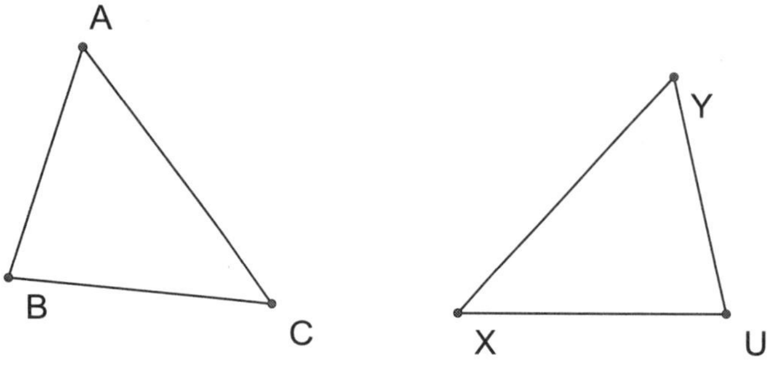

b. m∠A = m∠Y; m∠C = m∠X; CA = XY

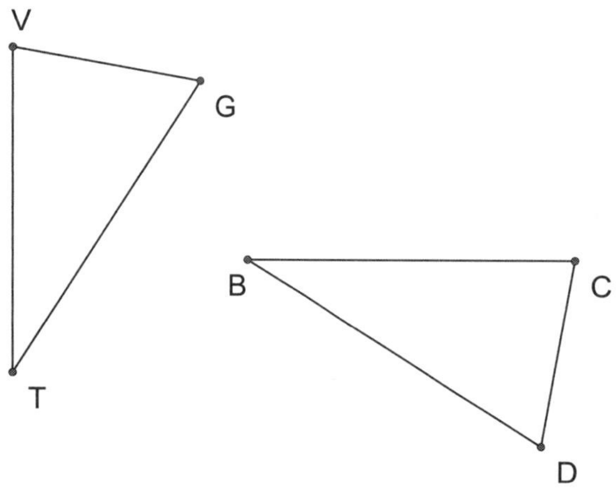

c. VT = CB; VG = CD; TG = BD

process: practice

Proving Triangles Congruent

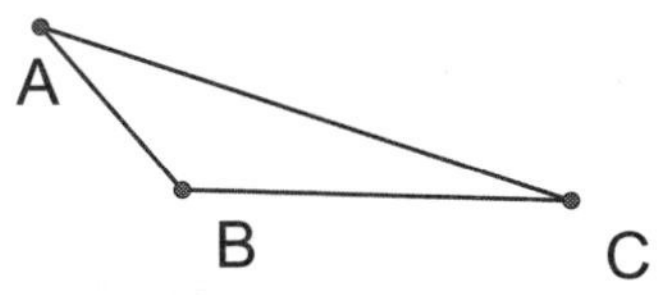

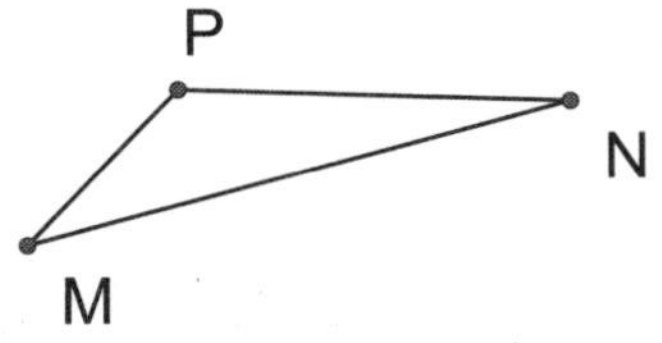

d. m∠A = m∠M; m∠B = m∠P; BC = PN

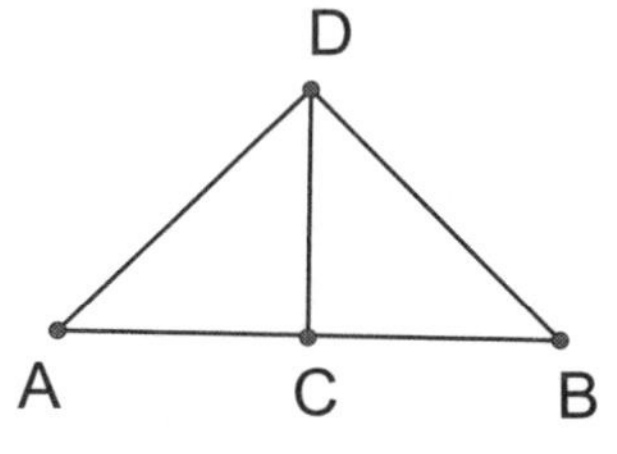

e. C is the midpoint of $\overline{AB}$; $\overline{DC} \perp \overline{AB}$

Note that in this problem, no parts are given as equal. Use what is given along with the diagram to deduce which parts of the triangle can be shown as equal. Use those equal parts to deduce how the triangle can be shown to be congruent.

process: practice

Proving Triangles Congruent

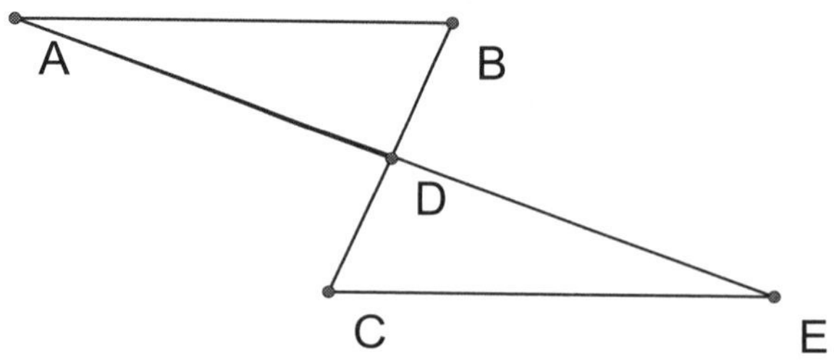

f. D is the midpoint of AE; AB // CE

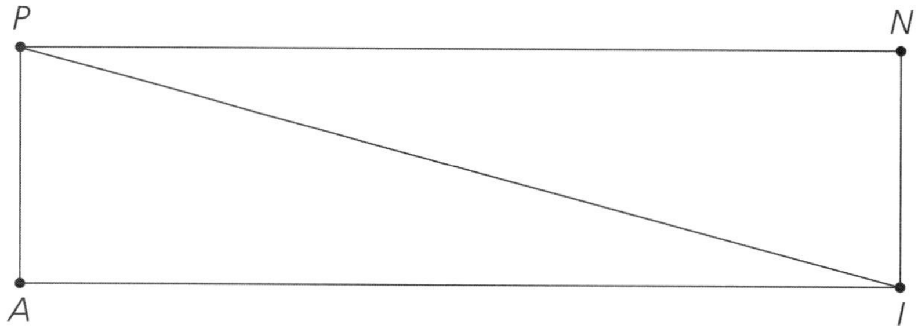

g. The figure above is a rectangle.

7. Create your own problem like those above. Exchange papers with another member of your group. Prove your partner's triangles are congruent using one of the methods your class has created.

extension

Proving Triangles Congruent

8. If all the corresponding angles of two quadrilaterals are congruent and all the corresponding sides are congruent, then the quadrilaterals are similar and congruent. Do you need to know this information about all eight parts of the quadrilaterals, or can you determine whether two quadrilaterals are similar or congruent using less information? If so, how much information is sufficient?

9. How many pairs of parts will be needed to prove that quadrilaterals are congruent?

10. Create at least two methods to prove quadrilaterals congruent as you created methods to prove triangles congruent in the last section.

input

Proving Triangles Similar

In the previous two sections, you dealt with congruent triangles. In order for triangles to be congruent, you know they must have the same *size* and the same *shape*.

Which parts must be the same for two figures to have the same size?

Which parts must be the same for two figures to have the same shape?

To show that triangles are similar, do you need to show that they have the same size *and* the same shape? Why or why not?

You have created methods to show that triangles have the same size and shape in order to prove them congruent. You found that the minimum number of parts needed to show triangles congruent is three parts, and you found that only a few groupings of different parts (SAS, ASA, and SSS) will give congruent triangles. In this section, you will try to create methods to show that triangles are similar.

analysis

Proving Triangles Similar

1. In order to show that triangles are similar to one another, what property needs to be preserved?

2. What parts of the triangle must be shown as equal in order to preserve that property?

Proving Triangles Similar

analysis

3. Can you show that two triangles are similar by showing that one pair of parts is equal? Why or why not?

4. Can you show that two triangles are similar by showing that one pair of sides and one pair of angles are equal? Explain.

5. Can you show that two triangles are similar by showing that two pairs of sides are equal? Why or why not?

6. If two pairs of angles are shown to be equal in a pair of triangles, what else can you deduce?

7. Can you show that two triangles are similar by showing that two pairs of angles are equal? Why or why not?

process: practice

Proving Triangles Similar

8. Can the following pairs of triangles be proven as similar? If so, explain how you know the triangles are similar. Use full sentences to create a complete and coherent explanation. If not, explain why the triangles cannot be shown as congruent.

a.

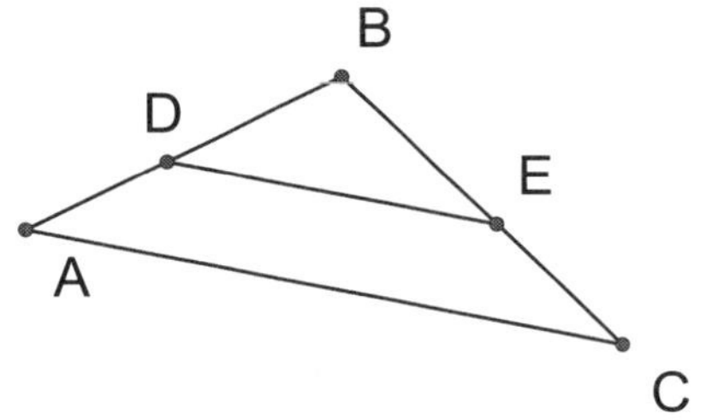

DE //AC

b.

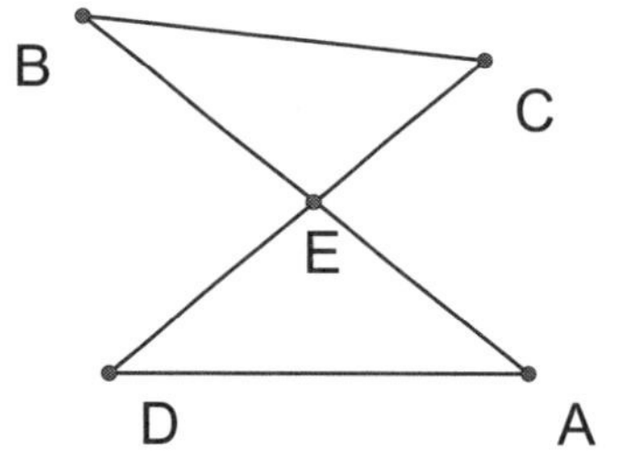

analysis

Proving Triangles Similar

9. Aside from showing that two pairs of angles are equal, you may be able to use the sides to show that triangles are similar. What is true about the relationship between the lengths of the sides in similar triangles?

10. f you do not know that any pairs of angles are equal, how many pairs of sides must be shown to be proportional in order to show that triangles are similar? Why?

11. Are there any conditions under which you need to show only that two pairs of sides are in proportion? How many pairs of angles do you need to know are equal in order to use only two pairs of sides?

12. There are two different explanations for why the following pair of triangles is similar. Provide at least one of these explanations.

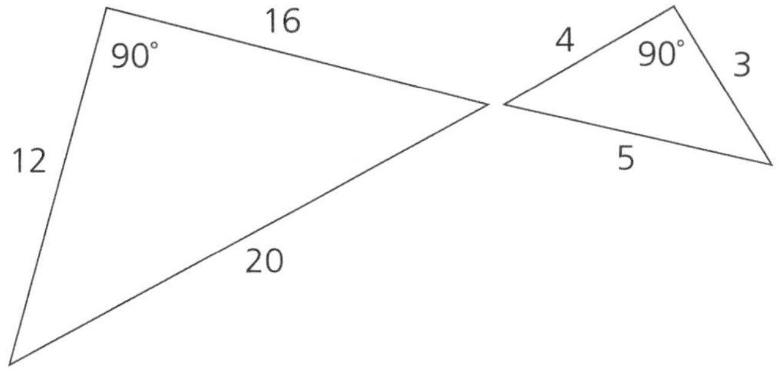

analysis

Proving Triangles Similar

13. The figure below contains two triangles.

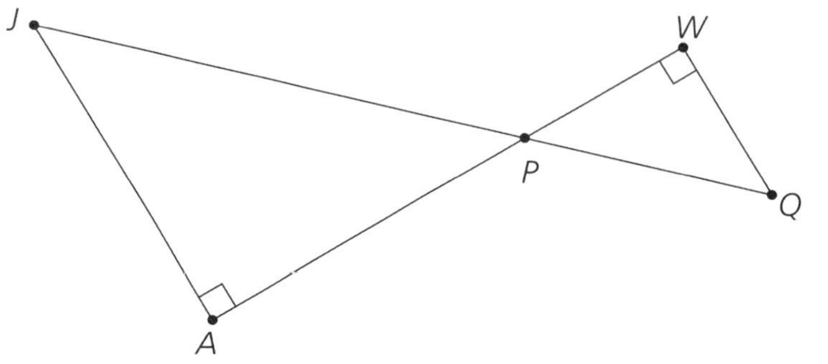

a. Are these triangles similar?

b. If they are similar, fill in this similarity statement about the triangles and set up a proportion comparing their sides.

Δ ________ ~ Δ ________

$$\frac{\square}{\square} = \frac{\square}{\square} = \frac{\square}{\square}$$

process: evaluate

Applications Involving Similar Triangles

1. Your heating bills this winter have been outrageous, so you decide to insulate the region between the ceiling of the second floor and the roof. This area is a very low attic, known as a crawl space. You need to know the maximum height of this space. You begin by standing on your knees at the opening to the crawl space, with your head touching the roof. In this position you are 120 centimeters tall, and the roof meets the floor 40 centimeters to the left. From where you are squatting to the center of the crawl space is 100 centimeters.

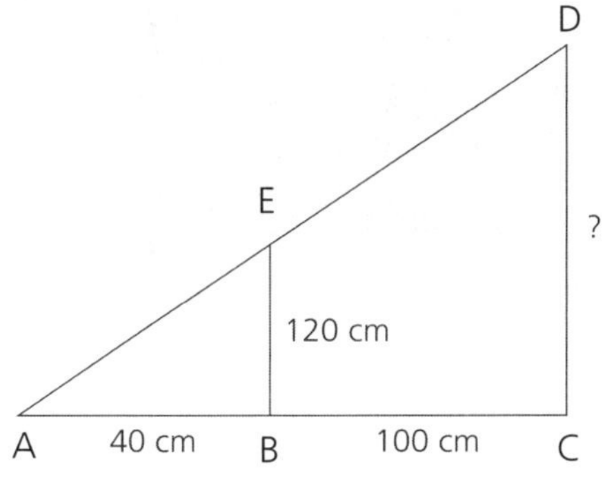

a. Explain in full sentences how you know these triangles are similar.

b. Explain how you use similar triangles to find the height of the crawl space.

c. Create a similarity statement between the two similar triangles.

d. Create ratios between the sides of the two similar triangles.

$$\frac{\square}{\square} = \frac{\square}{\square} = \frac{\square}{\square}$$

e. Replace the names of the sides of the triangles with known measurements and use this proportion to find the height of the crawl space.

process: evaluate

Applications Involving Similar Triangles

2. Lily is taking a picture of a mural painted on the side of a building. The focal length in a camera is the distance between the film and the lens. The width of the film in her camera is 35 millimeters and the focal length is 70 millimeters. The distance from the camera lens to the mural is 10 feet. Find the width of the mural.

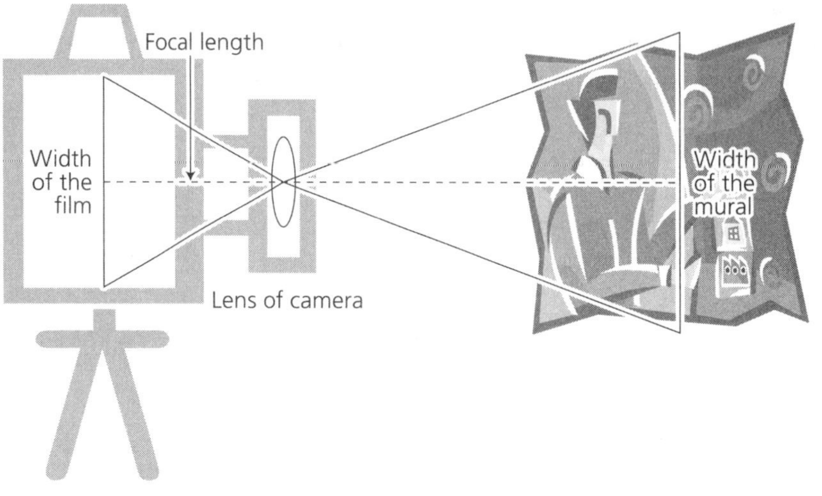

a. Provide a complete explanation of how you know the two triangles pictured above are similar.

b. Label the vertices of the triangles and create a similarity statement between them.

c. Create ratios between the sides of the two similar triangles.

$$\frac{\square}{\square} = \frac{\square}{\square} = \frac{\square}{\square}$$

d. Replace the segment names with known measurements. Use this proportion to find the width of the mural.

Applications Involving Similar Triangles

process: evaluate

3. Emma spikes a volleyball so that it bounces off the ground 4 feet from her. Her hand was 5 feet off the ground when she hit the ball, and the net is 6 feet beyond the position where the ball hit the ground. The ball bounced over the net, just grazing the top of it. How high is the top of the net?

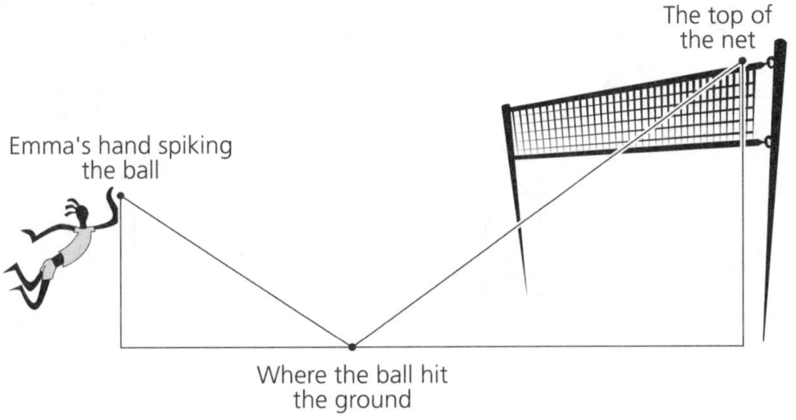

a. Provide a complete explanation of how you know these triangles are similar.

b. Label the vertices of the triangles and create a similarity statement between them.

c. Create ratios between the sides of the two similar triangles.

$$\frac{\square}{\square} = \frac{\square}{\square} = \frac{\square}{\square}$$

d. Replace the segment names with known measurements. Use the resulting proportion to find the height of the top of the net.

Applications Involving Similar Triangles

process: practice

4. In the figure below, the cue ball hits the black ball. The dashed line represents the black ball's path to the pocket. The distance from the black ball to the point at which it will bank off the left side of the table, represented by *AB*, is 50 cm. The perpendicular distance from the black ball to the left side of the table, represented by *TA*, is 40 cm. The width of the pool table, *EL*, is 200 cm. Find *AB* + *BL*, the total distance the black ball will travel before it sinks into the pocket.

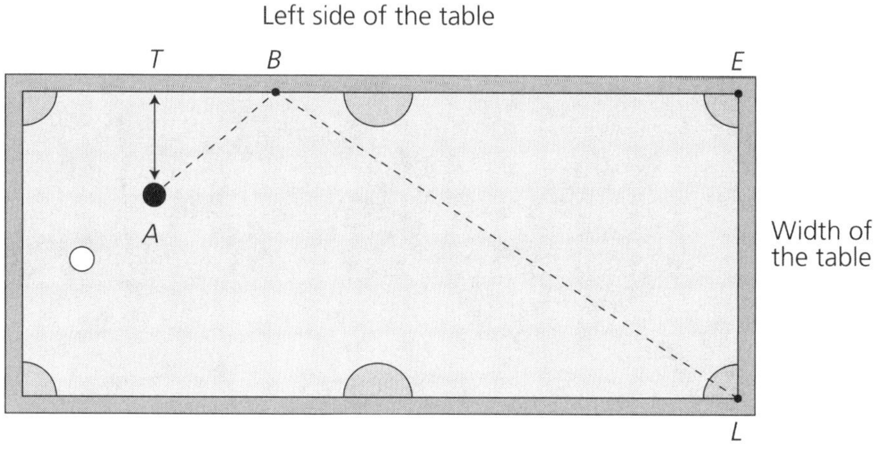

a. Can you use similar triangles to find the total distance the ball travels? Why or why not?

b. If you can use similar triangles to find the total distance traveled by the ball, do so below, making sure to show all of your work.

process: practice

Applications Involving Similar Triangles

5. Return to *The Washington Monument Problem* at the beginning of this unit and solve the problem again using what you know now about similar triangles. Draw a picture to illustrate this situation and show your work.

   a. Provide a complete explanation of how you solved this problem, justifying each step you took.

input

Indirect Measurement

There are situations in which it is physically undesirable or impossible to measure something directly. Your knowledge of similar triangles can be very helpful in these situations. For example, the head custodian at your school needs to purchase enough rope to string the flag up the newly installed flagpole. He does not know the height of the flagpole, and it is impractical for him to attempt to climb the flagpole to solve this problem. Your teacher has volunteered your class to help the head custodian determine the height of the flagpole.

process: model

Indirect Measurement

Your teacher will provide you with a mirror, a marker, and a tape measure to solve this problem.

1. Use the marker to make a dot near the center of the mirror. Face the flagpole and place the mirror on the ground in front of you. Keeping your eye on the mirror, carefully take a few steps backward. Stop when you see the top of the flagpole on the dot, as in the figure below. (This figure is not drawn to scale.) Note that in order to solve this problem correctly, the flagpole, the mirror, and the person sighting the top of the pole must stay in one straight line.

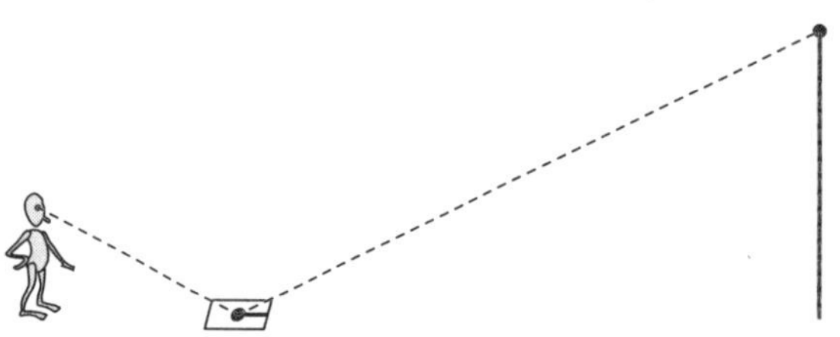

analysis

Indirect Measurement

2. Can you use similar triangles to find the height of the flagpole? Explain.

3. Which distances should your partner measure and record so that you can find the height of the flagpole?

process: evaluate

Indirect Measurement

4. Use the recorded distances and the figure to write a proportion. Solve this proportion for the height of the flagpole. Show your work.

analysis

Indirect Measurement

5. Compare your answer with the answers of other groups in the class and with the actual height of the flagpole, if it is known.
6. What are some possible sources of any error in your results? Discuss the possible reasons for these errors.

process: model

Midsegments of a Triangle

1. In triangle *MID*, draw the midpoint of $\overline{MI}$ and the midpoint of $\overline{MD}$. Label these points and connect them. You have just created a **midsegment** in triangle *MID*. Measure all the segments in the triangle and record their measurements on the figure.

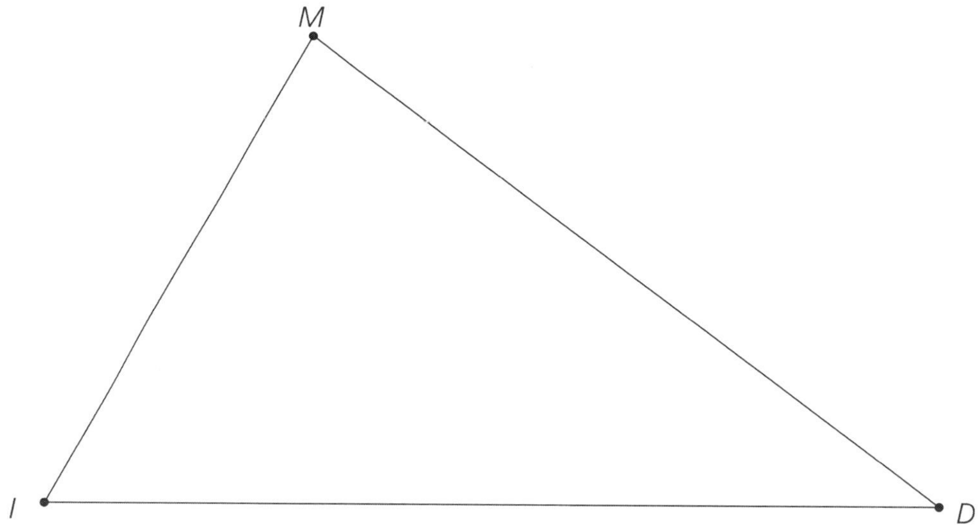

2. In triangle *MID*, draw the midpoint of $\overline{MI}$ and the midpoint of $\overline{ID}$. Label these points and connect them. You have just created another midsegment in triangle *MID*. Measure all the segments in the triangle and record their measurements on the figure.

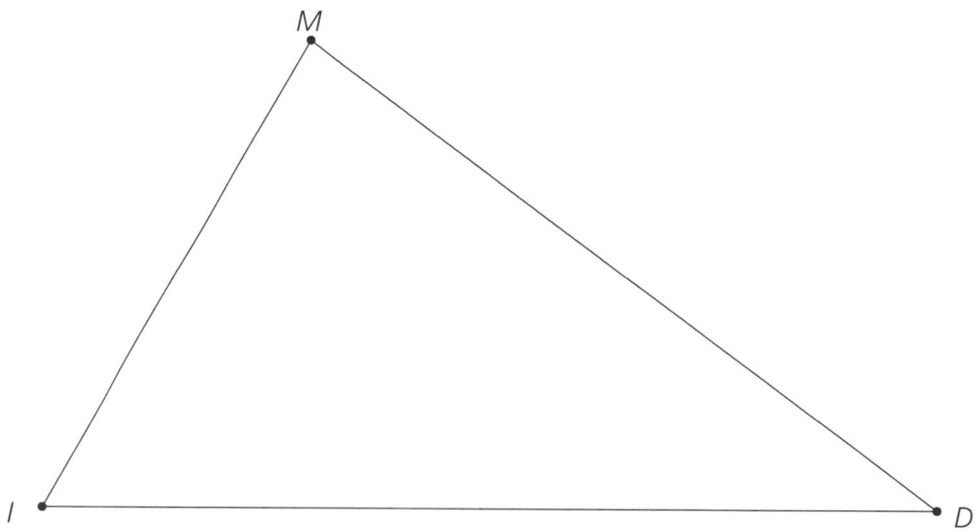

process: model

Midsegments of a Triangle

3. In triangle *MID*, draw the midpoint of $\overline{MD}$ and the midpoint of $\overline{ID}$. Label these points and connect them. You have just created another midsegment in triangle *MID*. Measure all the segments in the triangle and record their measurements on the figure.

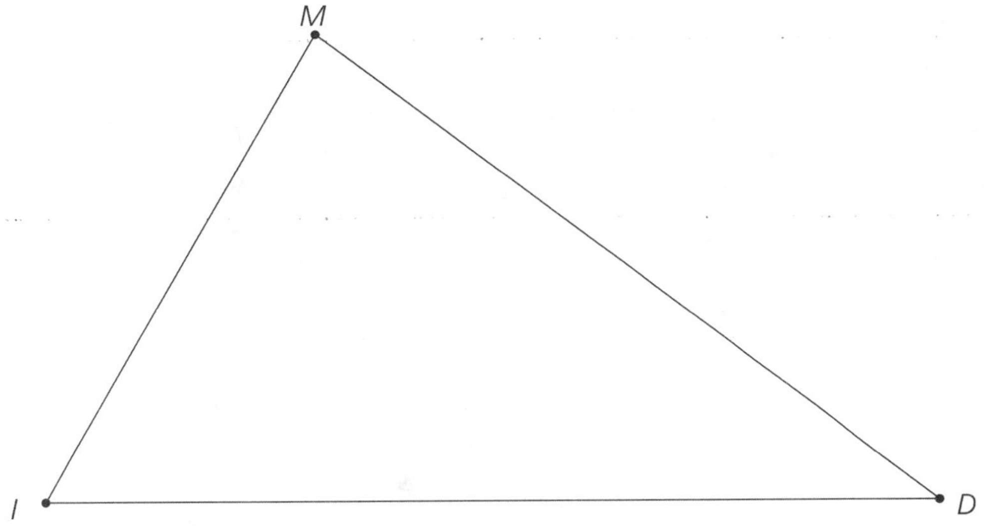

analysis

Midsegments of a Triangle

4. Why is the original triangle MID similar to the smaller triangle formed when the midsegment is created?

5. Create a similarity statement for each of the figures in questions 1 through 3.

Midsegments of a Triangle

analysis

6. How do the lengths of the sides in the smaller triangle relate to the lengths of the sides in the original triangle?

7. Based on this relationship, what conclusion can you make about the midsegment's length?

8. Does this property seem to make sense? Why or why not?

9. How do the measures of the angles in the smaller triangle relate to the measures of the angles in the original triangle?

10. Based on this relationship, what can you say about the midsegment's relation to its corresponding side in the original triangle?

11. Does this property seem to make sense? Why or why not?

process: practice

Midsegments of a Triangle

12. Use what you have just learned about midsegments to find all the missing measurements in triangle ABC with midsegment DE.

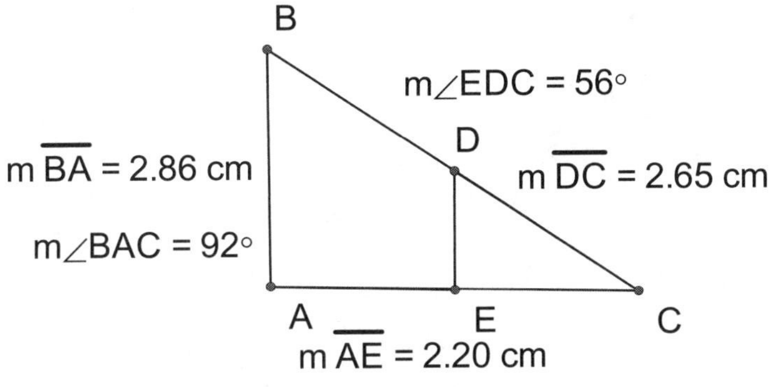

m∠CBA =

m∠C =

m∠CED =

BD =

BC =

EC =

AC =

DE =

process: evaluate

Similarity with Two- and Three-Dimensional Figures

1. Determine if the following triangles are similar. Record your measurements on the figures, and show any calculations you use to establish if the triangles are similar.

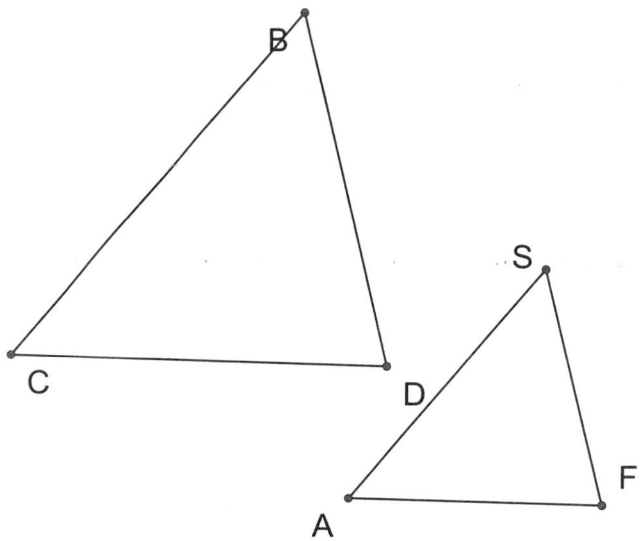

2. Create a segment perpendicular to the base of each triangle to form the height.

3. Measure each height and record your measurements.

4. Compute the ratio of the heights. Show your work.

Similarity with Two- and Three-Dimensional Figures

analysis

5. How does the ratio of the heights compare to the ratio of the sides in these triangles? Show any work.

6. Does this result make sense? Why or why not?

Similarity with Two- and Three-Dimensional Figures

process: evaluate

7. Calculate the area of each triangle in the space below. Show your work.

8. Compute the ratio of the areas. Show your work.

Similarity with Two- and Three-Dimensional Figures

analysis

9. How does the ratio of the areas compare to the scale factor?

10. How can you manipulate the scale factor to get the ratio of the areas?

11. Why is the ratio of the areas different from the scale factor? (Think about what you did to calculate the area and about the units on the area calculation.)

12. You and your best friend are going to the local home improvement store to purchase carpet for the living room of your friend's new house. The house is still under construction. You are using the blueprint to calculate the amount of carpet needed. On the blueprint 1 inch = 2 feet. The living room measures 6 inches by 8 inches on the blueprint. Your friend reasons that 6 inches by 8 inches is 48 square inches. With a scale factor of 1 inch to 2 feet, the room will need 96 square feet of carpet to cover the floor. Do you agree or disagree with your friend's reasoning? Explain.

    a. How is this problem related to the ideas of similar figures and scale factors?

process: evaluate

Similarity with Two- and Three-Dimensional Figures

13. The measurement of a side is given for each of the cubes pictured below. How do you know that the cubes are similar?

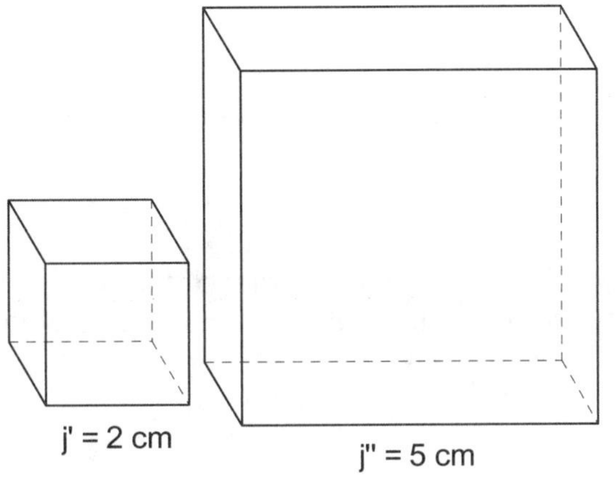

a. What is the scale factor between the two cubes?

b. What is the ratio between the areas of the front faces of the cubes? Is it the same as the scale factor? Why or why not?

c. Calculate the volume of each cube by multiplying the area of the front face by the depth. (Recall that all of the sides of a cube are the same length.)

d. What is the ratio of the volumes of the cubes?

e. Is it the same as the scale factor? Is it the same as the ratio of areas?

Similarity with Two- and Three-Dimensional Figures

analysis

f. How does the ratio of the volumes compare to the scale factor?

g. How can you manipulate the scale factor to get the ratio of the volumes?

h. Why is the ratio of the volumes different from the scale factor? (Think about what you did to calculate the volume and what the units were on the volume calculation.)

Similarity with Two- and Three-Dimensional Figures

process: summary

14. Summarize your findings about how the scale factor relates to two and three-dimensional measurements in similar figures.

Similarity with Two- and Three-Dimensional Figures

extension

15. If we could create a four-dimensional figure, theorize how the four-dimensional measurement would relate to the scale factor.

Koch Snowflake

input

The Koch Snowflake is an example of a fractal curve. Fractal curves make use of repetitive patterns operating on relatively simple figures in order to create very interesting designs. The Koch Snowflake begins with an equilateral triangle like the one pictured below. This equilateral triangle is called Stage 0 of the Koch Snowflake Curve because nothing has been done to the figure. For our purposes, each of the sides of this triangle will measure 1 unit.

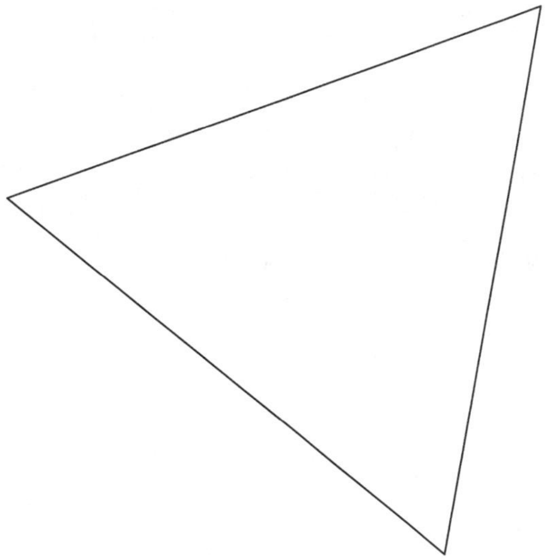

At each iteration, divide each side of the figure into thirds. Upon the middle, create a smaller equilateral triangle. The first iteration of the curve is shown below.

process: model

Koch Snowflake

1. Create the next iteration of the curve. Show this on the Stage 1 Snowflake Curve that is already pictured on the previous page.

2. Create one more iteration on the same curve, for a total of three complete iterations of the pattern.

analysis

Koch Snowflake

3. How do each of the new triangles created relate to *each other*? How do you know?

4. How do each of the new triangles created relate to *the triangles created in previous stages* of the curve? How do you know?

process: evaluate

Koch Snowflake

5. Complete the following table.

Stage	Number of sides	Length of each side	Perimeter of the curve
0	3		
1			
2			
3			
4			
n			

analysis

Koch Snowflake

6. Describe any patterns you see in the table above.

7. As the stage number increases, what happens to the number of sides? Does there seem to be a limit to the number of sides this curve can have? If so, what is it?

8. As the stage number increases, what happens to the length of each side? Does there seem to be a limit to the length of each side? If so, what is it?

analysis

Koch Snowflake

9. As the stage number increases, what happens to the perimeter of the Snowflake Curve? Does there seem to be a limit to the perimeter? If so, what is it?

10. Is there anything odd about what you have just found about the perimeter and the length of each side? If so, what is it?

process: evaluate

Koch Snowflake

11. Return to Stage 0 of the curve. Create an altitude in the original triangle. Calculate its length based upon trigonometric ratios.

12. Calculate the area of the original triangle.

process: evaluate

Koch Snowflake

13. Calculate the area of each of the smaller triangles in the first iteration.

14. Complete the following table.

Stage number	Number of new triangles	Area of each new triangle	Sum of the areas of the new triangles	Total area
0	1			
1	3			
2				
3				
4				
n				

analysis

Koch Snowflake

15. Describe any patterns you see in the table above.

16. As the stage number increases, what happens to the number of new triangles? Does there seem to be a limit to the number of new triangles that can be added? If so, what is it?

	Koch Snowflake
analysis	17. As the stage number increases, what happens to the area of each new triangle? Does there seem to be a limit to the area of each new triangle? If so, what is it? 18. As the stage number increases, what happens to the area of the Snowflake Curve? Does there seem to be a limit to the area? If so, what is it? 19. Is there anything odd about what you have just found about the perimeter and the area? If so, what is it?

	Koch Snowflake
process: summary	20. Analyze the following statement: *If your lawn was modeled by a Koch Snowflake Curve, you could mow the entire lawn but you could never fence it.*

Unit 7: Exponential and Logarithmic Functions

Contents

preview

The Homework Dilemma

Many students in your class are concerned about the amount of homework they receive each night. In response, your teacher makes this proposal on the first day of the month:

"Tonight I will give you one problem for homework. Every school day for the rest of the month, I will give you twice as many problems as the day before. On the first day of each month, we'll start over. Discuss the proposal among yourselves. I will ask one group to present in favor of the proposal and one group to present against it. The presentations will be followed by a vote."

1. To analyze the proposal, construct a table to record the number of homework problems assigned each day of the month and the total problems assigned up to and including that day. Assume there are 22 school days this month.

School Day of the Month	Homework Problems Assigned That Day	Homework Problems Assigned Up To That Day

2. Based on your analysis, would your group present for or against the proposal? What information would you include in your presentation?

input

Arithmetic Sequences and Series

Your group was chosen to present your analysis of the homework problem. Based on the information you provided, the class voted to decline the initial proposal. Expecting the offer to be rejected, your teacher makes a second proposal:

"Tonight I will give you 2 homework problems. Tomorrow I will give you 4 problems. Every day I will give you 2 more problems than the day before."

process: model

Arithmetic Sequences and Series

1. Write an arithmetic sequence to show the number of homework problems given to the class each day.

2. Why is it appropriate to model the second proposal using an *arithmetic* sequence?

3. Write a recursive formula to compute the number of homework problems given to the class on the nth day.

4. Write an explicit formula to compute the number of homework problems given to the class on the nth day.

5. Is it more appropriate to model the scenario using a sequence or a function? Why?

process: evaluate

Arithmetic Sequences and Series

6. Compute the following:
   a. The total homework problems given to the class by the 6th day, S_6.
   b. The total homework problems given to the class by the 22nd day, S_{22}.

input

Arithmetic Sequences and Series

Some students still hope for a better offer, so your teacher makes a third proposal to the class:

"I will give you 5 homework problems today, 6 homework problems tomorrow, and so on. Each day I will give you 1 more problem than the day before."

process: model

Arithmetic Sequences and Series

7. Write an arithmetic sequence to show the number of homework problems given to the class each day.
8. Write a recursive formula to compute the number of homework problems given to the class on the nth day.
9. Write an explicit formula to compute the number of homework problems given to the class on the nth day.
10. Is it more appropriate to model the scenario using a sequence or a function? Why?

process: evaluate

Arithmetic Sequences and Series

11. Compute the following:

   a. The total homework problems given to the class by the 6th day, S_6.

   b. The total homework problems given to the class by the 22nd day, S_{22}.

analysis

Arithmetic Sequences and Series

12. Which proposal would require the fewest problems to be done if this month has the full 22 school days? Explain.

13. Which proposal would require the fewest problems to be done if this month has 6 school days? Explain.

14. Knowing that the number of school days will vary by month, would you choose the second or third proposal? Explain.

Arithmetic Sequences and Series

extension

15. Examine your teacher's second and third proposal. What day of the month will the class receive the same total homework problems under each proposal?

process: model

Geometric Sequences and Series

Consider the original homework proposal made by your teacher – one problem on the first day, 2 on the second day, 4 on the third day, etc.

1. Write a sequence to show the number of homework problems given to the class each day.

2. How is the sequence above similar to the sequences you wrote for the other proposals? How is it different?

3. Is it appropriate to model the original proposal using an arithmetic sequence? Explain why or why not.

4. Write a recursive formula to compute the number of homework problems given to the class on the nth day.

5. How is the recursive formula above similar to the recursive formulas you wrote for the second and third proposals? How is it different?

6. Is it more appropriate to model the scenario using a sequence or a function? Explain.

input

Geometric Sequences and Series

The sequence you wrote to represent your teacher's original proposal is one example of a Geometric Sequence.

A sequence is called a **Geometric Sequence** if there is a number r, such that

$$g_n = r \cdot g_{n-1}$$

where g_n is the nth term in the sequence, g_{n-1} is the term before the nth term in the sequence, and r is the **common ratio**.

process: model

Geometric Sequences and Series

7. Complete the table below which displays the day, the number of homework problems received on that day, and how to calculate the number of homework problems received using multiplication and exponents.

Day	Homework problems received	Calculation using multiplication	Calculation using exponents
1			
2			
3			
4			
5			
6			
7			
8			
n			

Geometric Sequences and Series

process: model

8. Suppose the class received 10 homework problems on the first day and then the number of problems received doubled each day. Complete the table below to help generate the implicit formula to calculate the number of problems received on the nth day.

Day	Homework problems received	Calculation using multiplication	Calculation using exponents
1			
2			
3			
4			
5			
6			
7			
8			
n			

process: model

Geometric Sequences and Series

9. Suppose the class received 10 homework problems on the first day and then the number of problems received tripled each day. Complete the table below to help generate the implicit formula to calculate the number of problems received on the nth day.

Day	Homework problems received	Calculation using multiplication	Calculation using exponents
1			
2			
3			
4			
5			
6			
7			
8			
n			

analysis

Geometric Sequences and Series

10. Examine the formulas you generated in the last three questions. What do the three formulas have in common? What differences exist?

analysis

Geometric Sequences and Series

11. How is the common ratio indicated in the explicit formula to calculate the nth term in each of the last three examples?

12. How is the first term indicated in the explicit formula to calculate the nth term in each of the last three examples?

process: practice

Geometric Sequences and Series

13. Generate an explicit formula to calculate the nth term of each geometric sequence.
    a. The first term is 5 and the common ratio is 3.

    b. 40, 60, 90, 135, 202.5...

    c. -3, -6, -12, -24, -48...

    d. The first term is g_1 and the common ratio is r.

input

Geometric Sequences and Series

In general, the recursive formula for a geometric series can be written as:

In general, the explicit formula for a geometric sequence can be written as:

input

Chain Letters

Chain letters have been around for a long time and have become more common with the increased use of e-mail. Suppose a chain letter is sent to 10 people on the first day. The letter instructs the reader to send it to 5 additional people or they will have bad luck. Assume that everyone in the chain complies with the instructions.

process: model

Chain Letters

1. Write a recursive and an explicit formula to compute the number of people who receive the chain letter each day if everyone who receives it follows the instructions.

2. Is it more appropriate in this scenario to use a sequence or a function? Explain.

process: evaluate

Chain Letters

3. How many people will receive the chain letter on the:

   a. 5th day?

   b. 8th day?

   c. 15th day?

process: model

Chain Letters

4. Complete the table below to compute the total number of people who received the chain letter by each day.

Number of days	People who received the chain letter on that day	People who received the chain letter total
1		
2		
3		
4		
5		
6		
7		
8		

process: practice

Chain Letters

5. How long will it take for:

   a. 1000 people to receive the chain letter?

   b. 1,000,000 people to receive the chain letter?

input

Chain Letters

When discussing an arithmetic series, you used a formula to compute S_n, the sum of the first n terms of the arithmetic sequence. A formula also exists to compute the first n terms of a geometric sequence:

$$S_n = \frac{g_1\left(1-r^n\right)}{1-r} \text{ for } r \neq 1$$

process: evaluate

Chain Letters

6. Write a formula to compute S_n, the total number of people who received the chain letter by the nth day.

7. Compute the total number of people who would have received the chain letter:
   a. By the end of the fifth day.

   b. By the end of the 10th day.

   c. By the end of the 12th day.

input

Family Tree

As part of an English class assignment, you must research your family history and write a report. As you begin to gather information, you wonder how many ancestors you actually have. You decide to perform some calculations. To keep things simple, you decide to consider only direct ancestors such as parents, grandparents, great-grandparents, etc.

process: evaluate

Family Tree

1. Determine the number of direct ancestors you had for the given generation.
   a. At the generation one generation before yours

   b. At the generation two generations before yours

   c. At the generation three generations before yours

   d. At the generation five generations before yours

   e. At the generation seven generations before yours

   f. At the generation nine generations before yours

process: model

Family Tree

2. Summarize your data in the table below.

Quantity Name	**Number of generations before yours**	**Number of direct ancestors at that generation**	**Number of direct ancestors written as a power**
Unit	***Generations***	***Ancestors***	***Ancestors***
	0		
	1		
	2		
	3		
	5		
	7		
	9		
Expression			

3. Plot the points contained in your table and draw a smooth curve to connect them.

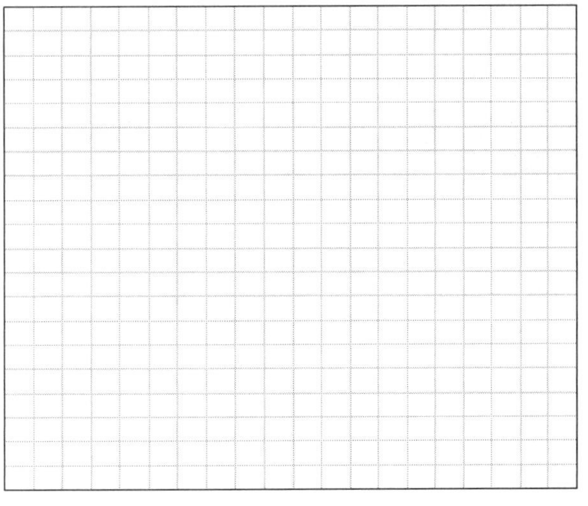

Family Tree

analysis

4. When would it be more appropriate to model the scenario using a sequence? Explain.

5. When would it be more appropriate to model the scenario using a function? Explain.

6. What is the y-intercept of the graph? Describe what this means in terms of the problem situation.

7. What is the x-intercept of the graph? Describe what this means in terms of the problem situation.

input

Computer Virus

One danger of the 'digital age' is the spread of computer viruses. Viruses are usually spread via e-mail. When a computer becomes infected, the virus is spread to everyone in the computer's address book. For one recent computer virus, every infected computer infected 5 new computers an hour after becoming infected.

process: evaluate

Computer Virus

1. Determine the number of newly infected computers during the:
   a. First hour

   b. Second hour

   c. Third hour

   d. Fifth hour

   e. Seventh hour

   f. Ninth hour

process: model

Computer Virus

2. Summarize your data in the table below.

	Time	Computers infected	Computers infected written as a power
Quantity Name			
Unit			
	0		
	1		
	2		
	3		
	5		
	7		
	9		
Expression			

3. Plot the points contained in your table and draw a smooth curve to connect them.

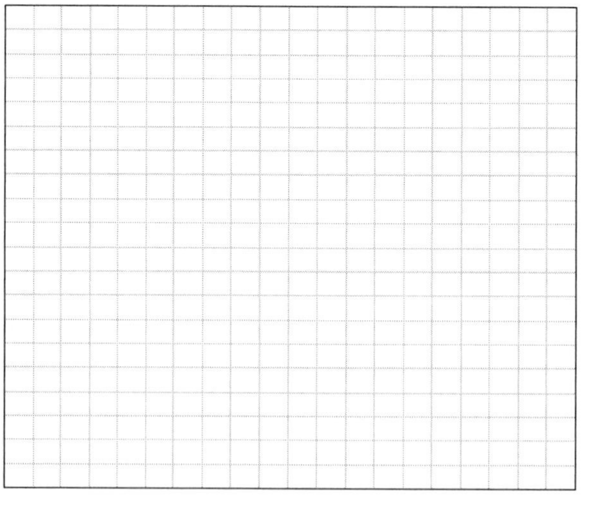

Computer Virus

analysis

4. Is it more appropriate to model the scenario using a sequence or a function? Explain.

5. If a scenario is modeled as a sequence, when might it be useful to model it using a function? Explain.

6. What is the y-intercept of the graph? Describe what this means in terms of the problem situation. Does this make sense in terms of the problem?

7. What is the x-intercept of the graph? Describe what this means in terms of the problem situation. Does this make sense in terms of the problem?

input

Pump Up the Base

The functions representing the number of direct ancestors and the number of infected computers are both examples of **exponential functions**. A parent exponential function is one in the form

$$f(x) = b^x$$

where b is the **base** and x is the **exponent** or **power**.

process: model

Pump Up the Base

1. Graph the functions $y = 2^x$, $y = 3^x$, $y = 4^x$ and $y = 5^x$ on the same set of axes.

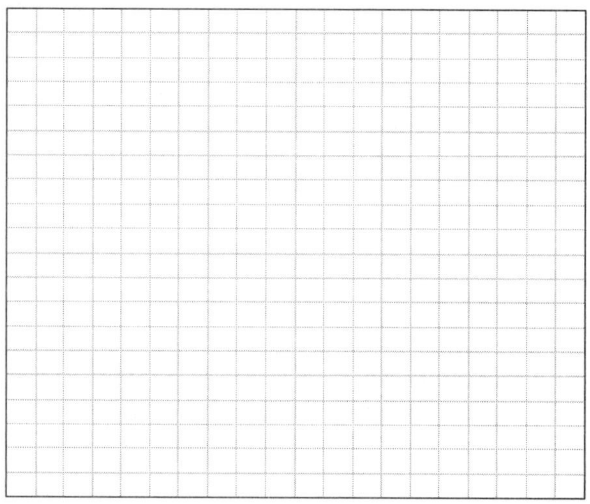

Pump Up the Base

analysis

2. How are the graphs of each exponential function similar? How are they different?

3. What happens to the graph of each exponential function as the value of the base increases? Why do you think this happens?

4. What do you notice about the y-intercept of each graph?

5. What do you notice about the x-intercept of each graph?

process: model

Pump Up the Base

6. Graph the functions $y=\left(\frac{1}{2}\right)^x$, $y=\left(\frac{1}{3}\right)^x$, $y=\left(\frac{1}{4}\right)^x$ and $y=\left(\frac{1}{5}\right)^x$ on the same set of axes.

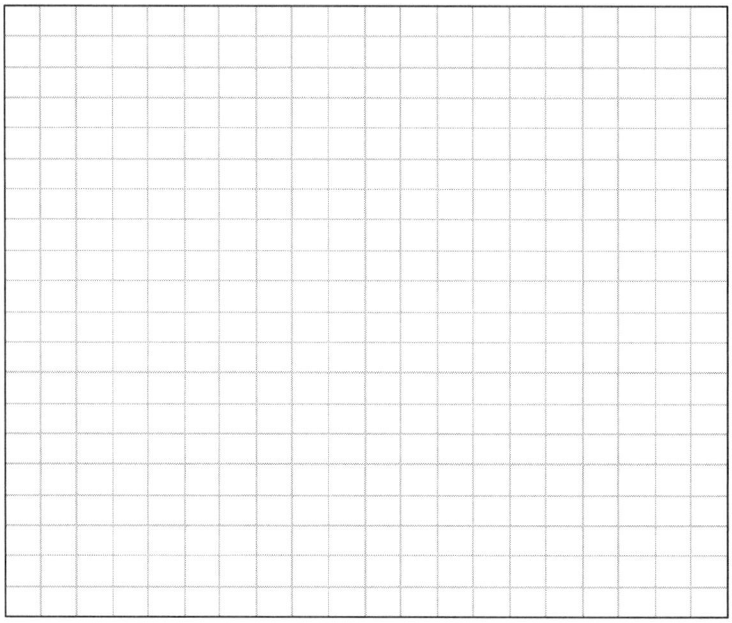

analysis

Pump Up the Base

7. How are the graphs of each exponential function similar? How are they different?

Pump Up the Base

analysis

8. What happens to the graph of each exponential function as the value of the base decreases? Why do you think this happens?

9. What do you notice about the y-intercept of each graph?

10. What do you notice about the x-intercept of each graph?

11. How is the shape of the graphs of $y = 2^x$, $y = 3^x$, $y = 4^x$ and $y = 5^x$ different than the shape of the graphs of $y = \left(\frac{1}{2}\right)^x$, $y = \left(\frac{1}{3}\right)^x$, $y = \left(\frac{1}{4}\right)^x$ and $y = \left(\frac{1}{5}\right)^x$? Why do you think this is the case?

12. What do you notice about the value of the base in each of the functions? Describe how the value of the base determines the shape of the graph of an exponential function.

process: summary

Pump Up the Base

13. Given the graph of any exponential function in the form $f(x) = b^x$, what can you say about the y-intercept?

$$f(x) = b^x$$

14. If $f(x) = b^x$, write an expression for $f(0)$.

15. If $f(x) = b^x$, what is the numerical value of $f(0)$?

16. What is the value of any number raised to the zero power?

17. Given the graph of any exponential function in the form $f(x) = b^x$, what can you say about the x-intercept?

input

Pump Up the Base

The graphs of all exponential functions have a horizontal **asymptote**, which is represented by a horizontal line. The graph gets infinitely close to the asymptote but never touches or intersects it.

For exponential functions of the form $f(x) = b^x$, the asymptote is the x-axis.

input

Double or Nothing

You recently tried out for and were chosen to appear on a game show called *Double or Nothing*. You have been extremely lucky on the show so far. As the current champion, you have $1000 as of today.

At the end of each episode, that day's champion decides whether to attempt the Double Dog Dare Question. If you decide to go for it and get the question right, your winnings double and you return for the next show. If you get it wrong, you go home with nothing. Of course, you may decide not to attempt the Double Dog Dare Question and go home with all the money you have won up to that point.

process: model

Double or Nothing

1. Complete the table below to compute your winnings if you keep answering the Double Dog Dare Questions correctly.

Number of episodes from today	Amount won	Amount won written as a power of 2
Episodes	Thousands of dollars	Thousands of dollars
0		
1		
2		
3		
4		

2. Write an equation to model the amount won. Be sure to define the variables.

process: evaluate

Double or Nothing

3. How much money did you win on the episode before today? Explain how you computed this value.

4. How much money did you win two episodes ago? Explain how you computed this value.

5. How much money did you win three episodes ago? Explain how you computed this value.

6. Complete the table below to compute your past winnings. Write the amount won using fractions.

Number of episodes from today	**Amount won**	**Amount won written as a power of 2**
Episodes	Thousands of dollars	Thousands of dollars
-4		
-3		
-2		
-1		
0		

Double or Nothing

analysis

7. Compare the amount (in thousands) you will win in one more episode to the amount (in thousands) you won one episode ago.

   a. What do you notice about the expressions used to calculate your winnings (in thousands) when written with exponents?

   b. What do you notice about the value of your winnings (in thousands) when written using fractions (or whole numbers)?

8. Compare the amount (in thousands) you will win in two more episodes to the amount (in thousands) you won two episodes ago.

   a. What do you notice about the expressions used to calculate your winnings (in thousands) when written with exponents?

   b. What do you notice about the value of your winnings (in thousands) when written using fractions (or whole numbers)?

analysis

Double or Nothing

9. Compare the amount (in thousands) you will win in four more episodes to the amount (in thousands) you won four episodes ago.

   a. What do you notice about the expressions used to calculate your winnings (in thousands) when written with exponents?

   b. What do you notice about the value of your winnings (in thousands) when written using fractions (or whole numbers)?

process: practice

Double or Nothing

10. Compute how much you would win in 7 more episodes. What do you think you would have won 7 episodes ago? Write your answer as a fraction.

11. Compute how much you would win in 10 more episodes. What do you think you would have won 10 episodes ago? Write your answer as a fraction.

process: summary

Double or Nothing

12. Summarize what you have discovered about how to evaluate expressions involving negative exponents.

    a. Based on your observations, how does the value of 300^{-956} compare to the value of 300^{956}?

    b. If you know that $q^{-5} = 75$, what does q^5 equal?

input

Double or Nothing

The new producers of *Double or Nothing* are considering changing the rules of the game because they find it cruel that the champions lose all their money if they get the Double Dog Dare Question wrong. They decide that if the champion gets the question wrong, only half of the money will be lost and the champion can come back the next episode.

Your friend has $1000 as of today and has been unable to answer the Double Dog Dare Question during his reign as champion.

process: model

Double or Nothing

13. Write an equation to model the amount (in thousands) won if your friend continues to answer the Double Dog Dare Question incorrectly. Be sure to define the variables.

Double or Nothing

process: model

14. Complete the table below to compute your friend's winnings. Write the amount won using fractions or whole numbers.

Number of episodes from today	Amount won	Amount won written as a power of $\frac{1}{2}$
Episodes	Thousands of dollars	Thousands of dollars
-3		
-2		
-1		
0		
1		
2		
3		

analysis

Double or Nothing

15. Consider your previous observations regarding negative exponents. If the base of an exponential expression is a fraction, do any of your observations change?

process: model

Double or Nothing

16. Graph the function to compute your friend's winnings.

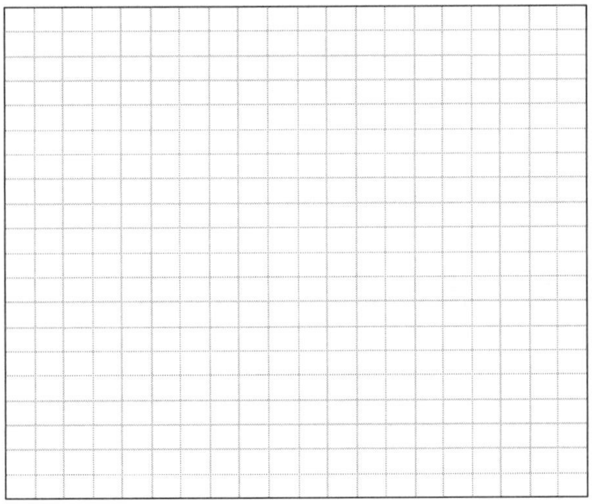

analysis

Double or Nothing

17. Once contestants lose all their winnings, they must leave the show. Based on the rules of the game, how many additional dare questions can your friend miss before he loses all his money? Use your graph to verify your answer.

18. How does the definition of an asymptote relate to your answer above?

19. Does the above answer make sense realistically?

process: practice

Double or Nothing

20. Write each expression using positive exponents and then evaluate each expression. Write your final answers as fractions.

    a. 4^{-5}

    b. $\left(\frac{1}{3}\right)^{-4}$

    c. $\left(\frac{2}{3}\right)^{-2}$

    d. 7^{-4}

Exponential Transformations

process: model

1. Graph the functions $f(x)=2^x$, $f(x)=3{\bullet}2^x$, $f(x)=5{\bullet}2^x$, $f(x)=\left(\frac{1}{2}\right){\bullet}2^x$ and $f(x)=\left(\frac{1}{3}\right){\bullet}2^x$ on the same coordinate plane.

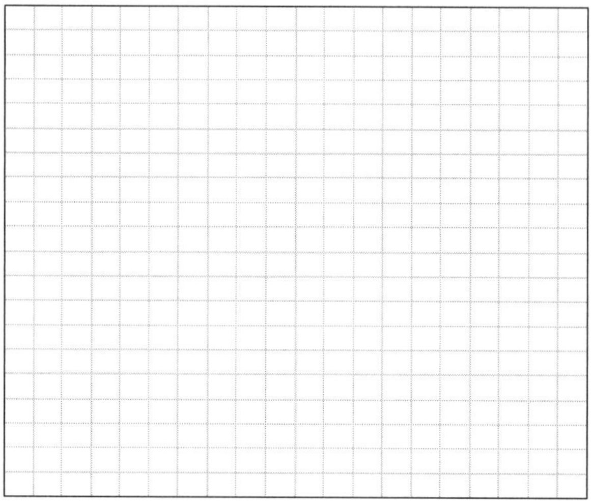

Exponential Transformations

analysis

2. What happens to the shape of the graph if you multiply the parent function by a constant greater than 1?

3. What happens to the shape of the graph if you multiply the parent function by a constant less than 1?

Exponential Transformations

analysis

4. What is the y-intercept of the graph of the function $f(x) = 2^x$?

5. What are the y-intercepts of the graphs of the other functions? What impact does multiplying the parent function by a constant have on the y-intercept?

6. What is the asymptote of the graph of the function $f(x) = 2^x$?

7. What are the asymptotes of the graphs of the other functions? What impact does multiplying the parent function by a constant have on the asymptote?

Exponential Transformations

process: model

8. Graph the functions $f(x) = 2^x$, $f(x) = 2^x + 1$, $f(x) = 2^x - 1$, $f(x) = 2^x + 4$ and $f(x) = 2^x - 4$ on the same coordinate plane.

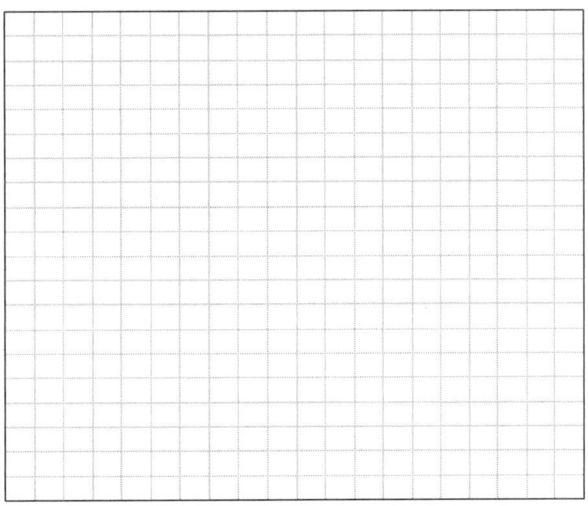

Exponential Transformations

analysis

9. What happens to the graph if you add a constant to the parent function?

10. What happens to the graph if you subtract a constant from the parent function?

11. What is the y-intercept of the graph of the function $f(x) = 2^x$?

12. What are the y-intercepts of the graphs of the other functions? What impact does adding or subtracting a constant have on the y-intercept?

13. What is the asymptote of the graph of the function $f(x) = 2^x$?

14. What are the asymptotes of the graphs of the other functions? What impact does adding or subtracting a constant have on the asymptote?

Exponential Transformations

process: model

15. Graph the functions $f(x) = 2^x$, $f(x) = -2^x$, $f(x) = 3 \cdot 2^x$ and $f(x) = -3 \cdot 2^x$ on the same coordinate plane

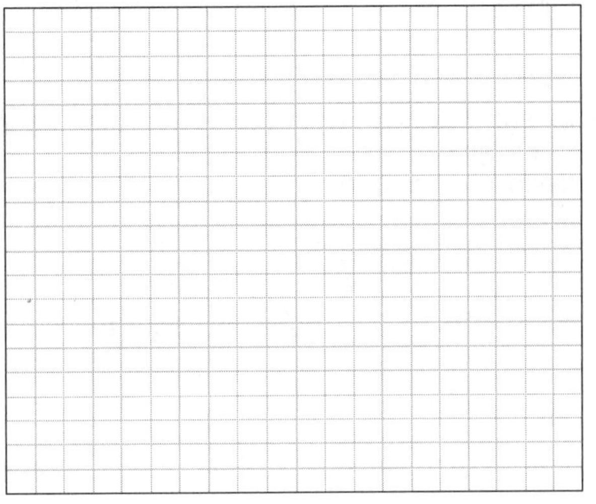

Exponential Transformations

analysis

16. What happens to the graph if you multiply the parent function by a negative number?

17. What impact does multiplying the parent function by a negative number have on the y-intercept?

18. What impact does multiplying the parent function by a negative number have on the asymptote?

process: practice

Exponential Transformations

19. Write the function that results when the following transformations are applied to the given parent function.

    a. The parent $f(x)=5^x$ is shifted up 3 units and stretched by a factor of 2.

    b. The parent $f(x)=3^x$ is shifted down 1 unit, reflected about the x-axis and stretched by a factor of $\frac{1}{2}$.

    c. The parent $f(x)=4^x$ is reflected about the x-axis and stretched by a factor of 10.

20. Describe the transformations applied to the parent function that result in each function below.

    a. $f(x)=-3 \cdot 2^x - 5$

    b. $f(x)=-3^x+6$

    c. $f(x)=\frac{2}{3}\left(\frac{1}{3}\right)^x - 4$

process: summary

Properties of Exponents

1. Evaluate the following expressions.

   a. $7^0 =$

   b. $y^0 =$

   c. $\left(3xy^2\right)^0 =$

   d. $(x+2)^0 =$

2. Summarize the process to evaluate a term raised to a zero exponent.

3. Write each expression using positive exponents and fractions.

   a. 2^{-3}

   b. $\left(\frac{1}{3}\right)^{-4}$

   c. $\left(\frac{2}{3}\right)^{-2}$

   d. x^{-5}

4. Summarize the process to convert an expression written using a negative exponent to an expression involving a positive exponent.

process: evaluate

Properties of Exponents

5. Write each expression in expanded form.

a. $2^3 =$

b. $5^4 =$

c. $x^6 =$

6. Write each expression in expanded form and simplify.

a. $x^2 \bullet x^3$

b. $2^3 \bullet 2^5$

c. $y^7 \bullet y$

analysis

Properties of Exponents

7. What is the relationship between the bases in each original expression and each final expression?

8. What is the relationship between the exponents in each original expression and each final expression?

9. Summarize the process to multiply two exponential expressions with the same base.

process: evaluate

Properties of Exponents

10. Write each expression in expanded form and simplify. Use negative exponents as appropriate to write each final answer without fractions.

a. $\frac{x^5}{x^2}$

b. $\frac{y^2}{y^6}$

c. $\frac{3^7}{3^2}$

analysis

Properties of Exponents

11. What is the relationship between the bases in each original expression and each final expression?

12. What is the relationship between the exponents in each original expression and each final expression?

13. Summarize the process to divide two exponential expressions with the same base.

extension

Properties of Exponents

Another way to demonstrate that $a^0 = 1$ for $a^1 \neq 0$ is to use the property of exponents you just developed.

14. Write an expression involving the quotient of two exponential expressions with the same exponent.

15. Use the exponential property you just learned to simplify the exponential expression. Write your final answer as a base raised to an exponent.

16. Use what you know about simplifying fractions with the same numerator and denominator to simplify the exponential expression.

17. What did you observe when you simplified the original exponential expression using each of the simplification methods?

process: evaluate

Properties of Exponents

18. Write each expression in expanded form and simplify. Use negative exponents as appropriate to write each final answer without fractions.

    a. $\left(x^3\right)^2$

    b. $\left(y^{-2}\right)^3$

    c. $\left(4^3\right)^{-5}$

analysis

Properties of Exponents

19. What is the relationship between the bases in each original expression and each final expression?

20. What is the relationship between the exponents in each original expression and each final expression?

21. Summarize the process to raise an exponential expression to a power.

Properties of Exponents

process: evaluate

22. Write each expression in expanded form and simplify. Use negative exponents as appropriate to write each final answer without fractions.

    a. $(xy)^3$

    b. $(3y)^4$

    c. $(2xy)^{-3}$

Properties of Exponents

process: summary

23. Summarize the process to raise the product of exponential expressions to a power.

process: evaluate

Properties of Exponents

24. Write each expression in expanded form and simplify.

a. $\left(\frac{x}{y}\right)^3$

b. $\left(\frac{3}{x}\right)^2$

c. $\left(\frac{x}{y}\right)^{-4}$

process: summary

Properties of Exponents

25. Summarize the process to raise the quotient of two exponential expressions to a power.

process: summary

Properties of Exponents

26. Below is a summary of the properties of exponents. Provide a written summary of each property.

a. $a^0 = 1, a \neq 0$

b. $a^{-n} = \dfrac{1}{a^n}$

c. $a^m a^n = a^{m+n}$

d. $\dfrac{a^m}{a^n} = a^{m-n}$

e. $\left(a^m\right)^n = a^{mn}$

f. $\left(ab\right)^n = a^n b^n$

g. $\left(\dfrac{a}{b}\right)^n = \dfrac{a^n}{b^n}$

process: practice

Properties of Exponents

27. Simplify the following expressions. Write the final answer using only positive exponents.

a. $\dfrac{8x^6}{2x^2}$

b. $\left(-2x^3y\right)^2$

c. $\left(\dfrac{2x}{3y^3}\right)^3$

d. $\dfrac{\left(-3x^2y\right)^3}{\left(-2x^3y^2\right)^0}$

e. $\left(\dfrac{-15xy^2}{20x^2y^3}\right)^3$

f. $\left(2ab^2\right)^3\left(-3b\right)^{-2}$

input

College Fund

When you were born, your grandparents on your mother's side began a college fund on your behalf. They deposited $1000 in a savings account with an interest rate of 10% compounded annually. This means every year, 10% of the amount in the account is added to the balance. No money is added to the $1000 and it is left to grow on its own.

process: model

College Fund

1. Complete the table below to show the principal (the amount in the account over time).

Your age	**Principal in the account**
Years	**Dollars**
0	
1	
2	
3	
4	
5	
6	
7	
8	
9	
10	

analysis

College Fund

2. Would it be appropriate to model this scenario using an arithmetic sequence? Explain why or why not.

3. Would it be appropriate to model this scenario using a geometric sequence? Explain why or why not.

4. What is the starting point (y-intercept) in this scenario? What does this mean in terms of the problem situation?

5. Does the sequence modeling the scenario contain a common difference or common ratio? Identify the value of this constant.

process: model

College Fund

6. Write a function to model the scenario.

7. What values for the independent variable make sense in this problem?

8. What values for the dependent variable make sense in this problem?

College Fund

process: model

9. Graph the function below.

College Fund

process: evaluate

10. What is the amount in the account when you are:
    a. 15 years old?
    b. 18 years old?

11. How old are you when the principal reaches:
    a. $3000?
    b. $6000?

input

College Fund

In general, the formula to compute the principal of an account compounded *annually* can be written as

$$P = P_0\left(1+r\right)^t$$

where P_0 is the original principal of the account, r is the interest rate written as a decimal and P is the principal after t years.

input

College Fund

Your grandparents on your father's side also began a college fund for you when you were born. They deposited $1500 in a savings account with an interest rate of 7.5% compounded annually.

process: model

College Fund

12. Write a function to compute the principal of the account started by your father's parents over time.

13. Graph the function below.

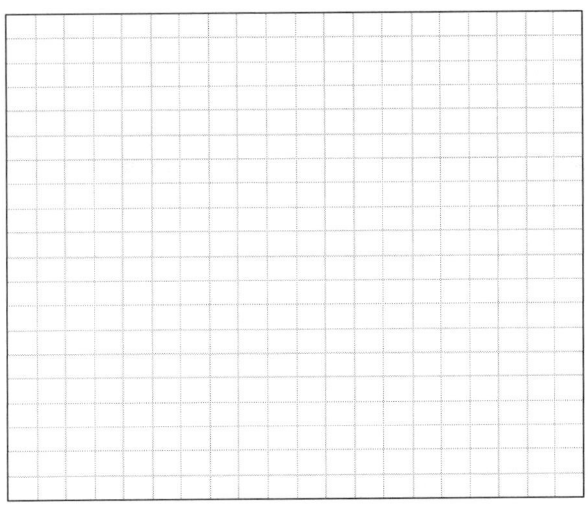

	College Fund
process: evaluate	14. What is the amount in the account when you are: a. 15 years old? b. 18 years old? 15. How old are you when the principal reaches: a. $3000? b. $6000?

	College Fund
process: evaluate	16. If you enter college at age 17, which account will have a larger balance? 17. If you enter college at age 18, which account will have a larger balance?

extension

College Fund

18. Your grandparents on your mother's side want to be able to cover the cost for your first semester of college completely. Assume when you enter college, the cost of your first semester's tuition will be $10,000. In order to cover your first semester costs completely, how much must be deposited in the account initially? What interest rate would be necessary?

input

Compounding the Problem

Some types of investments have their interest calculated more often than every year. For example, interest may be compound monthly, weekly, or even daily.

In general, the formula to compute the principal of an account with compounded interest can be written as

$$P = P_0\left(1+\frac{r}{n}\right)^{nt}$$

where P_0 is the original principal of the account, r is the interest rate written as a decimal, n is the number of times interest is calculated per year and P is the principal after t years.

The formula to compute principal for interest compounded *annually* is a simple case of the above formula with n = 1.

input

Compounding the Problem

As you continue to learn about how interest is calculated, you wonder how much more money you could have had for college if your grandparents had opened accounts in which the interest was compounded more frequently than annually. To find out, you construct a table to compute the principal in each account in which the interest rate and original principal remain the same, but the number of times per year the interest is compounded increases.

process: model

Compounding the Problem

1. Complete the following table to compute the principal of each account after 18 years.

   Remember that your grandparents on your mother's side opened an account with an initial principal of $1000 at 10% interest.

   Your grandparents on your father's side opened an account with an initial principal of $1500 at 7.5% interest.

process: model

Compounding the Problem

Type of compounded interest	Number of times interest is calculated per year	Principal in account from your mother's parents after 18 years	Principal in account from your father's parents after 18 years
		Dollars	Dollars
	n		
Annually			
Monthly			
Weekly			
Daily			
Hourly			
Minutely			

analysis

Compounding the Problem

2. What do you notice about the principal as the frequency of compounding interest increases over time?

3. What do you notice about the rate of increase in the principal as the frequency of compounding interest increases?

analysis

Compounding the Problem

4. As the frequency of compounding interest increases, does it appear that the principal is approaching a particular value? What behavior of the graphs of exponential functions can be used to describe this behavior?

input

Compounding the Problem

As the frequency of compounding interest continues to increase, the function is approaching what is known as **continuous growth**. In this scenario, the growth involves interest. This means that interest is constantly being calculated and adding to the balance.

Continuous growth is modeled using a special irrational number, represented by the letter 'e'. The mathematician Euler was the first to discover the importance of e during the 18th century.

The 'e' does not represent a variable, but rather a constant much like π represents the irrational number approximated by 3.14. It is possible to approximate a value for e using the formula to compute compound interest.

input

Compounding the Problem

While reading the newspaper, you see an advertisement for your bank's once in a lifetime Special 100% Interest Account. Anyone opening an account today with an initial principal of $100 will receive 100% interest for one year. The frequency of compounding is not given in the ad.

process: model

Compounding the Problem

5. Write an equation to calculate the balance of the account after one year based on the frequency that interest is calculated.

6. Complete the table below to show how the principal after one year changes as the frequency of compounded interest increases.

Type of compounded interest	**Number of times interest in calculated per year**	**Principal in account after one year**
		Dollars
Annually		
Monthly		
Weekly		
Daily		
Hourly		
Minutely		

analysis

Compounding the Problem

7. If the account offered continuous interest, what value will the balance approach?

8. Since the account began with an initial principal of $100, divide your estimate by 100 to approximate a value for e.

input

Compounding the Problem

While the constant e is irrational, it is usually approximated by the value 2.72. Many calculators also have an 'e' button that can be used when performing calculations involving e.

The formula to compute the principal of an account with continuous compound interest can be written as

$$P = P_0 e^{rt}$$

where P_0 is the original principal of the account, r is the rate written as a decimal and P is the principal after t years.

process: evaluate

Compounding the Problem

9. Compute the principal after 18 years for each account opened by your grandparents if the interest was calculated continuously.

Computer Hard Wear

input

You have been saving over the last year to buy a new computer. While researching different computers, you find an article in a leading magazine stating that computers depreciate at a rate of about 40% annually. This means a computer purchased one year ago is worth 60% of its purchase price today. The computer you were thinking about buying costs $3000.

Computer Hard Wear

process: model

1. Write a function to calculate the value of the computer over time. Be sure to define the variables used.

2. Create a graph of the function

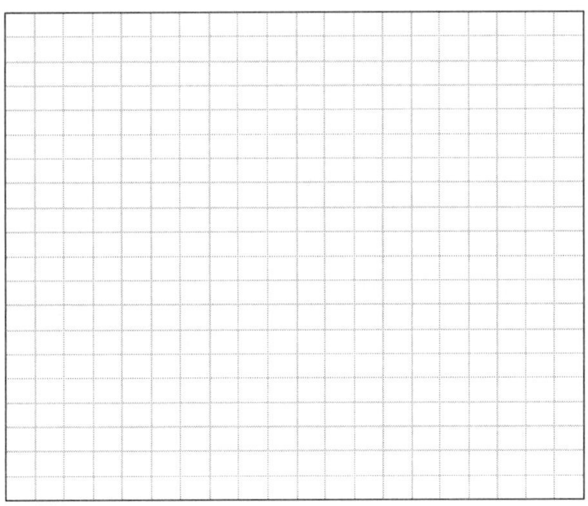

Computer Hard Wear

process: evaluate

3. What is the value of the computer after:
   a. 2 years?

   b. 5 years?

   c. 10 years?

4. Use the graph to determine when the value of the computer will be:
   a. $1500

   b. $750

   c. $100

5. Use the equation to determine algebraically when the value of the computer will be $2000.

Computer Hard Wear

analysis

6. Explain how you attempted to solve the problem algebraically. Did you encounter any difficulty?

Computer Hard Wear

extension

7. When will the computer be worth nothing?

8. Does your answer make sense? Why or why not?

input

Computer Hard Wear

So far, the only method you have used to solve an exponential equation has been to use a graph. However, it is also desirable to be able to solve exponential equations algebraically.

When solving a linear equation like $x - 4 = 7$, you add 4 to both sides since addition and subtraction are inverse operations. Similarly, solving an exponential equation like $2^x = 7$ requires the ability to "undo," or take the inverse of, the exponent when isolating the variable.

The inverse of the family of exponential functions is a family of functions called **logarithms**.

Consider an exponential function of the form $y = b^x$. The inverse of $y = b^x$ is $x = b^y$. This inverse can be written using logarithms to isolate the variable y.

The exponential equation $x = b^y$ can be rewritten with logarithms as $y = \log_b x$.

The logarithm $y = \log_b x$ is read "y equals the log base b of x."

process: practice

Computer Hard Wear

9. Convert the following exponential equations to logarithmic form. Write a sentence to demonstrate how the logarithmic equation would be spoken.

   a. $5^x = 14$

   b. $e^x = 23$

   c. $10^{3x+1} = 6$

process: practice

Computer Hard Wear

10. Write a sentence to demonstrate how each logarithmic equation would be spoken and convert each to exponential form.

    a. $x - 7 = \log_3 7.6$

    b. $2x = \log_5 34.9$

    c. $x = \log_8 6 + 1$

input

Computer Hard Wear

Some bases, such as 10 and e, are used more frequently than other bases. These two logarithms have special names and notations.

A logarithm of base 10 is called a **common logarithm** and can be written without a base. For example, $y = \log_{10} 6$ can be written as $y = \log 6$. When there is no explicit base, it is assumed to be base 10.

A logarithm of base e is called a **natural logarithm** and is written using 'ln' instead of 'log.' For example, $y = \log_e 6$ can be written as $y = \ln 6$.

Most scientific calculators only have buttons for the common and natural logs.

process: practice

Properties of Logarithms

1. Simplify the following exponential expressions.

   a. 5^0

   b. x^0

   c. $\left(\dfrac{5a}{b}\right)^0$

process: summary

Properties of Logarithms

2. What is the value of any base raised to the zero power? Write an exponential equation to demonstrate this.

3. Write this rule in logarithmic form.

4. Write a sentence to summarize a logarithmic rule corresponding to the exponential rule.

process: practice

Properties of Logarithms

5. Simplify the following exponential expressions.

   a. 16^1

   b. r^1

   c. $\left(\frac{x}{3y}\right)^1$

process: summary

Properties of Logarithms

6. What is the value of any base raised to the first power? Write an exponential equation to demonstrate this.

7. Write this rule in logarithmic form.

8. Write a sentence to summarize a logarithmic rule corresponding to the exponential rule.

process: practice

Properties of Logarithms

9. Simplify the following exponential expressions.

   a. $x^2 \bullet x^4$

   b. $s \bullet s^4$

   c. $y^3 \bullet y^5$

process: summary

Properties of Logarithms

10. Write a sentence to summarize the exponential property $a^m a^n = a^{m+n}$.

input

Properties of Logarithms

There is also a logarithmic property corresponding to the exponential property $a^m a^n = a^{m+n}$.

The next window will guide you through deriving this logarithmic property.

analysis

Properties of Logarithms

11. Let $x = a^m$ and $y = a^n$. Convert each of these exponential equations to logarithmic form.

12. Substitute x and y into the original exponential property to write an equation of the form a^{m+n} equals.

13. Convert the exponential equation above to logarithmic form.

14. Substitute the logarithmic expressions for m and n into the equation above.

15. Summarize the logarithmic property derived from the exponential property $a^m a^n = a^{m+n}$.

process: practice

Properties of Logarithms

16. Simplify the following exponential expressions.

a. $\dfrac{x^6}{x^4}$

b. $\dfrac{w^{10}}{w}$

c. $\dfrac{(x+1)^6}{(x+1)^3}$

process: summary

Properties of Logarithms

17. Write a sentence to summarize the exponential property $\dfrac{a^m}{a^n} = a^{m-n}$.

analysis

Properties of Logarithms

The next set of questions will guide you through deriving the logarithmic property corresponding to the exponential property $\frac{a^m}{a^n} = a^{m-n}$.

18. Let $x = a^m$ and $y = a^n$. Convert each of these exponential equations to logarithmic form.

19. Substitute x and y into the original exponential property to write an equation of the form a^{m-n} equals.

20. Convert the exponential equation above to logarithmic form.

21. Substitute the logarithmic expressions for m and n into the equation above.

22. Write a sentence to summarize the logarithmic property.

process: summary

Properties of Logarithms

Another logarithmic property can be expressed as $\log_b a^x = x\log_b a$.

23. Write a sentence to summarize this property.

extension

Properties of Logarithms

Derive the logarithmic property $\log_b a^x = x\log_b a$. (Hint: recall that a^x can be thought of as 'a' multiplied by itself x times.)

analysis

Properties of Logarithms

Most scientific calculators only provide the common log (base 10) and the natural log (base e). In order to evaluate $\log_5 17$ using a calculator, it is necessary to write the logarithmic expression using only common or natural logs.

24. Rewrite $\log_5 17$ using common logs by setting the logarithmic expression equal to a variable, converting the logarithmic equation to exponential form, taking the common log of both sides of the exponential equation, and solving for the variable using properties of logarithms as needed.

25. In general, how can $\log_b a$ be written using only common logs?

26. Rewrite the following logarithms using common logs and evaluate.

    a. $\log_4 6$

    b. $\log_8 145$

    c. $\log_3 18.96$

Properties of Logarithms

process: practice

27. Write the following logarithmic expressions in expanded form.

   a. $\log_4 6x^5$

   b. $\log_7 \dfrac{3y^4}{x^3}$

   c. $\ln 3xy^3$

28. Write the following logarithmic expressions using a single logarithm.

   a. $\log_2 10 + 3\log_2 x$

   b. $4\log 12 - 4\log 2$

   c. $3(\ln 3 - \ln x) + (\ln x - \ln 9)$

process: model

Graphs of Logarithms

1. Determine an equation for the inverse of the linear function $f(x)=3x+2$.

2. Complete the table of values for $f(x)$ and $f^{-1}(x)$. Describe the relationship between the table of values of a function and its inverse.

x	f(x)
-2	
-1	
0	
1	
2	

x	$f^{-1}(x)$
	-2
	-1
	0
	1
	2

3. Create a graph of $f(x)$, $f^{-1}(x)$ and the line y=x. Describe the relationship between the graph of a function and its inverse.

Graphs of Logarithms

process: model

4. Determine an equation for the inverse of the exponential function $f(x) = 10^x$.

5. Create a graph of $f(x)$, $f^{-1}(x)$ and the line y=x.

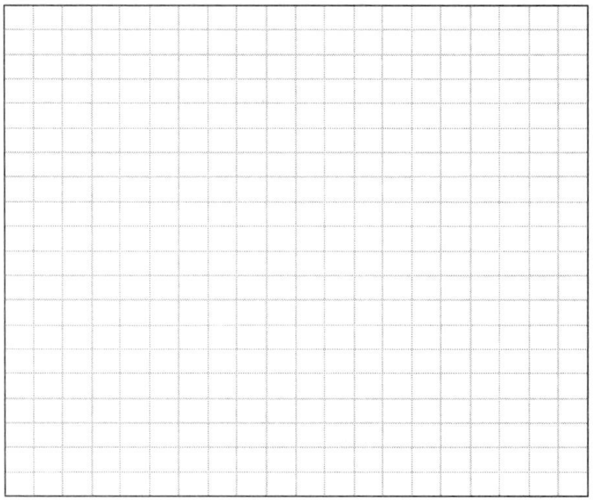

6. Complete the table of values below so that the relationship between $f(x) = 10^x$ and $f^{-1}(x)$ is apparent. Describe the relationship between the table of values of a function and its inverse.

x	f(x)
-2	
-1	
0	
1	
2	

x	$f^{-1}(x)$

Graphs of Logarithms

analysis

7. Based on your graph, determine the y-intercept of $f(x) = 10^x$. What is the x-intercept of $f^{-1}(x)$?

8. Are your graphical answers consistent with the definition of an inverse? Explain why or why not.

9. Based on your graph, determine the equation of the asymptote for $f(x) = 10^x$. What is the equation of the asymptote for $f^{-1}(x)$?

10. Are your graphical answers consistent with the definition of an inverse? Explain why or why not.

11. What similarities are there between the asymptote for an exponential function and the asymptote for a logarithmic function? What differences are there?

Graphs of Logarithms

process: model

12. Determine an equation for the inverse of the exponential function $f(x) = 2^x$.

13. Create a graph of $f(x)$, $f^{-1}(x)$ and the line y=x. You will need to perform a change of base conversion to enter the inverse function in your calculator.

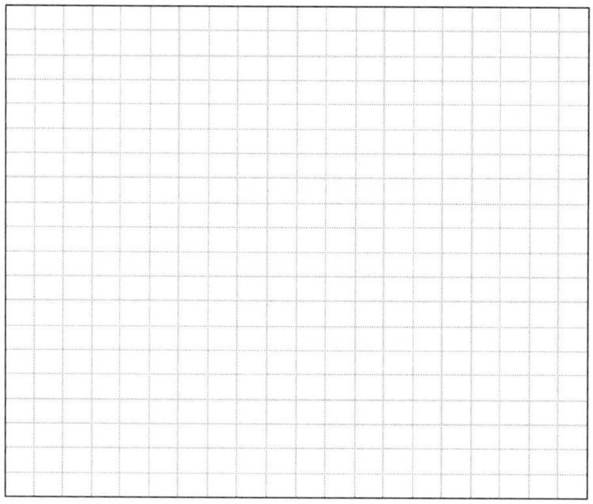

14. Complete the table of values below so that the relationship between $f(x) = 2^x$ and $f^{-1}(x)$ is apparent. Describe the relationship between the table of values of a function and its inverse.

x	f(x)
-2	
-1	
0	
1	
2	

x	$f^{-1}(x)$

analysis

Graphs of Logarithms

15. Based on your graph, determine the y-intercept of $f(x) = 2^x$. What is the x-intercept of $f^{-1}(x)$?

16. Are your graphical answers consistent with the definition of an inverse? Explain why or why not.

17. Based on your graph, determine the equation of the asymptote for $f(x) = 2^x$. What is the equation of the asymptote for $f^{-1}(x)$?

18. Are your graphical answers consistent with the definition of an inverse? Explain why or why not.

19. What similarities are there between the asymptote for an exponential function and the asymptote for a logarithmic function? What differences are there?

process: summary

Graphs of Logarithms

20. What observations can you make about the x-intercept of a logarithmic function in the form $y = \log_b a$?

21. What observations can you make about the y-intercept of a logarithmic function in the form $y = \log_b a$?

process: model

Logarithmic Transformations

1. Graph the parent logarithmic function $f(x) = \log x$ and the transformed functions $g(x) = \log x + 2$, $h(x) = \log x + 5$, $k(x) = \log x - 2$ and $p(x) = \log x - 5$ on the same coordinate plane.

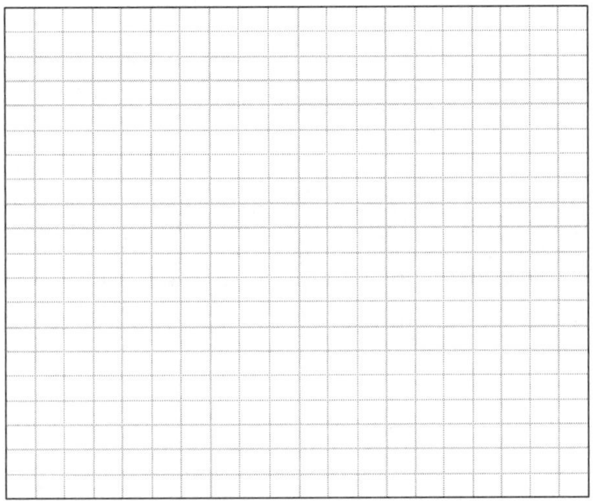

analysis

Logarithmic Transformations

2. What happens to the graph of the parent logarithmic function if you add a constant?

3. What happens to the graph of the parent logarithmic function if you subtract a constant?

4. What is the equation of the asymptote for $f(x) = \log x$?

5. What is the impact of adding or subtracting a constant to the parent function on the asymptote?

analysis

Logarithmic Transformations

6. What is the x-intercept of $f(x) = \log x$?

7. Determine the x-intercept of $g(x) = \log x + 2$ using an algebraic approach.

8. Write the x-intercept of $g(x) = \log x + 2$ as a power of 10. What is the relationship between the constant added to the parent function and the x-intercept?

9. Determine the x-intercept of $g(x) = \log x - 2$ using an algebraic approach.

10. Write the x-intercept of $g(x) = \log x - 2$ as a power of 10. What is the relationship between the constant added to the parent function and the x-intercept?

11. What is the x-intercept of $h(x) = \log x + 5$ and $p(x) = \log x - 5$? Verify algebraically.

process: model

Logarithmic Transformations

12. Graph the functions $f(x) = \log x$, $g(x) = \log(x + 2)$, $h(x) = \log(x + 5)$, $k(x) = \log(x - 2)$ and $p(x) = \log(x - 5)$ on the same coordinate plane.

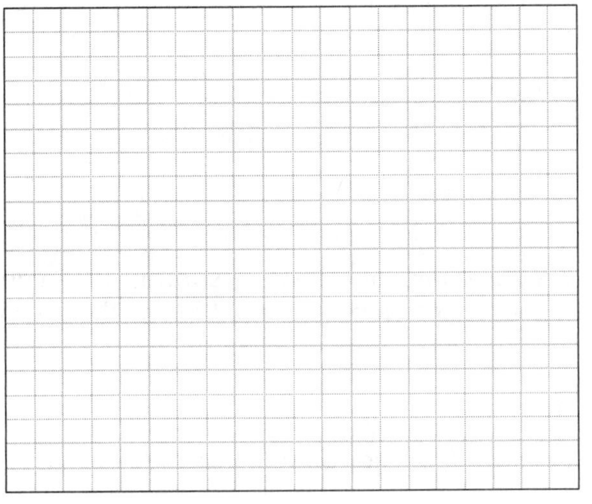

analysis

Logarithmic Transformations

13. What happens to the graph of the parent function if you add a constant within the logarithm?

14. What happens to the graph of the parent function if you subtract a constant within the logarithm?

Logarithmic Transformations

analysis

15. What is the equation of the asymptote for $f(x) = \log x$?

16. What is the equation of the asymptote for the four transformed functions?

17. What is the impact of adding or subtracting a constant to the parent function within the logarithm on the equation of the asymptote?

18. What is the x-intercept of $f(x) = \log x$?

19. What is the x-intercept of the four transformed functions?

20. What is the impact of adding or subtracting a constant to the parent function within the logarithm on the x-intercept?

Logarithmic Transformations

process: model

21. Graph the functions $f(x) = \log x$, $g(x) = -\log x$, $h(x) = 2\log x$, $j(x) = 5\log x$, $k(x) = -2\log x$ and $p(x) = -5\log x$ on the same coordinate plane.

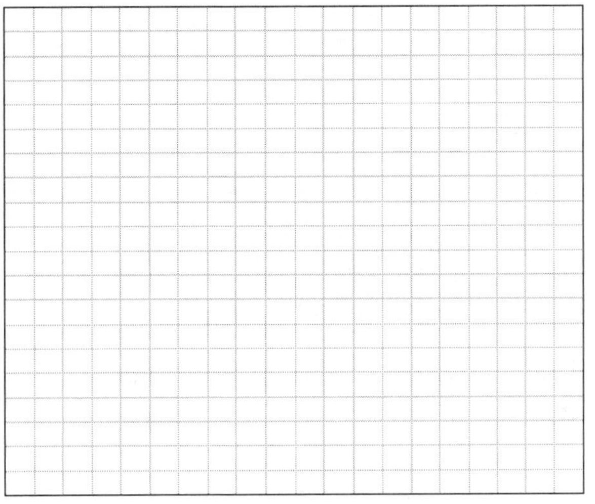

Logarithmic Transformations

analysis

22. What happens to the graph of the parent function if you multiply by a positive constant?

23. What happens to the graph of the parent function if you multiply by a negative constant?

Logarithmic Transformations

analysis

24. What is the equation of the vertical asymptote for $f(x) = \log x$?

25. What is the impact of multiplying the parent function by a constant on the asymptote?

26. What is the x-intercept of $f(x) = \log x$?

27. What is the impact of multiplying the parent function by a constant on the x-intercept?

Logarithmic Transformations

process: practice

28. Rewrite the following logarithms using only common logs.

    a. $g(x) = \log_2 x$

    b. $h(x) = \log_3 x$

    c. $k(x) = \log_5 x$

process: model

Logarithmic Transformations

29. Graph the functions $f(x) = \log x$, $g(x) = \log_2 x$, $h(x) = \log_3 x$, and $k(x) = \log_5 x$ on the same coordinate plane using the rewritten forms from the previous question.

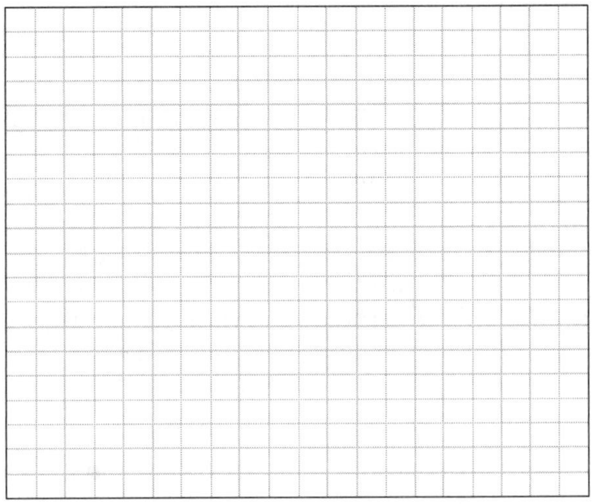

analysis

Logarithmic Transformations

30. What transformation does changing the base of the parent logarithms appear similar to?

input

Logarithmic Transformations

The rewritten form of $g(x) = \dfrac{\log x}{\log 2}$ can also be written as $g(x) = \left(\dfrac{1}{\log 2}\right)\log x$. When written in this form, it can be seen that changing the base is the same as multiplying the parent function by a constant.

process: practice

Logarithmic Transformations

31. Write the function that results when the following transformations are applied to the given parent function.

    a. The parent $f(x) = \log x$ is shifted up 3 units and stretched by a factor of 2.

    b. The parent $f(x) = \log x$ is shifted left 1 unit, reflected about the x-axis and stretched by a factor of $\frac{1}{2}$.

    c. The parent $f(x) = \log x$ is shifted up 3 units, shifted right 2 units, reflected about the x-axis and stretched by a factor of 10.

32. Describe the transformations applied to the parent function that result in each function below.

    a. $f(x) = 3\log(x - 1) - 4$

    b. $f(x) = -2\log x + 6$

    c. $f(x) = -\log(x + 5) + 2.5$

input

Richter Scale

The Richter Scale was developed by a scientist named Dr. Charles F. Richter in the 1930's to measure the strength of an earthquake from vibrations recorded with a machine called a seismograph.

A seismograph consists of a pendulum with a pen at the end. Paper scrolls beneath the pendulum. When there is no motion, the pen draws a straight line on the paper. When the earth shakes, the pen draws a squiggly line. The amplitude is the distance from the highest part to the lowest part of the squiggly line.

The Richter Scale is logarithmic, and is computed by taking the common logarithm of the amplitude, measured in millimeters.

A summary of intensity corresponding to each Richter Scale number appears below.

Richter Scale number	Intensity
1	Not detectable by humans
2	Hanging lamps sway
3	Vibrations begin to be felt by humans
4	Windows break
5	Tables and lamps topple
6	Wooden houses sustain damage
7	All buildings sustain damage
8	Catastrophic damage

process: model

Richter Scale

1. Write a function to calculate the Richter Scale magnitude from the amplitude measured from a seismograph.

process: model

Richter Scale

2. Complete the table of values to show the Richter Scale magnitude.

Amplitude measured using seismograph	Richter Scale magnitude
mm	Richter number
1	
10	
100	
1,000	
10,000	
100,000	
1,000,000	
10,000,000	
100,000,000	
1,000,000,000	
10,000,000,000	

3. Graph the function below.

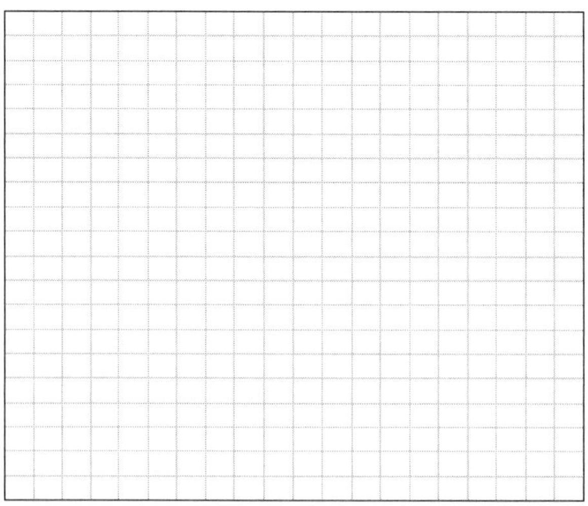

Richter Scale

process: evaluate

4. What is the Richter Scale magnitude if the seismograph measures an amplitude of:
   a. 1600 mm?

   b. 13,000,000 mm?

5. What is the amplitude of an earthquake measuring:
   a. 6.7 on the Richter Scale?

   b. 4.7 on the Richter Scale?

6. How much more powerful is an earthquake measuring 6.7 on the Richter Scale than an earthquake measuring 4.7 on the Richter Scale?

input

Carbon 14

Carbon-14 is a radioactive element used to date the age of plants and animals. It has a half-life of 5730 years.

When a plant or animal is alive, the amount of Carbon-14 present is constant, but when the plant or animal dies, the Carbon-14 begins to decay. To determine the length of time since the plant or animal died, you measure the percent of Carbon-14 remaining and apply the decay formula.

process: model

Carbon 14

1. Write an equation for the percentage of Carbon-14, using the value 100 as the initial amount present. Refer to the general equation for compounded interest to help in writing the equation.

analysis

Carbon 14

2. What is the y-intercept of the function? What does the y-intercept mean in terms of the problem situation?

3. What is the domain and range for this problem situation?

Carbon 14

process: model

4. Complete the table of values to show the percent of Carbon 14 remaining over time.

Time since the plant or animal died	Percent of Carbon 14 remaining
Years	**Percent**
0	
2500	
5000	
7500	
10000	
12500	
15000	
17500	
20000	
22500	
25000	

5. Graph the function below.

Carbon 14

process: evaluate

6. If the skeletal remains of an animal have been measured to have 70% of their Carbon-14 left, how long has it been since the animal died?

7. Determine the age of an animal's skeletal remains if 33% of their Carbon-14 is left.

input

Newton's Law of Cooling

The scientist Isaac Newton discovered that the cooling of a liquid can be modeled using an exponential function of the form $D = D_{room} + D_{diff}e^{Ct}$ where

- D is the temperature of the liquid in degrees Celsius.
- T is the time in minutes.
- D_{room} is the room temperature.
- D_{diff} is the difference between the room temperature and the starting temperature of the liquid.
- C is a cooling constant depending on the liquid.

You bought a cup of hot chocolate at lunch. The initial temperature of the hot chocolate is 200 degrees Celsius. The cafeteria is kept at 70 degrees Celsius. The cooling constant for hot chocolate is -0.04.

process: model

Newton's Law of Cooling

1. Write an equation to model the temperature of the hot chocolate over time.

analysis

Newton's Law of Cooling

2. Why is the base of e used to model the cooling of a liquid?

Newton's Law of Cooling

analysis

3. What is the y-intercept of the function? What does the y-intercept mean in terms of the problem situation?

4. What is the domain and range for this problem situation?

Newton's Law of Cooling

process: model

5. Graph the function below.

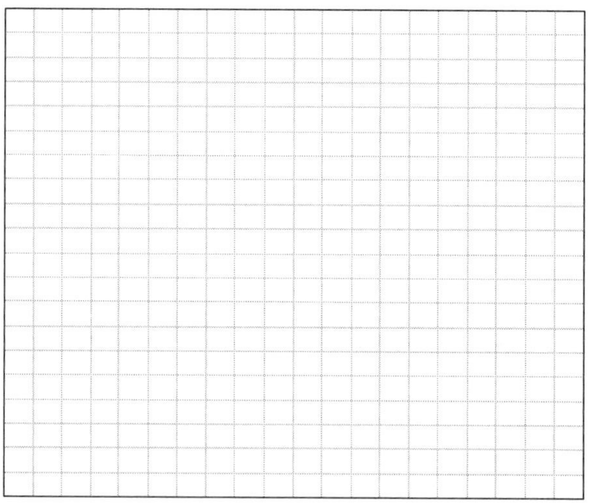

Newton's Law of Cooling

process: evaluate

6. What will be the temperature of the hot chocolate after 5 minutes?

7. What will be the temperature of the hot chocolate after 30 minutes?

8. How long will it take for the temperature of the hot chocolate to be 150 degrees Celsius?

9. How long will it take for the temperature of the hot chocolate to be 100 degrees Celsius?

input

Old Cars

In 1923, your great-grandparents purchased a brand new Ford Model T for $250. The car was inherited by your grandparents in 1948 and was appraised at $1000. When your parents inherited the car in 1980, it was valued at $7200. Knowing that one day you would inherit the car from your parents, you visited many internet auction sites. On one site, a similar model and condition Model T sold for $26,500 in 2002.

process: model

Old Cars

1. Using your calculator, find an equation of best-fit to model the data. Write the equation.

analysis

Old Cars

2. What is the y-intercept of the function? What does the y-intercept mean in terms of the problem situation?

3. What is the domain and range for this problem situation?

Old Cars

process: model

4. Using the equation of best-fit, create a table of values to show the value of the car over time.

5. Graph the best-fit equation below.

process: evaluate

Old Cars

6. What will be the value of the car if you inherit it on your 18th birthday?

7. What will be the value of the car when your children inherit it? Assume you keep the car for 30 years before passing it on to your children.

8. Are the previous two answers reasonable? Explain why or why not.

9. When will the car be worth $100,000?

10. When will the car be worth $1,000,000?

11. Are the previous two answers reasonable? Explain why or why not.

Unit 8: Quadratics

Contents

input

Animated gif Program

Last spring you created a home page for a client who, being a dog lover, wanted her page to open with a growing image of her dog. You designed the home page so that the dog's image would increase by one inch each second for a total of five seconds.

You recorded the information relating to the design in case adjustments needed to be made, but also for use with different projects in the future. The table below includes the data about the design and its animation process.

	Time	Image Length	Image Area
Labels			
Units			
	0	0	
	1	1	
	2		
	3		
	4		
	5		
Expression	x		

analysis

Animated gif Program

1. Draw the graph to show the relationship between time and image length.

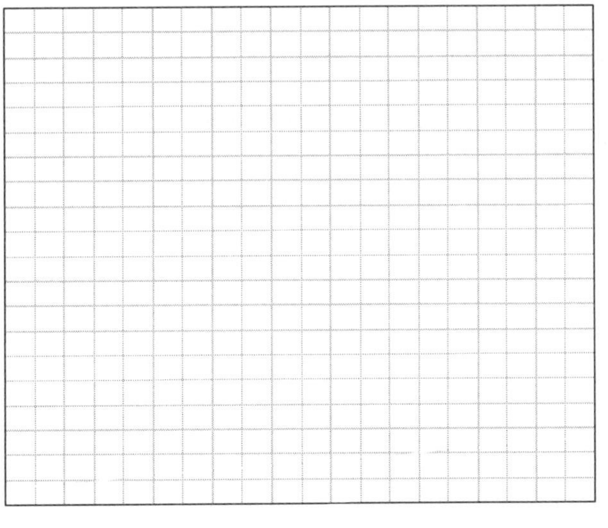

analysis

Animated gif Program

2. Draw the graph to show the relationship between time and image area.

3. Describe the growth patterns in the image length and the image area.

4. Use your graph to describe the rate of change for image length.

5. Use your graph to describe the rate of change for image area.

process: model

Animated gif Program

6. Based on the numerical data and the graph, write the function that describes the relationship between time and image length.

7. Write the function that describes the relationship for the image area with respect to time. (If you have difficulty describing this relationship, think about how you determined the actual value of the area given a specific value for the time).

analysis

Animated gif Program

8. How does the second graph describe the relationship between the *side length* of the image and the *area* of the image?

process: model

Animated gif Program

9. Suppose the image length started at 1 and was doubling every second for five seconds. Generate a table of values for the image length and image area.

Labels	**Time**	**Image Length**	**Image Area**
Units			
Expression			

Animated gif Program

process: model

10. Under these conditions, what is the algebraic equation that models the image length versus time?

11. What is the algebraic equation that models the image area with respect to time?

12. Graph the function that models image length versus time on the first grid, and graph that function that models image area versus time on the second grid.

analysis

Animated gif Program

13. Compare the graphs of the exponential growth pattern for the image length and image area with those in questions 1 and 2.

input

Animated gif Program: Adding a Banner

Your client decides she wants to modify her Web site by adding a rectangular banner of fixed size (2 in. by 5 in.) above the image as it grows.

To satisfy the client, you plan to modify the original program so that when the Web page first opens, a banner of fixed size appears. Then for each second that passes, the side length of the dog's image will increase by 1 inch, just as in the original program.

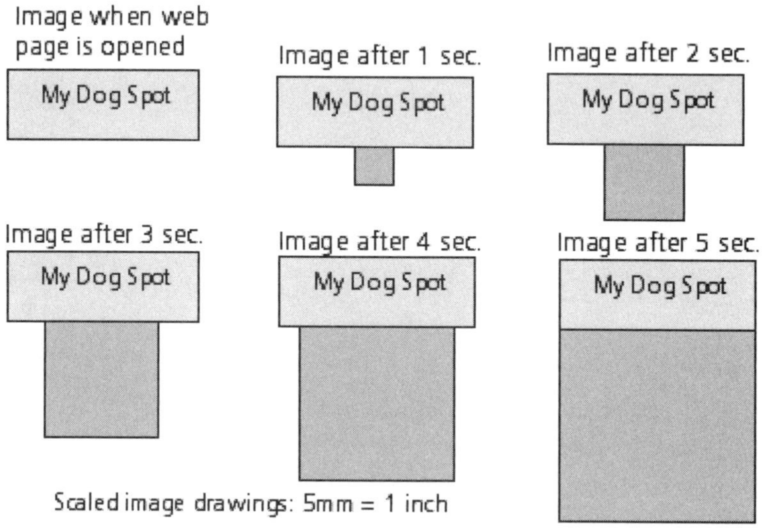

Scaled image drawings: 5mm = 1 inch

process: model

Animated gif Program: Adding a Banner

1. As always, you record information when you make the homepage.

	Time	Banner Area	Image Area	Total Area
Labels				
Units				
	0			
	1			
	2			
	3			
	4			
	5			
Expression				

process: model

Animated gif Program: Adding a Banner

2. To make sure that the design is working as it should, you verify that the growth of the image is smooth and continuous by sketching the graph for image area versus time.

3. Sketch the graph for total area versus time on this grid.

Animated gif Program: Adding a Banner

analysis

4. How is the graph of the image area the same as the graph of the total area?

5. How is the graph of the image area different from the graph of the total area?

6. To express the total area as a function of time, what change must be made to the image area function?

Animated gif Program: Adding a Banner

process: model

7. Express the *total area* of the banner and image combined as a function of time.

process: evaluate

Animated gif Program: Adding a Banner

8. As a final touch to the Web page with the banner, you will add a sound bite of a dog barking. The sound bite is to begin when the image area is 16.25 square inches. Use your graph to *estimate* how many seconds after the animation starts the sound bite will be heard. Explain how you estimated your answer.

process: model

Animated gif Program: Adding a Banner

9. Since precision is important, set up and solve an equation to determine the starting time of the sound bite.

process: summary

Animated gif Program: Adding a Banner

10. Write a paragraph in which you compare the representations (numerical, graphical and symbolic) for the original animated gif program and the new program with the banner added. Make sure to note the similarities and differences between each of the representations.

11. If you had to choose between the different representations, which do you think is generally most useful when solving problems? Why?

analysis

Parent Functions and Vertical Transformations

The graphs below represent the parent functions of:

$y = x$ $y = x^2$ $y = 2^x$

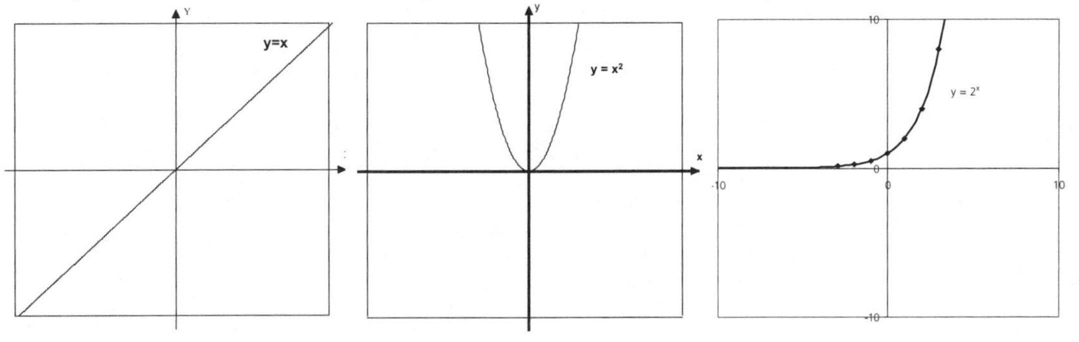

1. In your own words, describe how each of the above graphs is similar to each other and how they are different from each other.

Parent Functions and Vertical Transformations

process: model

2. Graph the following functions using your calculator. Then sketch the graph on the grids below.

a. $y = x + 3$

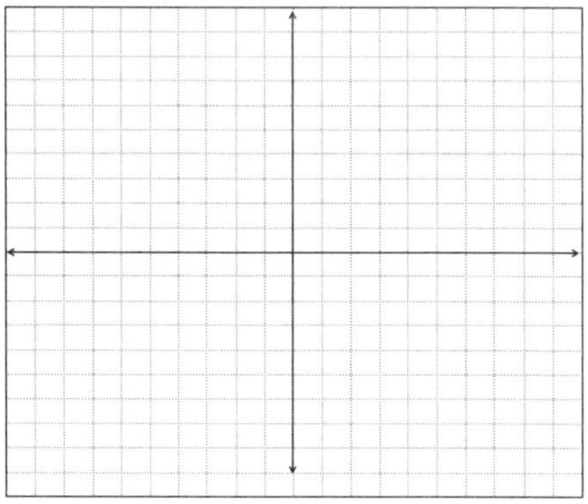

b. $y = x^2 + 3$

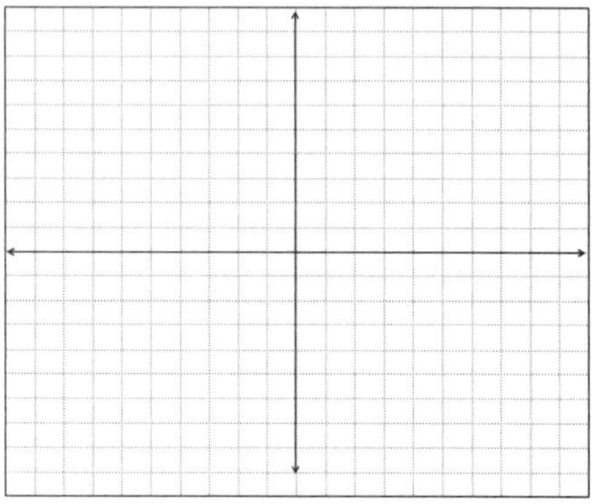

c. $y = 2^x + 3$

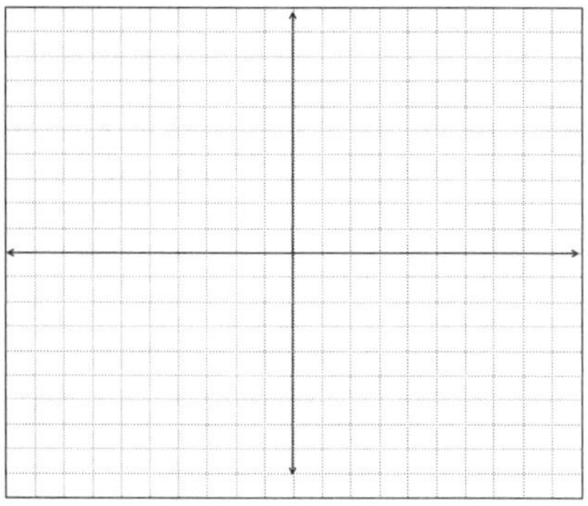

d. $y = x - 2$

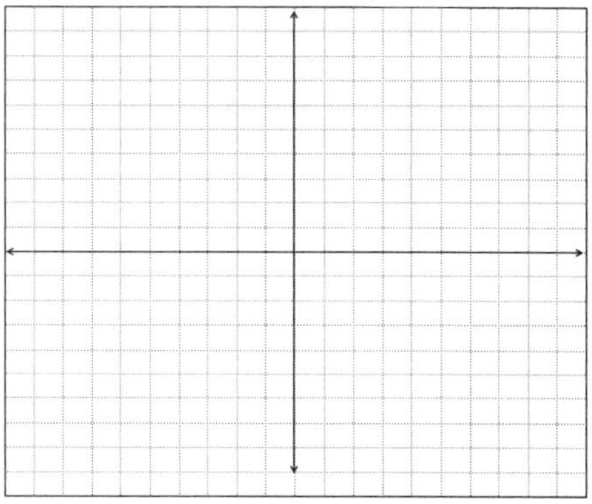

e. $y = x^2 - 2$

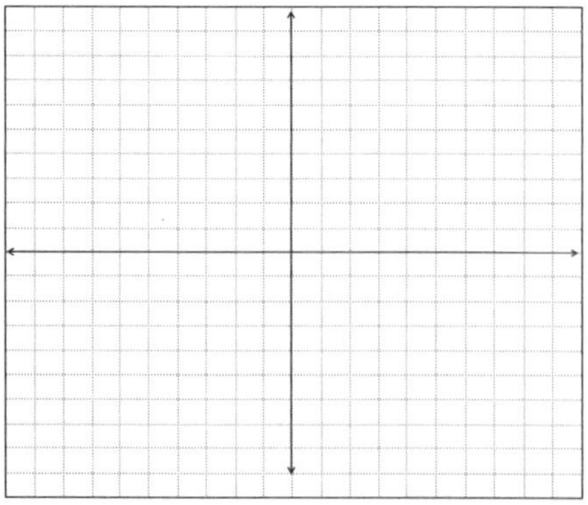

f. $y = 2^x - 2$

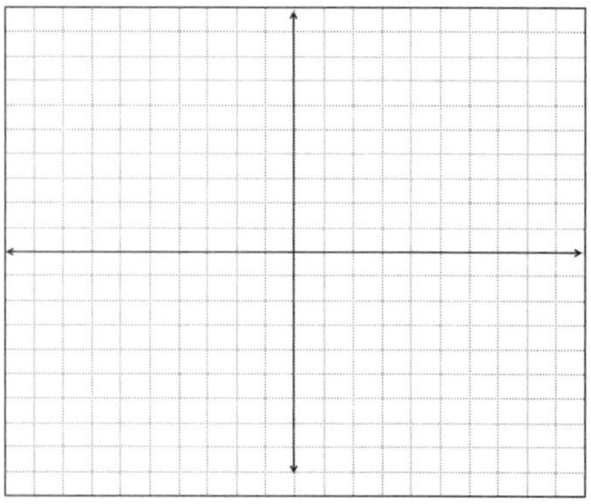

Parent Functions and Vertical Transformations

analysis

3. Describe each of the equations on the previous page in relation to their parent functions. How are they similar? How are they different?

Parent Functions and Vertical Transformations

extension

4. When the equation of a line is written in the form of $y = x + k$, how does its graph compare to the graph of $y = x$ if:

   a. $k > 0$

   b. $k < 0$

extension

Parent Functions and Vertical Transformations

5. When the equation of an exponential function is written in the form of $y = b^x + k$, how does its graph compare to the graph of $y = b^x$ if:

   a. $k > 0$

   b. $k < 0$

6. When the equation of a parabola is written in the form of $y = x^2 + k$, how does its graph compare to the graph of $y = x^2$ if:

   a. $k > 0$

   b. $k < 0$

Animated gif Program: Time Changes

input

You have a new design assignment. This time the client has a pet parrot. The assignment is similar to the work you did for the client who wanted an animated image of her dog on her website, but this time, you will be changing the timing of the animation in relation to the opening of the web page.

Your plan is to work with the programs you wrote by modifying the original one. Instead of the animation starting when the web page is opened, it will be delayed 2 seconds. Once the animation begins, the length of the image will still increase by 1 inch every second until the last image reaches a side length of 5 inches, just like with the image of the dog.

Animated gif Program: Time Changes

process: model

1. You want to study how the change in the start of the animation affects the length and area of the image. You record the information in a table to track the changes and understand the effect, if any, caused by the changes.

	Time	Image Length	Image Area
Labels			
Units			
	2	0	
	3		
	4		
	5		
	6		
	7		
Expression			

Animated gif Program: Time Changes

process: model

2. After examining the data, you want to model it algebraically. Define variables to represent the time, image length, and image area.

3. Write a function in which you express the image length as a function of time.

4. Write a function in which you express the image area as a function of time.

5. Compare the functions for image length versus time from the original animated gif program and the new animated gif program with the delayed appearance of the image by graphing both of them on the same set of axes.

process: model

Animated gif Program: Time Changes

6. Compare the original animated gif program to the new one with the delayed timing by sketching the graphs of image area versus time for each function on the same set of axes.

analysis

Animated gif Program: Time Changes

7. How do the graphs of the new program compare to the graphs of the original program?

8. Look at the positions of both graphs for the animated gif program with the delayed appearance of the image. How do the algebraic equations for each of these seem to conflict with the graphs?

process: evaluate

Animated gif Program: Time Changes

9. Using the new function for the area vs. time, determine how long it will take the area of the image to reach 20.25 square inches.

input

Animated gif Program: Time Changes

You have tested the 2-second delay, and now you would like to "push-up" the start of the animation by 1 second. The idea here is that when the Web page is opened, an image with a side length of 1 inch will already be displayed. It will increase by 1 inch every second until the last image has a side length of 5 inches.

process: model

Animated gif Program: Time Changes

10. Record all the relevant information in case further changes need to be made.

	Time	Image Length	Image Area
Labels			
Units	seconds	inches	inches^2
	0	1	
	1		
	2		
	3		
	4		
Expression			

process: model

Animated gif Program: Time Changes

11. Write an equation to represent image length as a function of time.

12. Express the image area as a function of time.

13. Sketch the original length vs. time graph and the new length vs. time graph for the "pushed-up" animation o the grid. Sketch the original area vs. time graph and the new area vs. time graph for the "pushed-up" animation on the second grid.

process: model

Animated gif Program: Time Changes

14. Plot a point on the new length vs. time graph showing when the length of the image would have been 0 (as if going back in time).
15. Plot a point on the new image area vs. time graph showing when the area of the image would have been 0 (as if going back in time).
16. Compare the original graphs to the graphs for the "pushed-up" animation. How have the original graphs been transformed to create the new graphs?

process: evaluate

Animated gif Program: Time Changes

17. How long it will take the area of the "pushed-up" image to reach 20.25 square inches? Use the graph to approximate the time, and confirm your answer by solving an algebraic equation.

analysis

Animated gif Program: Time Changes

18. Look at the positions of both graphs for the animated gif program with the "pushed-up" appearance of the image. How do the algebraic equations for each of these seem to conflict with the graphs?

process: model

Parent Functions and Horizontal Transformations

1. Using your graphing calculator, graph each of the functions. Sketch the graph below.

a. $y = x^2$

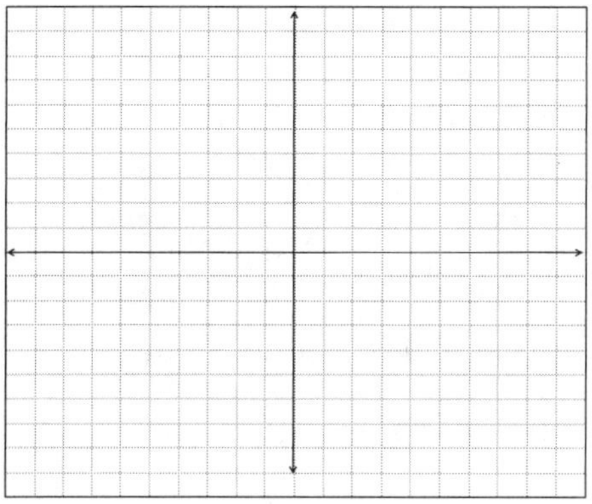

b. $y = (x - 1)^2$

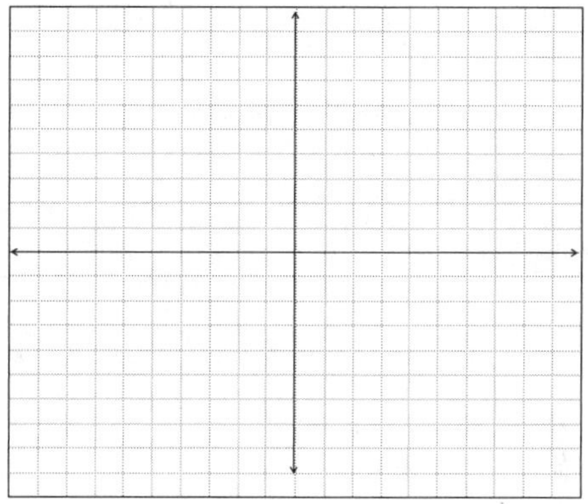

c. $y = (x + 2)^2$

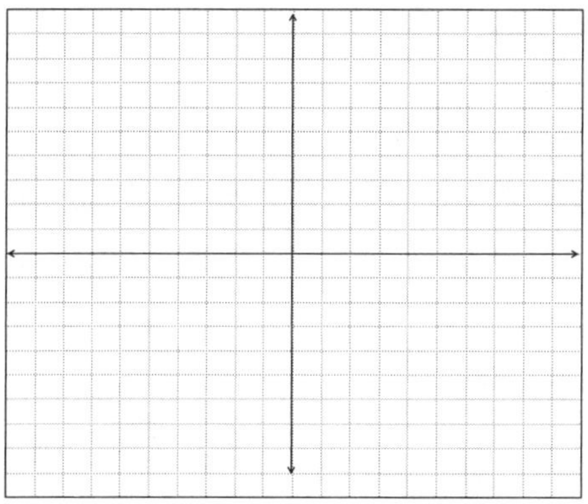

d. $y = 2^x$

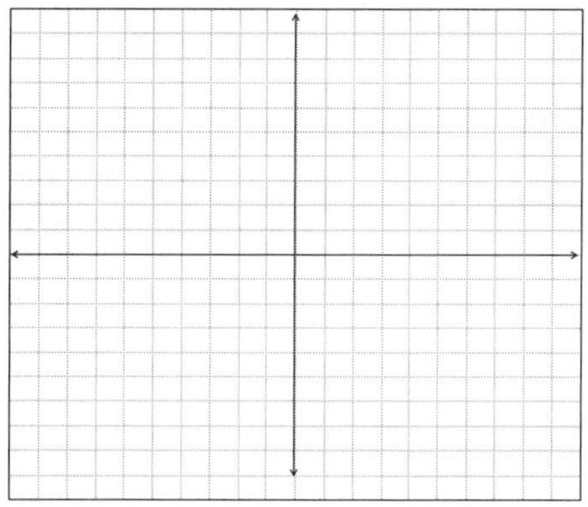

e. $y = 2^{x - 1}$

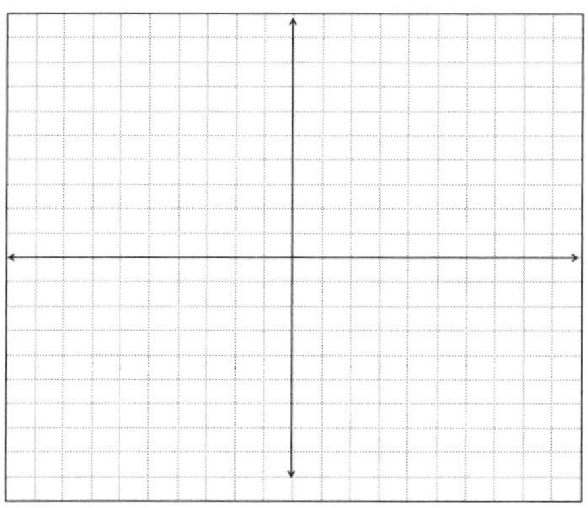

f. $y = 2^{x + 2}$

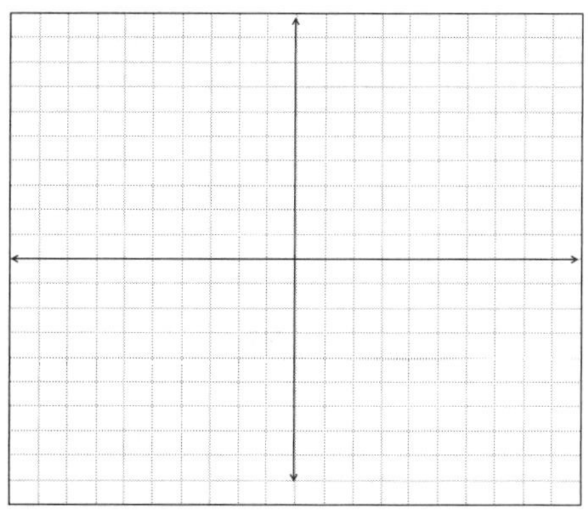

Parent Functions and Horizontal Transformations

analysis

2. Cite at least one similarity and at least one difference among the graphs of the three quadratic functions on the previous page.

3. How do the graphs of the three exponential equations compare to one another?

4. In general, what can you say about the effect of adding or subtracting a number from x in either a quadratic or an exponential function?

5. If the independent variable (x) is time, how do you interpret a horizontal translation to the right? How do you interpret a horizontal translation to the left?

process: practice

Parent Functions and Horizontal Transformations

6. Sketch the graphs of the following functions, which are members of the family of quadratic functions. Label each graph with its equation.

   a. $y = (x - 5)^2$

   b. $y = (x + 2)^2$

   c. $y = (x + 5)^2$

   d. $y = (x - 1)^2$

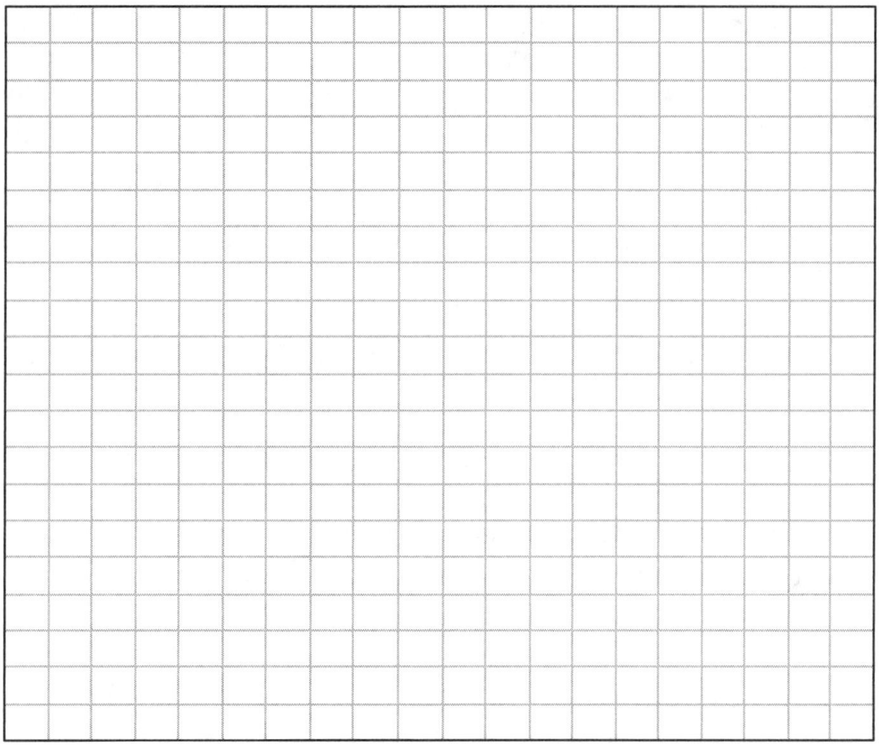

7. Plot a point on each graph showing the zero(s), which are the x-intercepts.
8. List the x-intercept(s) for each of the four quadratic functions.

extension

Parent Functions and Horizontal Transformations

9. When the equation of a parabola is written in the form of $y = (x - h)^2$, how does its graph compare to the graph of $y = x^2$, if:

   a. $h > 0$

   b. $h < 0$

10. What can you say about the line of symmetry for $y = (x - h)^2$ when h is positive? …when h is negative?

11. When an exponential equation is written in the form of $y = b^{x - h}$, how does it compare to the graph of $y = b^x$ if:

    a. $h > 0$

    b. $h < 0$

input

Animated gif Program: Size Changes

Just when you thought you were done with Web sites with animated animal images, you meet Leslie the cat lover. She wants an image to appear and grow when the Web page opens, just like the original dog-owner's Web page did, except she wants the side length of the image of her cat to grow by 1.5 inches every second for a total of 5 seconds. (Recall that in the original program, the image grew 1 inch every second).

What will the maximum length of the cat's image be after the 5 seconds have passed?

process: model

Animated gif Program: Size Changes

1. Gather and record information about the cat-lover's animation in the table.

Labels	Time	Image Length	Image Area
Units			
	0	0	0
	1	1.5	
	2		
	3		
	4		
	5		
Expression			

process: evaluate

Animated gif Program: Size Changes

2. By what factor does the new program change the side length of the image in the animation when compared to the side length in the very first gif program? (To answer this question, you may want to add a column to your table showing the side lengths from the original program.)

	Animated gif Program: Size Changes
process: model	3. Use the dilation factor (scale factor) to create an equation for the side length of the cat's image as a function of time. Make sure to define your variables and constants.

	Animated gif Program: Size Changes
process: evaluate	4. What is the image area after 5 seconds? 5. By what factor does this new program change the area of each image in the animation when compared to the area in the very first gif program? (To answer this question, you may want to add a column to your table showing the areas from the original program.)

	Animated gif Program: Size Changes
process: model	6. Create an equation for the area of the cat's image as a function of time.

process: model

Animated gif Program: Size Changes

7. Sketch the original length vs. time graph for the dog's animation, and sketch the new length vs. time graph for the cat's animation on the grid.

8. Sketch the original area vs. time graph for the dog's animation, and sketch the new area vs. time graph for the cat's animation on this grid.

analysis

Animated gif Program: Size Changes

9. Compare the graphs for image area in the original and dilated programs. Indicate at least two similarities and two differences between the original and dilated functions.

process: evaluate

Animated gif Program: Size Changes

10. You are thinking it might be fun to include the sound bite of a cat meowing along with the animation. If the sound bite is to begin when the image area is 20.25 square inches, how many seconds after the animation starts will the sound bite be heard?

input

Animated gif Program: Size Changes

The final thing you want to check is how the animation would look if you set the maximum image length to be 3.5 inches in 5 seconds, and the maximum image area to be 12.25 square inches after five seconds.

In this situation, how much is the side length changing each second?

process: model

Animated gif Program: Size Changes

11. For purposes of comparison, gather and record information in the table.

Labels	**Time**	**Image Length**	**Image Area**
Units			
	0		
	1		
	2		
	3		
	4		
	5		
Expression			

12. Sketch the original length vs. time graph for the dog's animation, and sketch the newest length vs. time graph for the cat's animation on the grid.

Animated gif Program: Size Changes

process: model

13. Sketch the original area vs. time graph for the dog's animation, and sketch the newest area vs. time graph for the cat's animation on the grid.

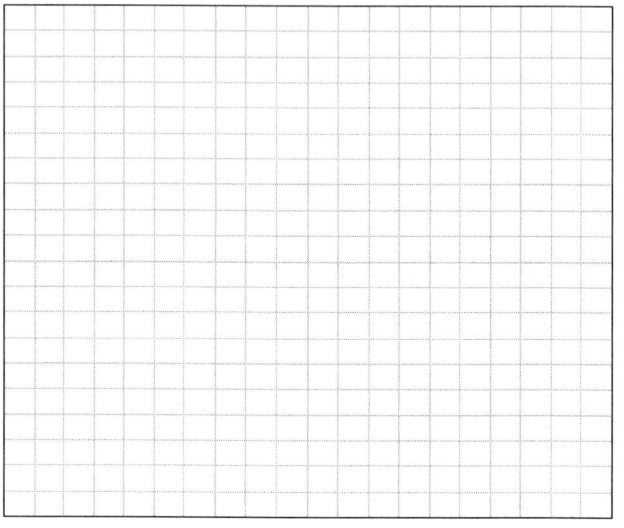

14. Using this dilation factor, express the length of the new images as a function of time and the area of the new images as a function of time.

Animated gif Program: Size Changes

process: evaluate

15. You want the maximum image area to be 12.25 square inches. For the original program, this area was a maximum of 25 square inches. By what factor does this new program change the size of each image in the animation?

process: summary

Animated gif Program: Size Changes

16. Compare the second dilation to the original program. What are the similarities and differences?

Squares, and Squares, and More Squares

process: model

1. Create a square centimeter. Use complete sentences to tell how you know that what you have constructed is a square centimeter.

Squares, and Squares, and More Squares

process: evaluate

2. What is the length of each side?

3. What is the perimeter?

4. What is the area?

Squares, and Squares, and More Squares

process: model

5. Create the next largest square, which has full centimeters for the measurements of its sides.

process: evaluate

Squares, and Squares, and More Squares

6. What is the length of each side?

7. How does this length compare to the length of each side in the previous square?

8. Write a ratio to compare these two lengths.

9. What is the perimeter of your new square?

10. How does this perimeter compare to the perimeter of the previous square?

11. Write a ratio to compare the two perimeters.

12. How does this ratio of perimeters compare to the scale factor?

13. What is the area of the new square?

14. How does this area compare to the area of the previous square?

15. Use a ratio to compare the two areas.

16. How does the ratio of areas compare to the scale factor?

process: model

Squares, and Squares, and More Squares

17. Create the next largest square that has full centimeters for the measurements of its sides.

process: evaluate

Squares, and Squares, and More Squares

18. What is the length of each side?

19. How does this length compare to the length of each side in the previous square?

20. Write a ratio to compare these two lengths.

21. What is the mathematical name given to this ratio?

22. How does the length of each side in the newest square compare to the length of each side in the *first square*? Write a ratio to compare these lengths.

process: evaluate

Squares, and Squares, and More Squares

23. What is the perimeter of your newest square?

24. How does this perimeter compare to the perimeter of the *previous square*? Write a ratio to compare the two perimeters.

25. How does the perimeter of the newest square compare to the perimeter of the *first square*? Write a ratio to compare the two perimeters.

26. How does this ratio of perimeters compare to the scale factor?

27. Explain why the ratio of perimeters compares to the scale factor in this way.

28. What is the area of the newest square?

29. How does this area compare to the area of the *previous square*? Use a ratio to compare the two areas.

30. How does the area of the newest square compare to the area of the *first square*? Use a ratio to compare the two areas.

31. How does the ratio of areas compare to the scale factor in each case?

Squares, and Squares, and More Squares

analysis

32. Why might the ratio of areas compare to the scale factor in this way?

33. If we continue to make squares with full-centimeter sides, what would be the length of each side in the next largest square?

34. What would the scale factor between the fourth square and the first square be?

35. What would the perimeter of the fourth square be?

36. What is the ratio of the perimeter of the fourth square to the perimeter of the first square?

37. How many square centimeters would be needed to build the fourth square?

38. What is the ratio of the area of the fourth square to the area of the first square?

39. Why is the ratio of perimeters the same as the scale factor, while the ratio of areas is not the same?

40. Are these squares dilations of one another? Explain. If so, where is the center of dilation? If not, what type of transformation has occurred?

process: model

Parent Functions and Dilations and Reflections

1. For problems a – d, first graph the parent function, y = x. Then, graph each transformed function. Use your graphing calculator for help if you need it.

a. Graph $y = x$ and $y = 2x$.

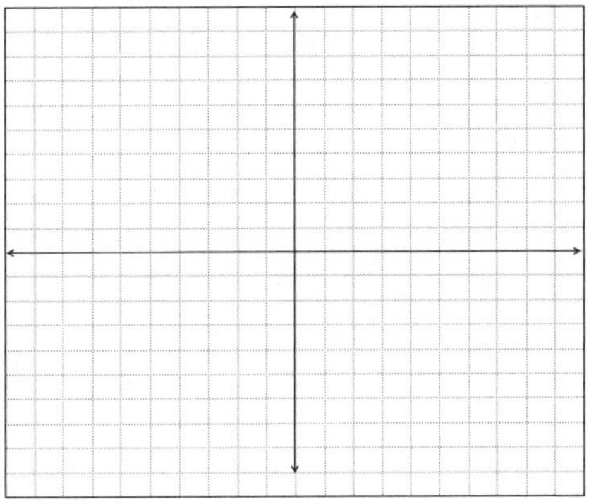

b. Graph $y = x$ and $y = \frac{x}{2}$.

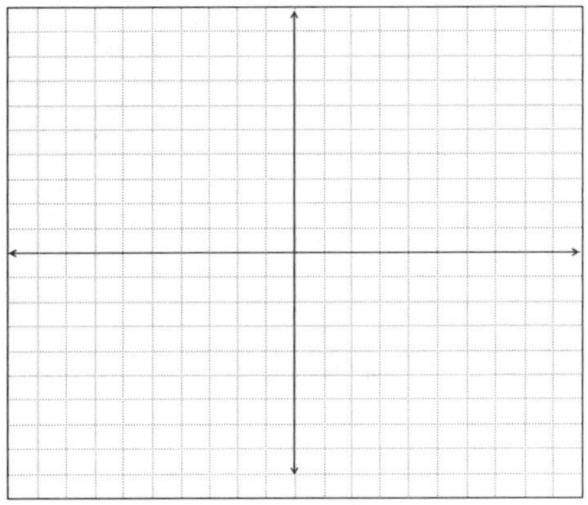

c. Graph $y = x$ and $y = -x$.

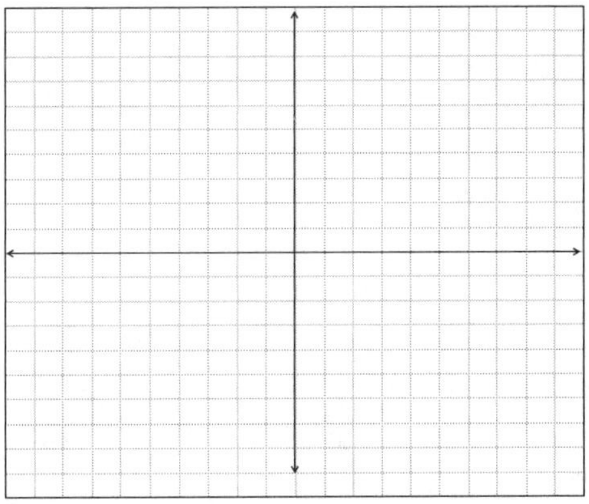

d. Graph $y = x$ and $y = \frac{-x}{2}$.

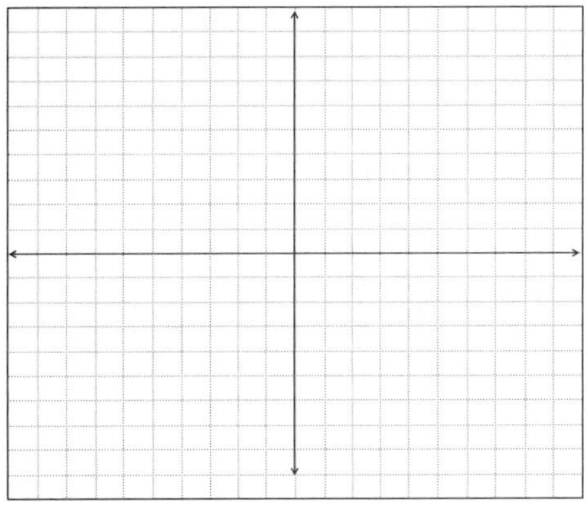

2. Describe how the graph of the transformed function compares to the graph of its parent function.

process: model

Parent Functions and Dilations and Reflections

3. For problems a – d, first graph the parent function, $y = x^2$. Then, graph each transformed function. Use your graphing calculator for help if you need it.

a. Graph $y = x^2$ and $y = 2x^2$.

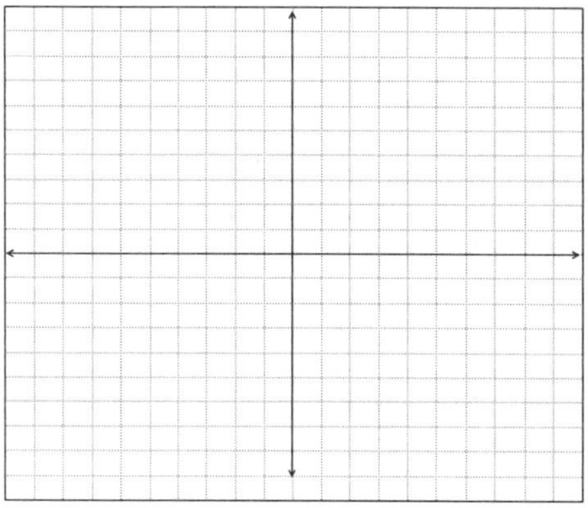

b. Graph $y = x^2$ and $y = \frac{x^2}{2}$.

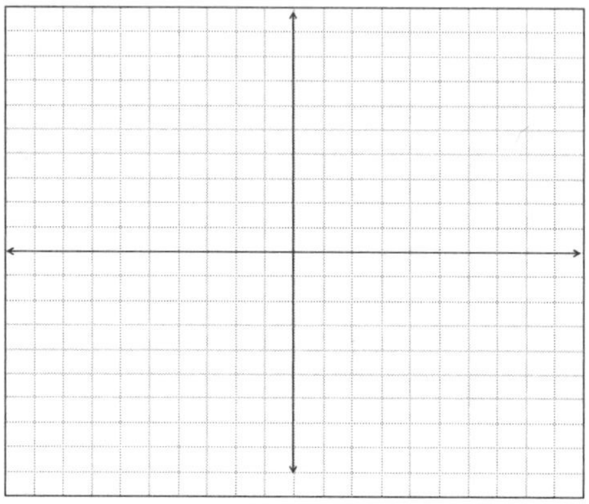

c. Graph $y = x^2$ and $y = -x^2$.

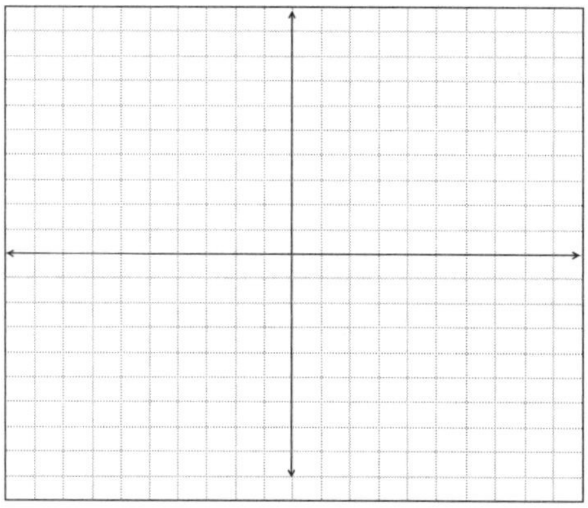

d. Graph $y = x^2$ and $y = \frac{-x^2}{2}$.

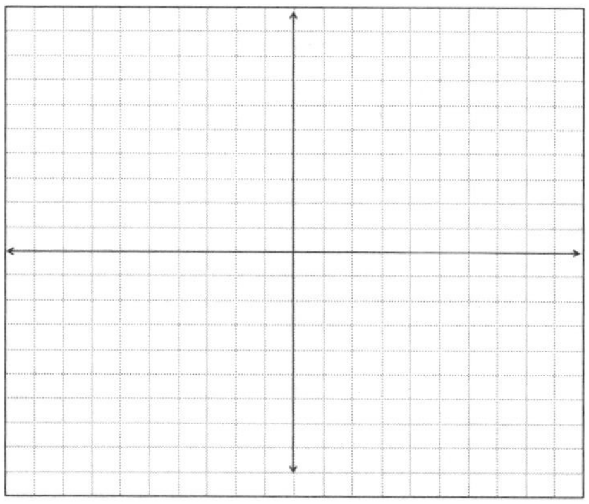

4. Describe how the graph of the transformed function compares to the graph of its parent function.

process: model

Parent Functions and Dilations and Reflections

5. For problems a – d, first graph the parent function, $y = 2^x$. Then, graph each transformed function. Use your graphing calculator for help if you need it.

a. Graph $y = 2^x$ and $y = 2\,(2^x)$.

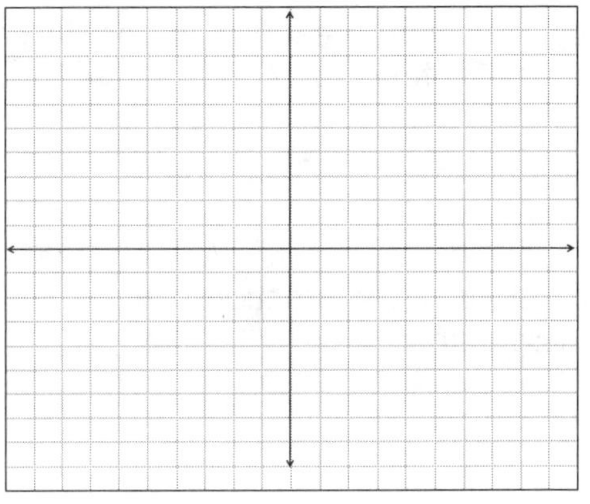

b. Graph $y = 2^x$ and $y = \frac{2^x}{2}$.

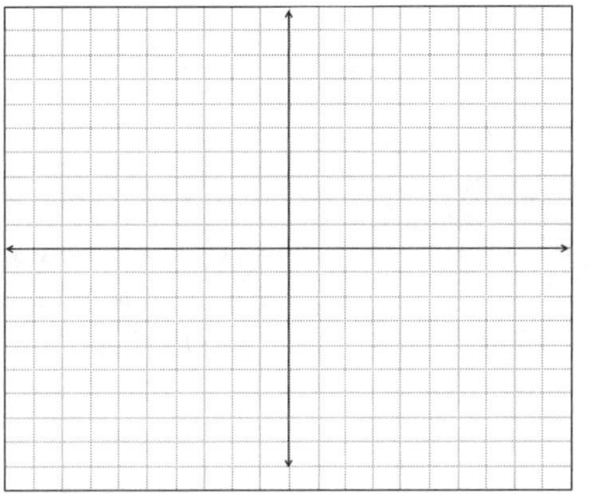

c. Graph $y = 2^x$ and $y = -(2^x)$.

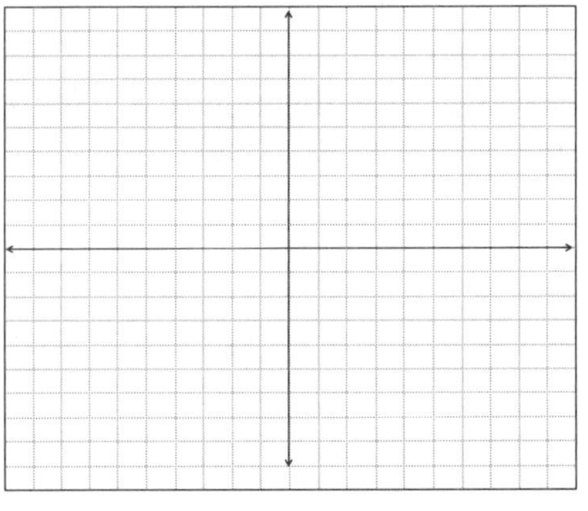

d. Graph $y = 2^x$ and $y = \frac{-(2^x)}{2}$.

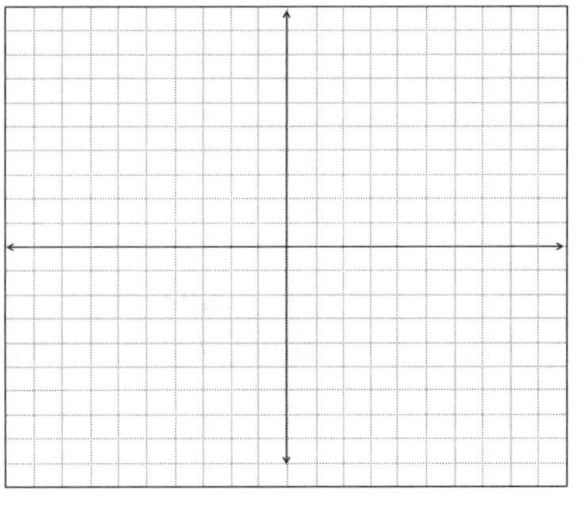

6. Describe how the graph of the transformed function compares to the graph of its parent function.

Putting it All Together

analysis

1. What do you observe about the graphs of the functions $y = x$, $y = x^2$, and $y = b^x$ when multiplying by a constant greater than one?

   a. Is the effect of multiplying by a constant greater than one the same or different in linear, quadratic, and exponential functions? Why?

2. What do you observe about the graphs of the functions $y = x$, $y = x^2$, and $y = b^x$ when multiplying by a constant less than zero?

   a. Is the effect of multiplying by a constant less than zero the same or different in linear, quadratic, and exponential functions? Why?

3. What do you observe about the graphs of the functions $y = x$, $y = x^2$, and $y = b^x$ when multiplying by a constant between zero and one?

   a. Is the effect the same or different in linear, quadratic, and exponential functions? Why?

4. What do you observe about the graphs of the functions $y = x$, $y = x^2$, and $y = b^x$ when adding or subtracting a constant from the variable in the expression.

   a. Is the effect the same or different in linear, quadratic, and exponential functions? Why?

Putting it All Together

analysis

5. What do you observe about the graphs of the functions $y = x$, $y = x^2$, and $y = b^x$ when adding or subtracting a constant from the entire expression.

   a. Is the effect the same or different in linear, quadratic, and exponential functions? Why?

6. Summarize the effects that transformations can have on graphs, and summarize the effects that transformations have on algebraic equations.

7. Given the made up function $f(x) = \Re \nabla x$:

   a. How would the graph of $f(x) = \Re \nabla x + 5$ compare to the original graph, whatever it looks like?

   b. How would the graph of $f(x) = -(\Re \nabla x)$ compare to the original graph?

   c. How would the graph of $f(x) = 3(\Re \nabla x)$ compare to the original graph?

   d. How would the graph of $f(x) = \Re \nabla (x + 5)$ compare to the original graph?

Writing Quadratic Equations

analysis

1. For each of the animated gif program activities following the first one (which was the parent function), describe the type of graphical transformation required.

   Adding a Banner

   Time Changes

   Size Changes

Writing Quadratic Equations

process: model

2. In the animated gif examples, you were introduced to the equation of a quadratic function in the vertex form. Here is an example of a quadratic equation in which two transformations are performed. What information can you gather from reading the equation in symbolic form?

$$f(x) = 0.5x^2 + 3$$

   a. Sketch a graph of this function based on the information you gathered above.

Writing Quadratic Equations

process: model

3. Here is another example of a quadratic equation with two transformations. What information can you gather from reading the equation in symbolic form?

$$g(x) = (x - 4)^2 - 2$$

   a. Sketch a graph of this function based on the information you gathered above.

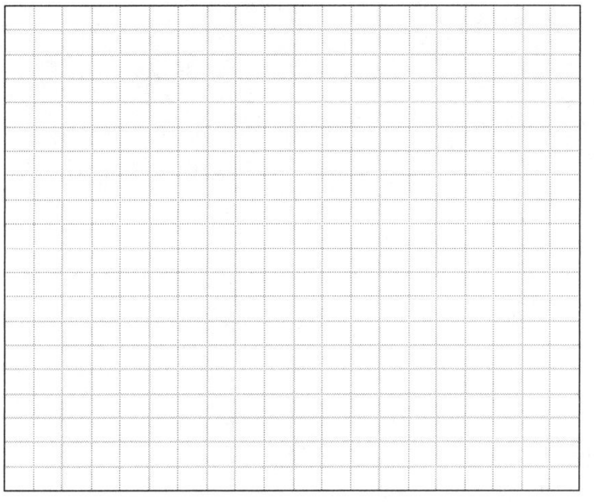

4. Here is another example of a quadratic equation with more than one transformation. What information can you gather from the equation?

$$h(x) = -2(x + 3)^2$$

   a. Sketch a graph of this function based on the information you gathered above.

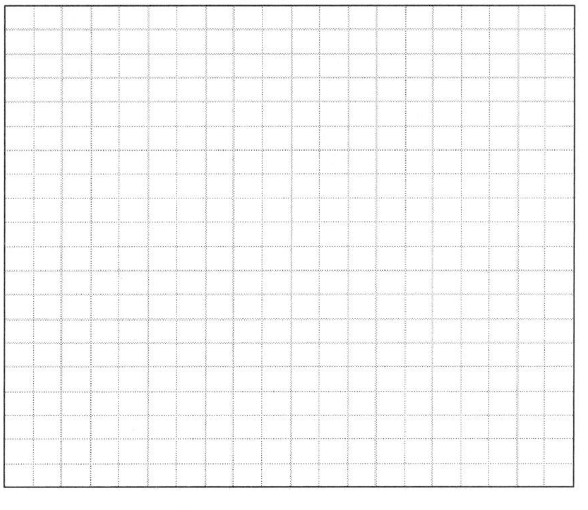

Writing Quadratic Equations

analysis

5. Given the function in the vertex form: $f(x) = 4(x - 1)^2 + 2$, what does the "2" tell you about the graph?

   a. What does the "1" tell you about the graph?

   b. What does the "4" tell you about the graph?

   c. Can you tell what the vertex of the graph is from examining the equation? If so, what is it? If not, why not?

   d. Can you tell what the axis of symmetry is from examining the equation? If so, what is it? If not, why not?

   e. Can you tell what the x-intercept(s) of the graph is/are from examining the equation? If so, what is it? If not, why not?

   f. Can you tell what the y-intercept of the graph is from examining the equation? If so, what is it? If not, why not?

6. Why is the form of the equation you have been working with called "vertex form?"

Writing Quadratic Equations

analysis

7. Given the following graph, use the vertex form to write the function it represents (assume the coefficient of x^2 is 1).

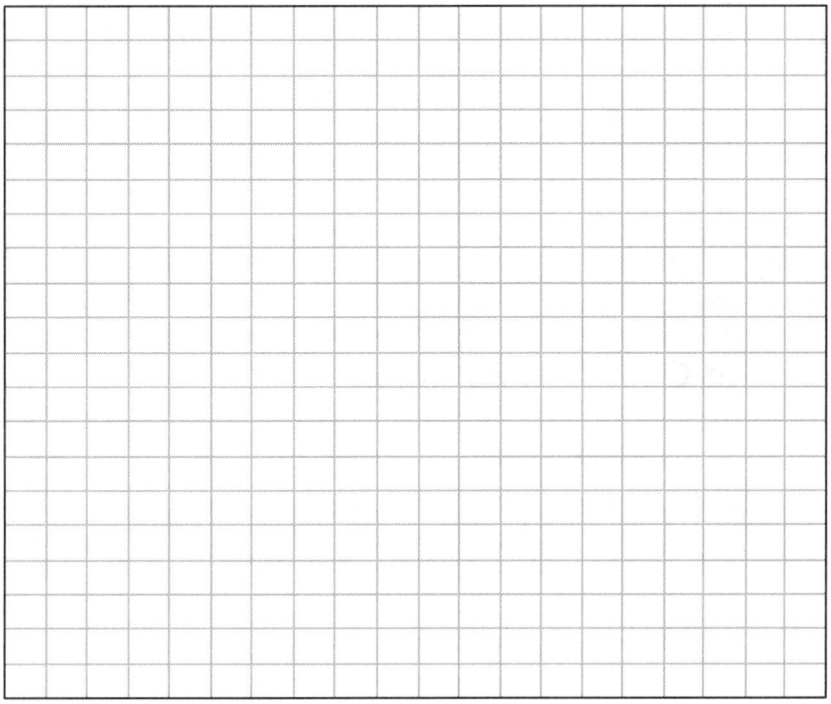

   a. What is the vertex of the function? Is the vertex a maximum or minimum?

   b. What is the equation of the line of symmetry for this function?

   c. What is/are the x-intercept(s) (zeros or roots)?

   d. What is/are the y- intercept(s)?

Writing Quadratic Equations

process: model

8. If you dilated the function graphed on the previous page by a factor of one-half, how would you represent the function symbolically? Sketch a graph of the transformed function on the graph on the previous page.

Writing Quadratic Equations

process: evaluate

9. Based on the dilated function in question 8, determine the:
   a. Zeroes of the transformed function

   b. y–intercept(s) of the transformed function

   c. Line of symmetry for the transformed function

   d. Maximum or minimum value of the function

process: model

Writing Quadratic Equations

10. If you reflected the dilated function from question 8, how would you represent it symbolically? Write the equation for the function. Sketch a graph of the new function on the same set of axes on page 8-47.

process: evaluate

Writing Quadratic Equations

11. Based on the reflected and dilated function in question 10, determine the:

    a. Zeroes of the transformed function.

    b. y–intercept(s) of the transformed function.

    c. Line of symmetry for the transformed function.

    d. Maximum or minimum value of the function

process: model

Writing Quadratic Equations

12. Provide an example of a quadratic equation showing a vertical shift, a horizontal shift, a dilation, and a reflection.

13. Graph the quadratic equation you created above.

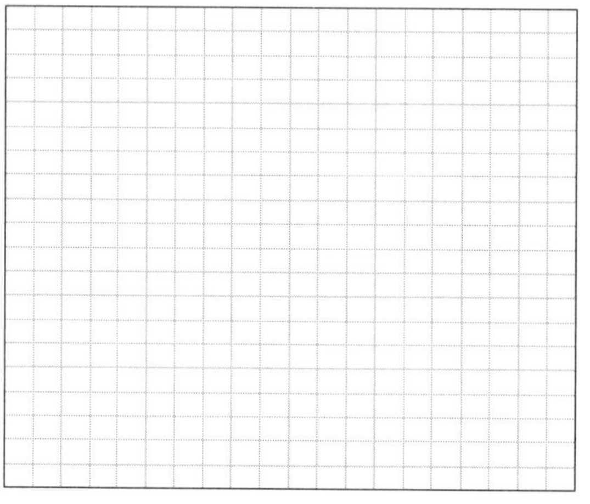

14. What information can you gather from reading the equation in symbolic form?

15. Begin with the parent function for your example. Provide a step-by-step written description of how to use the transformations to graph this quadratic function.

input

Writing Quadratic Equations

As you know, the form of quadratic function that you generated is called the **vertex** form. In general, the vertex form can be represented by:

$$y = a\,(x - h)^2 + k$$

where a is the dilation factor, h tells about the vertical shift, and k tells about the horizontal shift.

Animated gif: the Final Program

input

The dog owner for whom you created the first animated gif program heard about all the variations you offered to your other clients and got very jealous. She returned to you and demanded that you alter her Web page so that it had all of the following modifications.

- Add the 2-inch by 5-inch banner.
- Delay the start time by one second.
- Make the size of the image grow by 2 inches each second.

You are so sick of doing these web pages that you declare this will be the ***final program***.

Animated gif: the Final Program

process: model

1. Write the algebraic equation that models the area of the new animated gif over time.

2. Graph the function that models the area of the new animated gif with respect to time.

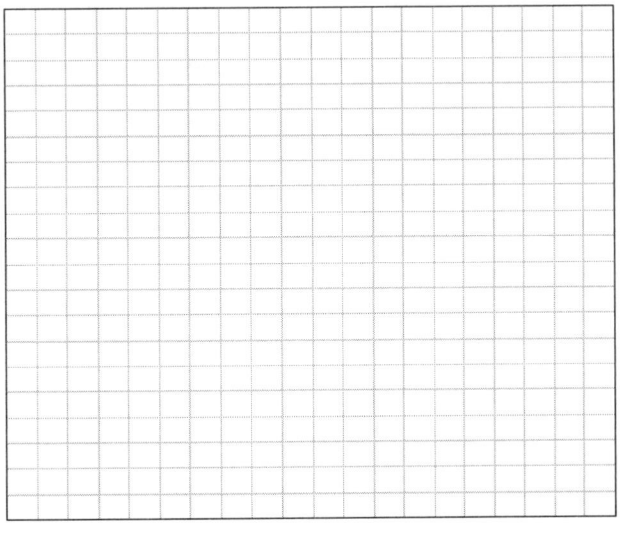

analysis

Animated gif: the Final Program

3. The graph of the function for the final program does not cross the x-axis. In other cases, the graph of the functions had y- and x-intercepts, but this graph has only a y-intercept. What does that mean? How would you explain this to someone who thinks that every quadratic function should cross both the x- and y-axes?

4. How do you interpret the minimum value for the function for the animated gif program?

5. What are the possible values for the time (domain) and area (range) of the function used to model the final gif program?

   a. If you were considering the quadratic function out of its context, what is its domain and range?

preview

Solving Quadratic Equations

When you generated the initial gif program, the graph was represented by a parabola of the form $y = x^2$. By looking at the graph, it was easy to tell that the x- and y-intercepts were at the origin, so applying an algebraic technique to determine those values was not necessary. However, for determining the zeros for most other functions, this will not be the case. In the upcoming section, algebraic techniques will be identified and applied.

analysis

Solving Quadratic Equations

1. A quadratic equation written as a product is said to be in **factored form**. Graph the equation $y = (x - 3)(x + 5)$.

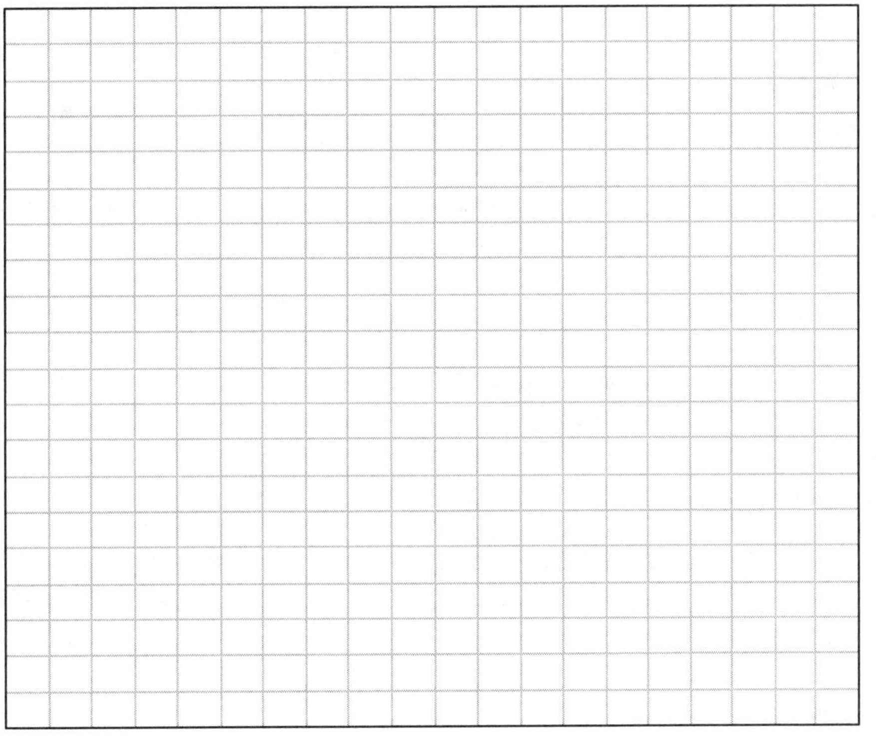

   a. Can you tell what the x-intercepts of the graph are from examining the equation? If so, what are they? If not, why not?

Solving Quadratic Equations

analysis

b. Can you tell what the y-intercept of the graph is from examining the equation? If so, what is it? If not, why not?

c. Can you tell if the graph is dilated from examining the equation? If so, what is the dilation factor? If not, why not?

d. Can you tell if the graph is reflected from examining the equation? Why or why not?

e. Can you tell if there is a vertical shift from examining the equation? If so, what is it? If not, why not?

f. Can you tell if there is a horizontal shift from examining the equation? If so, what is it? If not, why not?

g. Can you find the axis of symmetry from examining the equation? If so, what is it? If not, why not?

h. Can you find the vertex of the graph from examining the equation? If so, what is it? If not, why not?

analysis

Solving Quadratic Equations

2. A quadratic equation written so that the powers of x are descending from one term to the next is said to be in **standard form**. Graph the equation $y = x^2 + 2x + 1$.

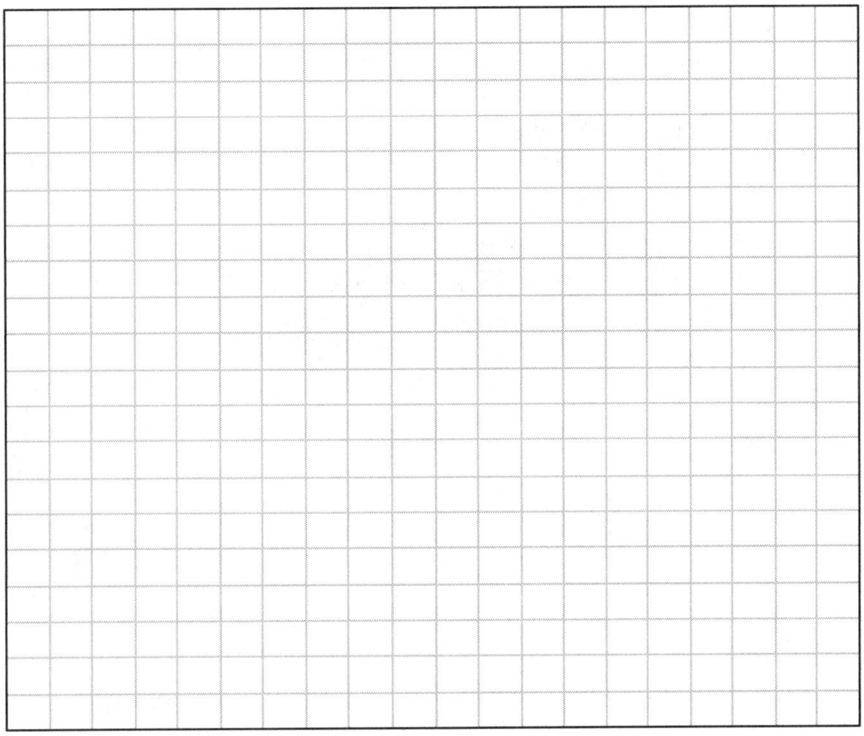

a. Can you tell what the x-intercepts of the graph are from examining the equation? If so, what are they? If not, why not?

b. Can you tell what the y-intercept of the graph is from examining the equation? If so, what is it? If not, why not?

c. Can you tell if the graph is dilated from examining the equation? If so, what is the dilation factor? If not, why not?

Solving Quadratic Equations

analysis

d. Can you tell if the graph is reflected from examining the equation? Why or why not?

e. Can you tell if there is a vertical shift from examining the equation? If so, what is it? If not, why not?

f. Can you tell if there is a horizontal shift from examining the equation? If so, what is it? If not, why not?

g. Can you find the axis of symmetry from examining the equation? If so, what is it? If not, why not?

h. Can you find the vertex of the graph is from examining the equation? If so, what is it? If not, why not?

3. Given equations of the form $y = x - 4$ and $y = x^2 - 4$, are the processes different for finding the y-intercepts? If so, explain.

Solving Quadratic Equations

analysis

4. Given equations of the form $y = x - 4$ and $y = x^2 - 4$, are the processes different for finding the x-intercept(s)? If so, explain. Can you actually solve the equation? If you can, explain how you would do it.

5. Given equations of the form $y = x - 1$ and $y = (x - 1)(x + 3)$, are the processes different for finding the x-intercept(s)? ...for finding the y-intercepts? If so, explain.

6. Given a quadratic equation in the form $y = (x - h)^2$, without the benefit of graphing it, how do you think you might determine the zero(s)? Explain.

7. Given a quadratic equation in the form $y = (x - h)^2 + k$, without the benefit of graphing it, how do you think you might determine the zeros? Explain.

input

Solving Quadratic Equations

To find the zeros of a quadratic equation algebraically, there are several methods to from which to choose. Your choice is likely to depend on the form in which the equation is written. In general, the objective is given $y = 0$, you will want to solve for x. For example, given an equation of the form $y = x^2 - k$, you will set $y = 0$ and solve for x.

General Example	Specific Example with k > 0	Specific Example with k < 0
$y = x^2 - k$ $0 = x^2 - k$ $x^2 = k$ $x = \pm\sqrt{k}$	$y = x^2 - 16$ $0 = x^2 - 16$ $x^2 = 16$ $x = \pm\sqrt{16} = \pm 4$	$y = x^2 + 16$ $0 = x^2 + 16$ $x^2 = -16$ $x = \pm\sqrt{-16}$ No answer is possible

Use your graphing calculator to confirm that 4 and -4 are the roots of $y = x^2 - 16$ and to confirm that there are no zeros for $y = x^2 + 16$.

analysis

Solving Quadratic Equations

8. Based on the examples above, describe the process for finding the zeros in an equation of the form $y = x^2 - k$, justifying each of the steps in the solution process.

9. How many zeros are possible with an equation of the form $y = x^2 - k$ if $k > 0$? …if $k = 0$? …if $k < 0$? Defend your answers based on what you know about transformations.

input

Solving Quadratic Equations

When you have an equation of the form $y = (x - h)^2$, there are two options for solving and finding the zeros. One option is to write $y = (x - h)(x- h)$ (factored form) and solve for x by setting each linear factor equal to zero.

General Example	Specific Example
$y = (x - h)^2$	$y = (x - 5)^2$
$0 = (x - h)(x - h)$	$0 = (x - 5)(x - 5)$
$0 = x - h$	$0 = x - 5$
$x = h$	$x = 5$

The other option is to set y = 0 and take the square root of both sides.

General Example	Specific Example
$y = (x - h)^2$	$y = (x - 5)^2$
$0 = (x - h)^2$	$0 = (x - 5)^2$
$\sqrt{0} = \pm\sqrt{(x - h)^2}$	$\sqrt{0} = \pm\sqrt{(x - 5)^2}$
$0 = \pm(x - h)$	$0 = \pm(x - 5)$
$0 = x - h$	$0 = x - 5$
$x = h$, and	$x = 5$, and
$0 = -x + h$	$0 = -x + 5$
$x = h$	$x = 5$

Use your graphing calculator to confirm that 5 is the only root for $y = (x - 5)^2$.

analysis

Solving Quadratic Equations

10. Based upon your experiences with transformations, why will an equation of the form $y = (x - h)^2$ have only one root, namely *h*?

analysis	**Solving Quadratic Equations**
	11. Based on the examples above, describe the process for finding the zeros in an equation of the form $y = (x - h)^2$, justifying each of the steps in the solution process.

process: evaluate	**Solving Quadratic Equations**
	12. Use algebra to find the zeros for the following quadratic equations:
	a. $y = 4(x + 2)^2$
	b. $y = \frac{x^2}{2} + 4$
	c. $y = (x + 1)^2 + 3$
	d. $y = 8(x - 1)^2 + 2$

input

Solving Quadratic Equations

The previous problems you examined to find the zeros are all in vertex form, but many quadratic equations are given in standard form. If an equation is in standard form, there are a few different possible ways to solve for the roots of the equation.

You might try factoring the equation and finding the x-intercepts from the factors. However, many quadratic equations cannot be easily factored.

You might try performing a process called "completing the square," which will change the quadratic equation into the vertex form. From there, you can use the techniques above to find the roots. However, "completing the square" is fairly complicated.

You might try using the quadratic formula. The quadratic formula can be applied to any quadratic equation in the form $ax^2 + bx + c = y$ where $y = 0$.

Recall from Integrated Math I that the quadratic formula states, given:

$$ax^2 + bx + c = 0 \quad \text{where } a \neq 0$$

all solutions can be found by solving:

$$x = \frac{-b \pm \sqrt{b^2 - 4ac}}{2a}$$

process: practice

Solving Quadratic Equations

13. Find the roots or zeros of the equations using the quadratic formula. Check your result by graphing the equation using your graphing calculator. Sketch the graph.

a. $y = x^2 + 9x + 14$

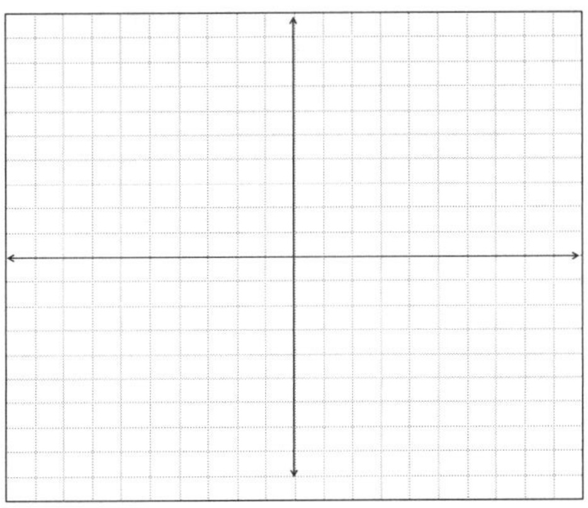

b. $y = -2x^2 + 5x - 2$

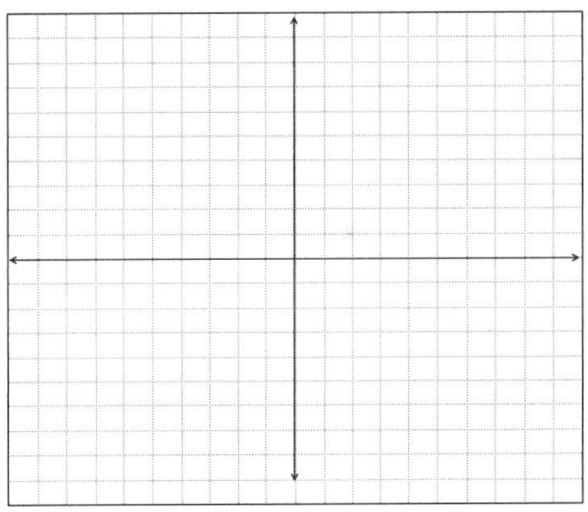

process: practice

Solving Quadratic Equations

c. $y = x^2 - 6x + 9$

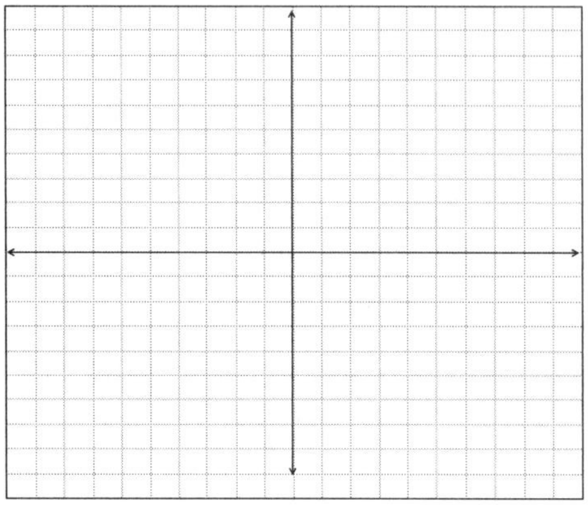

d. $y = 4x^2 - 4x - 3$

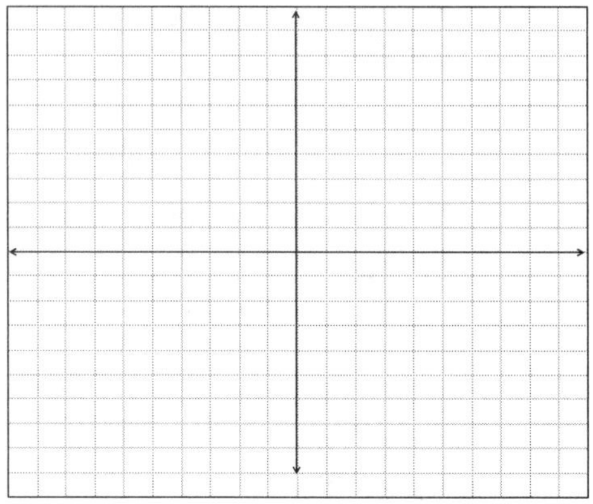

input

Solving Quadratic Equations

You can verify the number of zeros or x-intercepts to a quadratic equation by examining the **discriminant**. The **discriminant** is the portion of the quadratic formula that is contained within the square root symbol ($b^2 - 4ac$). If you determine the value of the discriminant, you will know the number and type of roots.

For example: $y = x^2 + 5x + 6$

$a = 1$, $b = 5$, and $c = 6$

The discriminant is $5^2 - 4(1)(6) = 25 - 24 = 1$.

$b^2 - 4ac > 0$ and therefore there are two real roots.

analysis

Solving Quadratic Equations

14. If the discriminant under the square root symbol is positive as in the previous example, why are there always two x-intercepts?

15. If the discriminant is 0, meaning nothing is added and subtracted from –b in the numerator of the quadratic formula, how many x-intercepts will the equation have?

16. The last possibility is that the discriminant is negative. In this case, how many x-intercepts will the equation have? Explain why.

process: evaluate

Solving Quadratic Equations

17. For each of the following quadratic functions, find the number of x-intercepts by using the discriminant portion of the quadratic formula. Verify the number of zeros by using your graphing calculator.

    a. $y = x^2 + 2x + 1$

process: evaluate

Solving Quadratic Equations

b. $y = x^2 + 4x + 5$

c. $y = x^2 - 15$

input

Solving Quadratic Equations

The intercepts are important to find in order to accurately sketch a quadratic function. However, without knowing the location of the vertex, you cannot make an accurate sketch.

Recall from Integrated Math I that the vertex is the highest or lowest point on the graph of a quadratic function, and that the coordinates of the vertex are defined to be $((\frac{-b}{2a}), f(\frac{-b}{2a}))$.

Based upon the general coordinates for the vertex given above, what will be the general equation for the axis of symmetry for a quadratic function? How do you know?

process: practice

Solving Quadratic Equations

18. Find the zeros, axes of symmetry, and vertices for the following quadratic equations.

    a. $y = 8x^2 - 32$

    b. $y = x^2 + 12x + 36$

    c. $y = 4x^2 + 5x - 6$

    d. $y = -5x^2 - x + 1$

Solving Quadratic Inequalities

extension

1. Graph $y = x^2 - x - 2$.

2. Based on the graph, determine the values for which $x^2 - x - 2 > 0$.

3. Based on the graph, determine the values for which $x^2 - x - 2 < 0$.

extension

Solving Quadratic Inequalities

4. Find the set of all values such that $4x^2 + 8x + 4 \le 6$.

5. Describe how you found the values that make the inequalities true in questions 2 through 4.

extension

Solving Quadratic Inequalities

6. Last July 4th you entered a contest offered through your local newspaper, and you won four tickets to an amazing concert. To win you determined the range of time it would take to project a canister of fireworks to a height greater than or equal to 200 feet, but less than or equal to 250 feet, from a platform 5 feet off the ground with an upward velocity of 120 feet per second. Present your winning solution below.

7. Using these constraints from your winning solution, what is the maximum height of the fireworks? How long will it take the fireworks to reach the ground?

extension

Solving Quadratic Inequalities

8. You work in a leather goods store in the mall. A shipment of 100 new jackets just came in for the fall. The suggested retail price for the jackets is $350. The store manager estimates that 12 jackets will be sold per week.

   a. If twelve jackets are sold, what is the revenue?

   b. If the revenue for the week is $12,250, how many jackets were sold?

9. Knowing that putting merchandise on sale increases sales, the manager estimates that if the jackets are reduced by $50, the number of sales per week will increase by 4. What should the price per jacket be if the manager wants to maximize the revenue?

   a. Write an expression for the income from the jackets.

   b. Write an expression representing the number of jackets sold.

   c. Write an expression representing the revenue.

   d. Find the price per jacket that will maximize revenue?

process: practice

Solving Quadratic Inequalities

10. Solve the following inequalities:

a. $x^2 - 9 > 0$

b. $x^2 + 2x + 1 \leq 0$

c. $3x^2 + 10x + 8 \geq 0$

d. $3x^2 - 6x \leq 2$

input

Functions and Inverse Functions

So far in Integrated Math I and Integrated Math II, you have examined relationships that are linear, quadratic, and exponential. All of these relationships are called functions. **Functions** are relationships for which every input value (x-value) is paired with exactly one output value (y-value). In other words, every element of the domain is paired with only one element in the range.

A correspondence between variable quantities that does not satisfy the definition of a function is a **relation**. A relation is defined as a correspondence between the elements of the domain and the elements of the range that does not carry the restriction that for each element of the domain there is only one element in the range.

For example, the relationship between the set of all people in the world and the set of possible birthdays is a functional relationship, because every person has only one birthday. In this example, the domain (set of inputs) is considered to be all the people in the world and the range (set of outputs) is considered to be all dates on the calendar.

However, examining this relationship in reverse does not give a functional relationship. If the domain is the set of all birthdays and the range is the set of all people in the world, it is NOT true that each calendar date is only one person's birthday. In fact, you may know someone who has the exact same birthday as you, so that date has at least two (and probably many, many more) person's birthdays associated with it.

analysis

Functions and Inverse Functions

1. Is the following a function? Every holder of a driver's license is assigned an identification number. Let the domain be the set of all the *names of license holders* and the range be all the *assigned identification numbers*. Explain your reasoning.

2. Is the reverse relationship a function: the correspondence of *license identification numbers* to *names of holders*? Explain your reasoning.

analysis

Functions and Inverse Functions

3. The domain is the set of all possible ages and the range is the set of all people in the world. Is this a functional relationship or merely a relation? Explain.

    a. What is the domain in the reverse relationship? What is the range of the reverse relationship? Is the reverse relationship a function, or merely a relation?

4. The domain is the set of all people in the world. The range is the set of possible DNA structures. Is this a function or a relation? Explain.

    a. What is the domain in the reverse relationship? What is the range of the reverse relationship? Is the reverse relationship a function, or merely a relation?

input

Functions and Inverse Functions

Functions that "work both ways" are called **one-to-one** functions. These are functions for which it is true that each element in the domain corresponds to one element in the range, and for which each element in the range corresponds to one element in the domain.

Instead of calling the "backwards" relationship the *reverse*, the functions that are one-to-one are said to have an inverse that is also a function. Recall that an inverse consists of all ordered pairs (y, x) or (f(x), x), where the pair (x, y) or (x, f(x)) belongs to the original function. (The input and output are simply switched to form the inverse.)

Recall that the notation you have been using for functions is f(x) notation, meaning *function of x*. The notation for the inverse of f(x) is $f^{-1}(x)$, read *"f inverse of x"* or *"the inverse of the function of x."* Note that the inverse in a one-to-one relationship is also a function.

Functions and Inverse Functions

5. Why is $f(x) = 2x + 3$ defined as a function?

   a. Does the inverse of $f(x)$ exist; in other words, is the backwards relationship a function? If so, what is $f^{-1}(x)$?

   b. Is $f(x) = 2x + 3$ a one-to-one function? Explain.

   c. Graph $f(x)$ and $f^{-1}(x)$. What relationship exists between the two graphs?

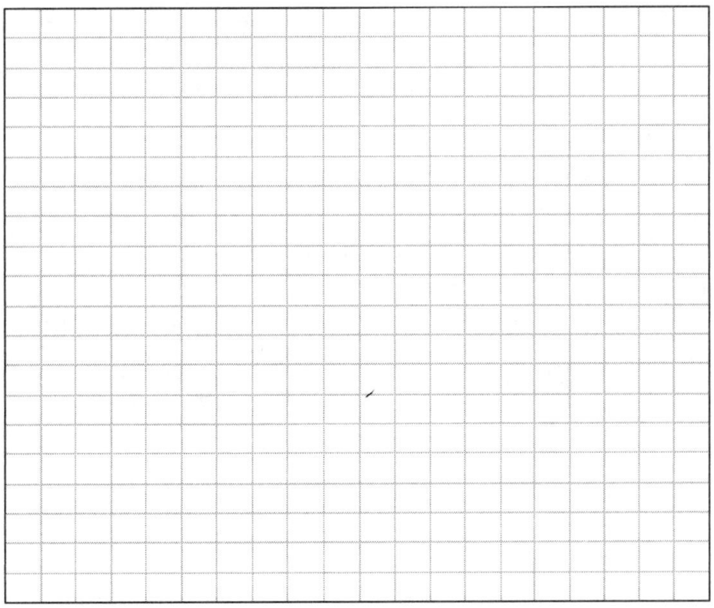

Functions and Inverse Functions

analysis

6. Why is $y = 2x^2 + 2x + 1$ defined as a function?

   a. Does the inverse of y exist? If so, what is y^{-1}?

   b. Is $y = 2x^2 + 2x + 1$ a one-to-one function? Explain.

   c. Graph the function y and the relation that is its inverse. What relationship exists between the two graphs?

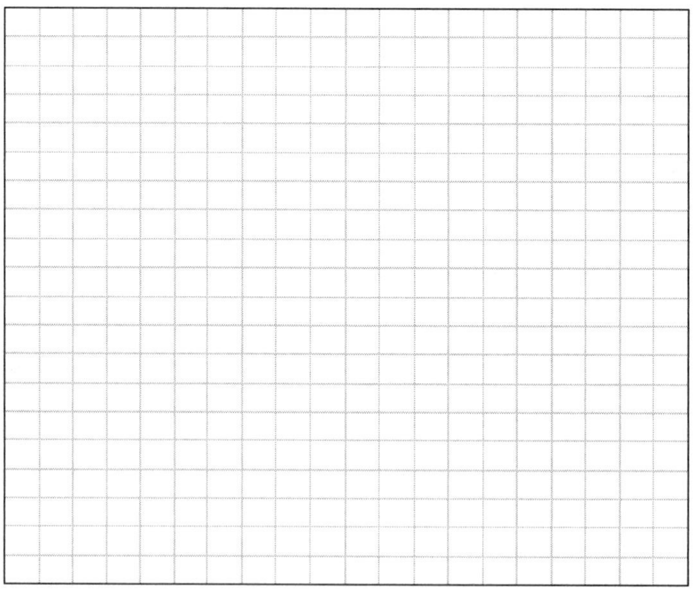

Functions and Inverse Functions

analysis

7. Why is $g(x) = 2^x$ defined as a function?

   a. Does the inverse of g(x) exist? If so, what is $g^{-1}(x)$?

   b. Is g(x) a one-to-one function? Explain.

   c. Graph the function and its inverse. What relationship exists between the two graphs?

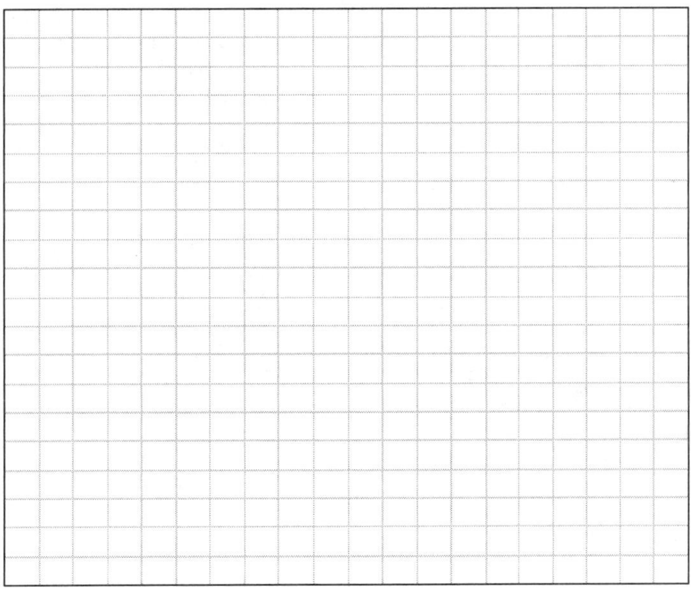

extension

Functions and Inverse Functions

8. Return to the graphs of each of the quadratic relationships on page 8-74. Draw vertical lines through each of the graphs.

   a. What is the maximum number of times that any vertical line will cross the graph of $y = 2x^2 + 2x + 1$?

   b. What is the maximum number of times that a vertical line will cross the relation that is the inverse of y?

9. How can the number of intersections of a vertical line and a graph tell whether the graph is a function?

10. Assuming that a graph does represent a function, what can a horizontal line tell you about the function? Test your conjecture on the graphs on pages 8-73 to 8-75.

input

Functions and Inverse Functions

The technique you performed, using vertical lines drawn through the graph to determine if the graph represents a function, is called the **vertical line test**. The vertical line test is used to determine if a relation is a function. If the vertical lines cross the graph at only one point, the relation is a function.

If a horizontal line is drawn through the graph and crosses it at most once, the function is said to be one-to-one.

Car Speed

input

One of the car manufacturers recently presented data on a test run for a new highly energy efficient car model. The car has a standard transmission (a stick shift), so it had to be put in gear by shifting three times during the 15 seconds in which the speed was being measured. The accelerator was being depressed at a constant rate during the 15-second period and then released at the same rate until the pedal was no longer being depressed.

Time	Speed
0	0
1	6
2	14
3	20
4	30
5	35.5
6	40
7	48
8	52
9	54
10	57
11	61
12	65.5
13	66.5
14	67
15	68.2
16	65

process: model

Car Speed

1. Model the data using a scatterplot and determine the equation of best fit.

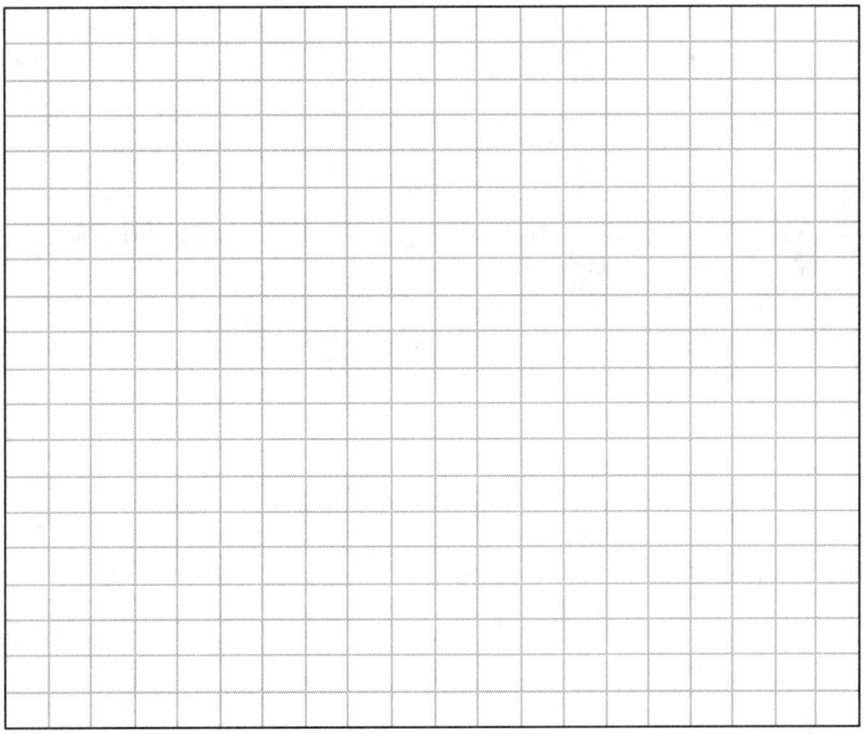

process: evaluate

Car Speed

2. Based on your model, when will the car come to a full stop?

analysis

Car Speed

3. What is the vertex of this quadratic graph? What meaning does the vertex have in this scenario?

4. What is the y-intercept for the graph? What meaning does it have in this scenario?

5. What are the x-intercepts for this graph? What do they mean in this scenario?

6. Based on the x-intercepts, what is the factored form of this quadratic function?

7. Multiply these factors to get the equation in standard form. Does the standard form of the equation approximate the equation of best fit? Explain.

Federal Budget

input

The annual Receipts and Outlays of the Federal Budget represent the money our government collects in taxes (Receipts) and the money our government spends on things that range from health care to the military (Outlays). The table below represents annual Receipts and Outlays for the Federal Budget between 1990 and 2000.

Year	Receipts ($trillions)	Outlays ($trillions)
1990	1.03	1.25
1991	1.05	1.32
1992	1.09	1.38
1993	1.15	1.41
1994	1.25	1.46
1995	1.35	1.51
1996	1.45	1.56
1997	1.58	1.60
1998	1.72	1.65
1999	1.83	1.70
2000	1.96	1.78

Federal Budget

process: model

1. Use a scatterplot to model the data for the receipts as a function of the number of years after 1990. Use your graphing calculator to determine the line of best fit and the correlation coefficient. Sketch the line of best fit.

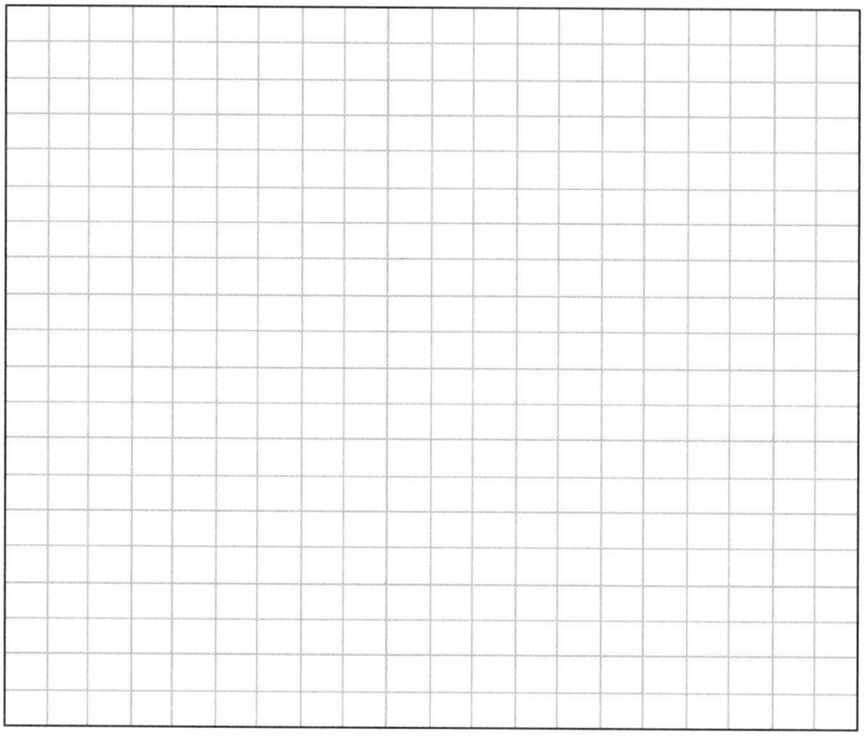

2. Use a scatterplot to model the data for the outlays as a function of the number of years after 1990 on the same set of axes. Use your graphing calculator to determine the line of best fit and the correlation coefficient. Sketch the line of best fit.

process: model

Federal Budget

3. Return to your scatterplot on the previous page. Use your graphing calculator to determine the *quadratic* equation of best fit and the correlation coefficient for receipts versus years after 1990. Sketch the quadratic equation of best fit on the grid using a different color pencil or dashes instead of a solid curve.

4. Return to your scatterplot and use your graphing calculator to determine the quadratic equation of best fit and the correlation coefficient for outlays as a function of time after 1990. Sketch the equation of best fit.

5. On the same scatterplot on the previous page, use your graphing calculator to determine the *exponential* equation of best fit and the correlation coefficient for receipts versus years after 1990. Sketch the exponential equation of best fit on the grid using a different color pencil or dots instead of a solid curve.

6. This time use your graphing calculator to determine the exponential equation of best fit and the correlation coefficient for outlays as a function of time after 1990. Sketch the equation of best fit.

analysis

Federal Budget

7. Which of the three types of functions (linear, quadratic, or exponential) provides the best mathematical fit for the data? How do you know?

Federal Budget

analysis

8. Which of the three types of functions (linear, quadratic, or exponential) do you think provides the best model of reality? Why? Be sure to take trends of the past as well as prospects for the future into consideration when responding to this question.

9. Describe the relationship between the money the government has collected and the money it has spent over the period of 1990 – 2000.

10. Is there ever a point in time when the receipts equal the outlays according to any of the three models you have used?

Federal Budget

extension

11. If there is a deficit (more outlays than receipts), in what area do you think the government should reduce spending and why?

12. If there is a surplus in receipts, what do you think the government should do with the excess dollars and why?

process: summary

Federal Budget

13. Summarize what you know about linear functions.

14. Summarize what you know about quadratic functions.

15. Summarize the similarities and differences between these two types of functions.

Unit 9: Higher Order Polynomial and Rational Functions

Contents

input

Saving Up

Your older sister is the type of person who plans everything. Currently she is saving for a vacation she plans on taking three years from now. She will need $1,400 for her airline tickets. Based on her current earnings and expenses, she figures she can deposit $400 per year into her vacation fund.

The neighborhood bank is offering 5% yearly interest compounded annually. Before opening a savings account for the vacation fund, your sister wants you to help determine if this interest rate is sufficient for her to reach her goal.

process: evaluate

Saving Up

1. Determine the balance after the first year if the first deposit was $400.

2. Determine the balance after the first year including a second deposit of $400. Show your work.

3. Determine the balance after the second year, but before the third deposit is made.

4. Determine the balance after a third deposit of $400 is made. Show your work.

5. Write an expression to determine the balance at the end of three years.

6. Will the 5% annual interest rate allow your sister to reach her goal?

process: model

Saving Up

7. Write the balance at the end of the three years in terms of each of the prior years.

input

Saving Up

The expression that represents the balance after three years is called a **cubic equation** or **third degree polynomial**.

Saving Up

process: model

8. Graph the equation using a graphing calculator and sketch it below. Use the following window setting:

   Xmin = -10

   Xmax = 10

   Xscl = 1

   Ymin = -5000

   Ymax = 5000

   Yscl = 100

   How would you describe the graph of this function?

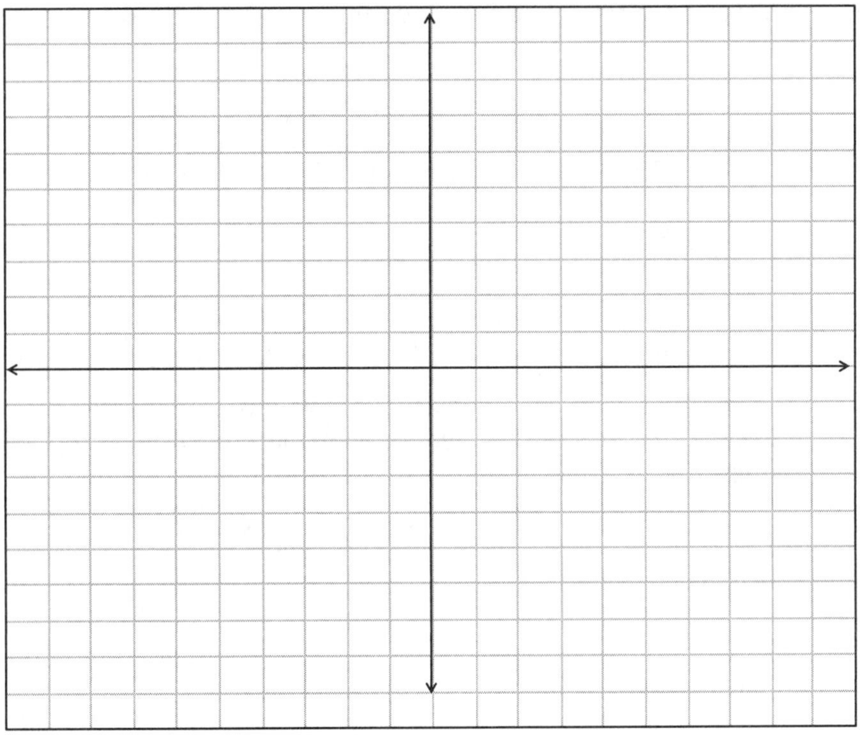

Saving Up

process: evaluate

9. Determine the growth factor or interest rate necessary to reach the goal of $1,400 after three years.

10. How many roots or solutions are there? What are the roots?

11. If your sister had invested $1,200 initially over three years at 5%, would this strategy have been better? How much would she have earned at the end of the time?

12. Use the cubic equation to model the situation and to answer the question:

    You put $1,000 in a savings account every year for three years and want to have $3,800 at the end of that period. What interest rate, compounded annually, is required?

    a. Is this answer realistic?

extension

Saving Up

13. Conjecture, how many roots do you think a cubic equation may have? Explain.

14. If you put $1,000 in a savings account every year that offers interest compounded annually for four years, how would you represent the balance at the end of this period with an algebraic model?

process: model

Transforming Cubic Functions

1. If $f(x) = x^3$ is the parent function for the cubic, based on what you know about transforming linear, quadratic, and exponential functions, provide an example of a cubic function that has undergone a:
   a. Vertical shift upward

   b. Vertical shift downward

2. Using your graphing calculator, verify your answers by graphing the parent function and the transformed functions.

3. Using the generalizations you developed in Unit 8, provide a general equation representing a cubic function that has undergone a:
   a. Vertical shift upward

   b. Vertical shift downward

4. Provide an example of a cubic function that has undergone a:
   a. Horizontal shift to the right

   b. Horizontal shift to the left

process: model

Transforming Cubic Functions

5. Provide a general equation representing a:
   a. Horizontal shift to the right

   b. Horizontal shift to the left

6. Provide an example of a cubic function that has been dilated by a factor:
   a. Between 0 and 1

   b. Greater than 1

7. Provide a general equation representing a dilation.

process: model

Transforming Cubic Functions

8. Provide an example of a cubic function that has been reflected.

9. Provide a general equation representing a reflection.

analysis

Transforming Cubic Functions

10. Write everything you know about the function $y = 3(x - 1)^3 + 2$ in relation to the parent function $y = x^3$.

11. Is the effect of the transformations on cubic functions the same or different than transformation performed on other polynomial functions (such as linear and quadratic functions)? Explain.

12. Is the effect of the transformations on cubic functions the same or different when they are performed on exponential functions?

Investigating Cubic Functions

process: model

1. If l represents the length of each edge, what does l^2 represent relative to the figures below? What does l^3 represent?

2. You want to organize your room, so you buy plastic cubes to hold some of the things you have scattered about. The cubes stack nicely and they will help you to create some space. In the empty corner of the room, you set five cubes as follows:

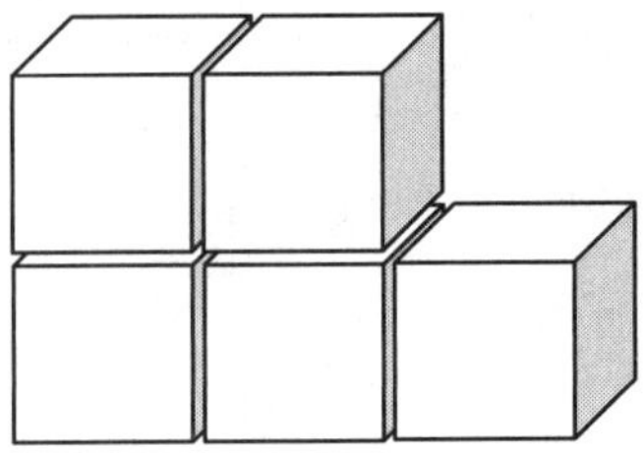

   a. Each cube has an edge of length x. Write an expression that represents the total volume and an expression that represents the total surface area of the stack of cubes.

process: model

Investigating Cubic Functions

3. You have made an open-top box from a rectangular piece of cardboard measuring 18 inches by 22 inches. To make the box, you cut and folded square flaps measuring x inches at each of the corners. Draw the diagram representing the work plan for creating the box. Write the volume function for the open-top box.

   a. Write the volume function as a polynomial with the terms written in descending order (this is called **standard form**).

process: practice

Investigating Cubic Functions

4. Write the following polynomial expressions in standard form:

   a. $(x - 4)(x + 2) =$

   b. $3(w - 1)(w - 3) =$

   c. $(-x + 5)(x - 6) =$

   d. $(r + s)(r - s) =$

   e. $(2t - 1)(3t + 2) =$

process: practice

Investigating Cubic Functions

5. Factor the following polynomial expressions:

   a. $x^2 + 2x - 15 =$

   b. $4x^2 - 16y^2 =$

   c. $a^3 + b^3 =$

   d. $2x^2 + 7x - 4 =$

   e. $x^2 + 2xy + y^2 =$

   f. $x^3 - 27 =$

   g. $x^4 + 2x^3 + x^2 =$

input

Investigating Cubic Functions

As you may have noticed in the previous problem set, you were asked to factor cubic polynomials of the form:

$$a^3 + b^3 \text{ and } a^3 - b^3$$

Based on what you know about factoring, it is likely that you said the equations were non-factorable. Polynomials of this form are actually factorable. Look at $a^3 + b^3$. If one factor is $(a + b)$, find the second factor.

$$a^3 + b^3 = (a + b)(\qquad\qquad)$$

Similarly for $a^3 - b^3$, if $(a - b)$ is one factor, find the second factor.

$$a^3 - b^3 = (a - b)(\qquad\qquad)$$

process: practice

Investigating Cubic Functions

6. Factor the following polynomials expressions:

   a. $x^3 - 1 =$

   b. $a^3 + 1 =$

   c. $3x^3 - 3y^3 =$

   d. $x^6 + 8 =$

process: model

Investigating Cubic Functions

7. Your friend saw what you did with the stacking cubes and decides to do something similar. She buys a cube of the same height x, but three inches longer and four inches wider.

   a. Determine the volume of each of the boxes.

   b. Write an expression to represent the total volume of the box.

   c. Graph the function representing the total volume using a graphing calculator.

Investigating Cubic Functions

analysis

8. Based on your earlier analyses of functions, what can you determine from the graph in question 7-b?

9. How would you compare this function to the graphs of a(n):

   a. Linear function   b. Quadratic function   c. Exponential function

Investigating Cubic Functions

process: model

10. Using your graphing calculator, graph the parent function $f(x) = x^3$ along with:

    a. $(x) = x^3 + 3x^2$ b. $f(x) = x^3 + 3x^2 - 12$

    and make a sketch of the graphs below.

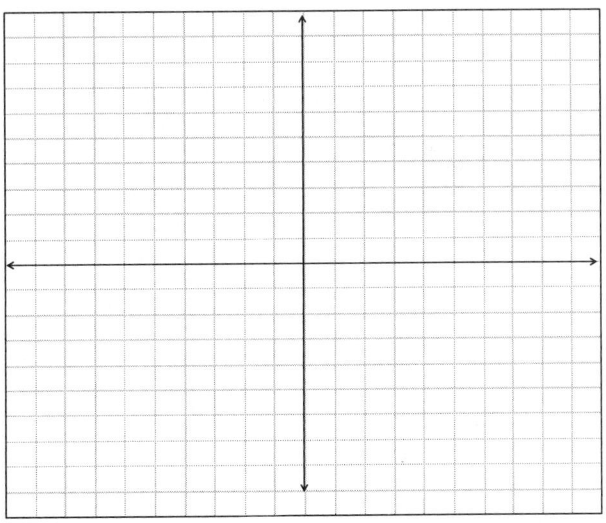

analysis

Investigating Cubic Functions

11. Compare the graphs in question 10 and indicate how they are similar and different.

12. How would you interpret the peaks and valleys of the cubic equations?

13. What happens to the function when it hits a peak or a valley?

input

Investigating Cubic Functions

The peaks and valley of functions that are of degree 2 or higher are referred to as **local extrema** or more specifically **local maximum** and **local minimum**.

analysis

Investigating Cubic Functions

14. Would you say that the maximum or minimum of a quadratic function is a local maximum or minimum? Explain your reasoning.

15. Using your graphing calculator, draw the graphs of:

    a. $y = x^2 + 6x + 8$

    b. $y = 2x^3 + x^2 + 3x$

    one at a time. Reading the graph from left to right, determine the intervals where each of the functions may be increasing or decreasing.

input

Investigating Cubic Functions

An **interval of increase** refers to the interval where the value of the independent variable corresponds to an increase in the dependent variable values. The **interval of decrease** refers to the interval where the value of the independent variable corresponds to a decrease in the dependent variable value.

analysis

Investigating Cubic Functions

16. How would you describe the behavior of a linear, quadratic, and cubic polynomial function on the far left and far right ends of the x-axis. That is, what happens to y as x gets very small or very large as you move from the origin in either direction from the origin?

    a. Linear

    b. Quadratic

    c. Cubic

input

Investigating Cubic Functions

The behavior of a function at its extreme right and left ends is called the **end behavior.**

analysis

Investigating Cubic Functions

17. Is $f(x) = x^3$ a one-to-one function? Does the inverse function ($f^{-1}(x)$) of $f(x) = x^3$ exist? Is it a function? Explain.

process: model

Investigating Cubic Functions

18. Describe the process of finding the inverse function. Find the inverse of the following functions:

    a. $y = x^3 - 4$

    b. $f(x) = x^3 + 2$

    c. $y = \sqrt[3]{x + 3}$

Polynomial Functions

input

The data below represents a recent analysis of the fluctuations in the market value of the stock of an internationally held conglomerate over a 12-month period.

Month	Change in the Number of Points in the Market Value of the Stock
1	-19
2	-8.125
3	-1
4	+3.125
5	+5
6	+5.375
7	+5
8	+4.625
9	+5
10	+6.875
11	+11
12	+18.128

process: model

Polynomial Functions

1. Conjecture, by looking at the data, whether a linear, quadratic, exponential, or cubic model will provide the best fit. Indicate your choice and explain why

2. Verify your conjecture by using a graphing calculator to draw a scatterplot to determine the equation of best fit. Was your conjecture correct?

analysis

Polynomial Functions

3. What do the zeros represent with respect to the situation?

4. Are there local maxima or minima? If so, where do they occur? How would you interpret these values in the given situation?

5. What are the intervals of increase or decrease? How would you interpret these intervals in the given situation?

6. In analyzing the model, does it makes sense to discuss the end behavior? Why or why not?

analysis

Polynomial Functions

7. Based on the information gathered in the previous questions, how would you describe the economic health of the company?

8. What might have influenced the fluctuations in the company's stock?

input

Polynomial Functions

These functions are all members of the family of polynomial functions. The degree of each polynomial function is defined by the highest power of the term.

A **polynomial function** is represented by the sum of terms. A term may contain a constant, a variable, or the product of constants and variables. In general, a polynomial function may be represented as:

$$y= f(x) = a_n x^n + a_{n-1} x^{n-1} + \ldots + a_1 x^1 + a_0$$

where a_n represents the coefficient associated to the variable of degree n. An example of a second degree polynomial function is $y = 5x^2 - 3x + 2$. What are a_2, a_1, a_0?

If a fourth degree polynomial function is written as:

$$y= f(x) = a_4 x^4 + a_3 x^3 + a_2 x^2 + a_1 x^1 + a_0$$

give an example of one and indicate the values for a_4, a_3, a_2, a_1, and a_0.

analysis

Polynomial Functions

9. Using your graphing calculator, graph each of the following polynomial functions and draw a sketch. Indicate the number of zeros, if local maxima or minima exist, and describe the end behavior.

   a. $f(x) = -2x^2 - 5x + 3$

   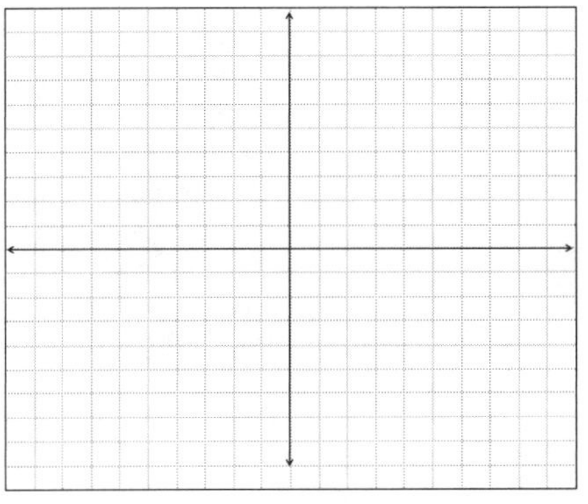

   b. $f(x) = 4x^3 + 16$

   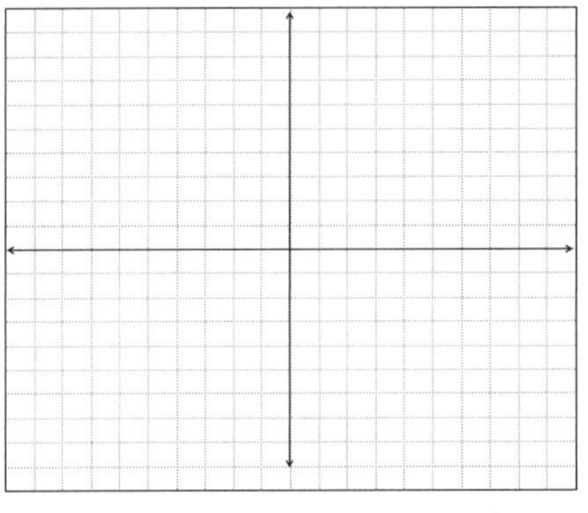

   c. $f(x) = x^4 - 4x^3 - 2x^2 + 15x - 8$

   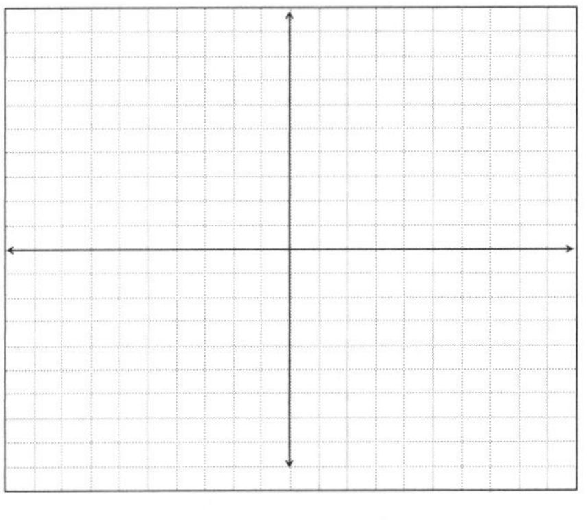

Polynomial Functions

analysis

d. $f(x) = \dfrac{-x^3}{4} + x^2 + 5$

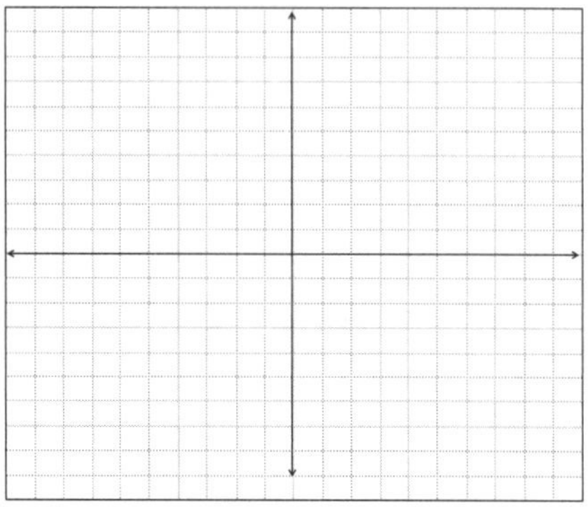

e. $f(x) = x^5 - 2x^4 + 3x^3 - x^2 + 2x + 3$

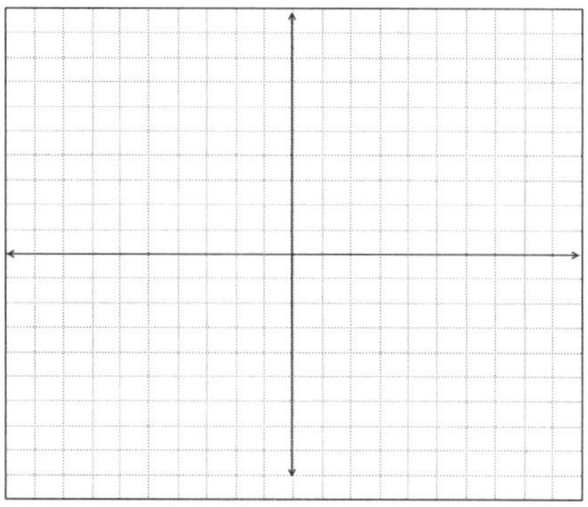

f. $f(x) = (x - 2)^4 + 1$

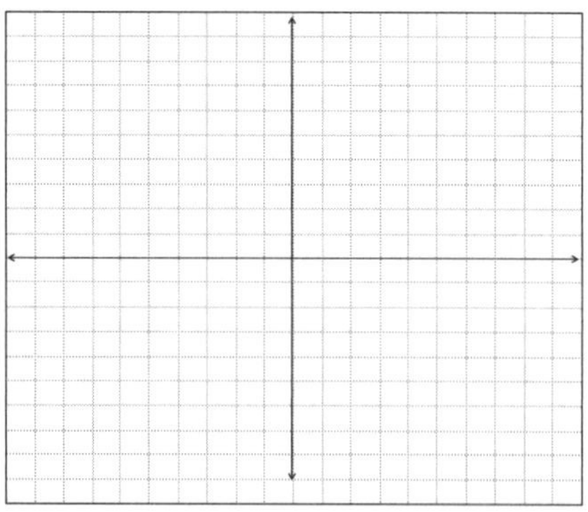

analysis

Polynomial Functions

10. What one characteristic do all the functions have in common?

11. Is it true to say that the number of zeros of a polynomial is less than or equal to its degree? Explain your reasoning.

12. Look at the graphs of the following polynomial functions. Determine if there are any patterns for the odd-degree functions and any patterns for the even-degree functions by analyzing and describing the behavior of each of the functions

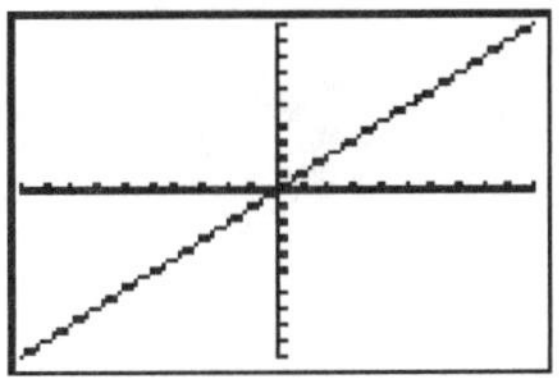

1st degree

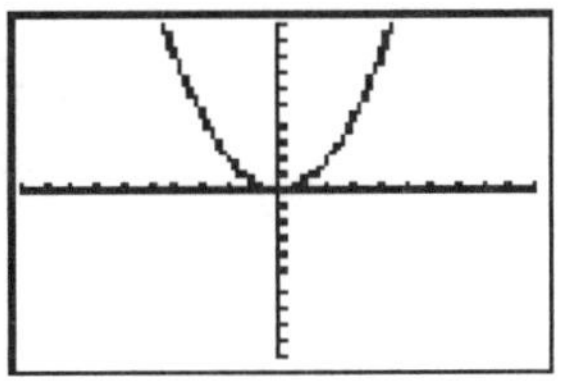

2nd degree

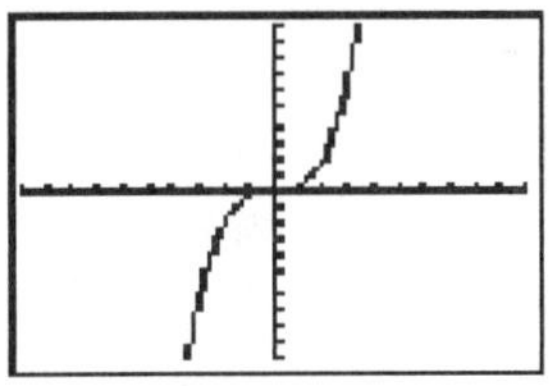

3rd degree

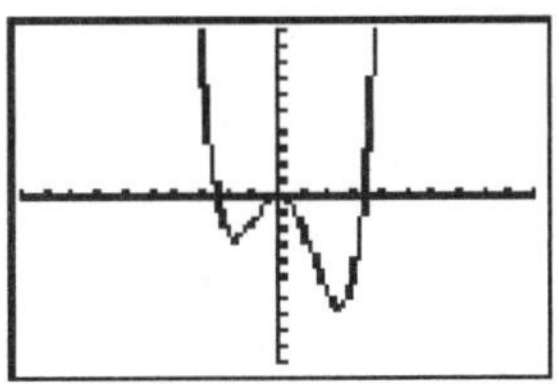

4th degree

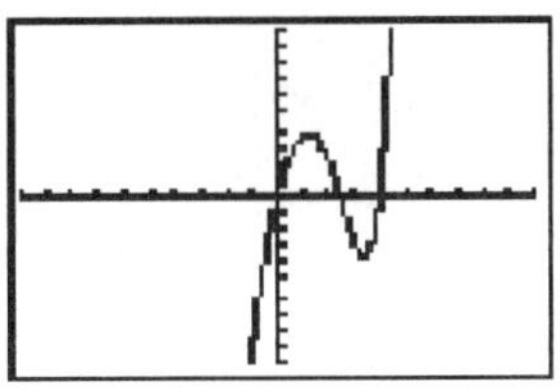

5th degree

Polynomial Functions

extension

13. Read the following statement and indicate whether it is true or false. Justify your answer.

    Every odd-degree function will cross the x-axis at least once, whereas an even-degree function may or may not, depending on its location in the coordinate plane.

input

Modeling with Polynomials: Spawning Salmon

The graph below represents data from an environmental study that examines the relationship the number of salmon swimming upstream to spawn in the Columbia River and the water temperature of the river.

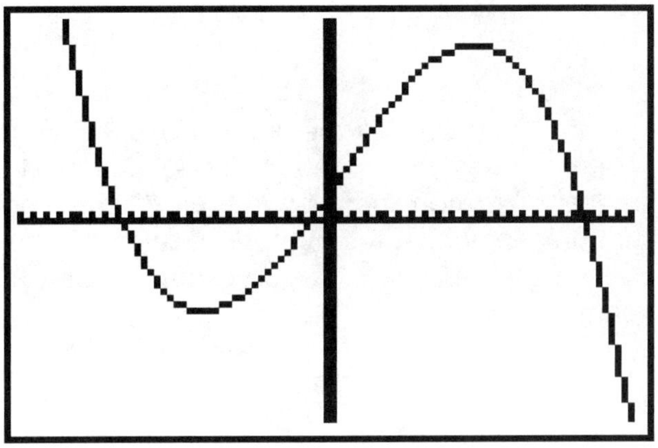

The function modeling this relationship may be written as:

$$N(t) = -t^3 + 3t^2 + 360t + 500$$

N represents the number of spawning salmon in the river and *t* is the water temperature in degrees Celsius.

Modeling with Polynomials: Spawning Salmon

analysis

1. Analyze the function and determine the range of values for "t" appropriate to the situation.

2. Describe the fluctuation in the number of spawning salmon with respect to the temperature that is appropriate to model the situation. In your description, include the number of salmon spawned at the extreme ranges of temperature, the temperatures for which the number of salmon increases and/or decreases, and the temperatures that produce the largest migration of salmon along with the maximum number.

3. Analyze the complete graph of the function over the domain of all real numbers and write a description of the function that describes the intervals of increase and decrease, the local extrema, the zeros, and end behavior.

input

Modeling with Polynomials: Recycling

For your science fair project you are going to do an analysis of recycling trends in the United States. To begin, you conducted a bit of research on the internet by looking at the U.S. Census Bureau Statistical Abstract of the United States 2002. You found an interesting data set on recycling, from which you extracted the following data that is in millions of tons.

Item and Material	1980	1990	1993	1994	1995	1996	1997	1998	1999	2000
Total Waste Pro-duced	151.5	205.2	211.8	214.2	211.4	209.2	216.4	223.0	229.9	231.9
Total Material Re-covered	14.5	33.6	43.8	50.8	54.9	57.3	59.4	61.6	63.9	69.9

process: model

Modeling with Polynomials: Recycling

1. Model the Total Materials Recovered data, first as a cubic function, and second as a quartic (fourth degree) function.

analysis

Modeling with Polynomials: Recycling

2. Determine which model is a better fit in terms of the data and explain your reasoning.

3. Based on the model you have chosen, write a report detailing the trends in recycling. Include in the report the intervals when recycling was on the increase or decrease; when recycling was at a relative high or low; and a commentary on the future of recycling as suggested by the model.

Modeling with Polynomials: Deer Population

input

During the early part of the 20th Century, the deer population of the Kaibab Plateau in Arizona was monitored by wildlife officers. For the period from 1905 to 1930, it was found that the deer population could be approximated by the function:

$$D(x) = -0.125x^5 + 3.125x^4 + 4000$$

Modeling with Polynomials: Deer Population

analysis

1. Using your graphing calculator, sketch the graph of the function.

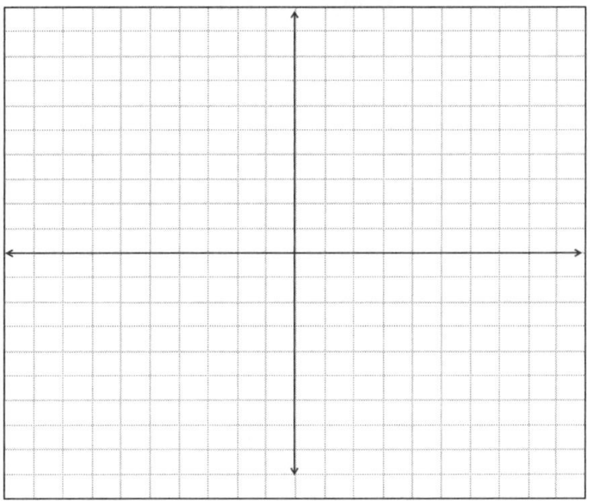

2. Write a summary report describing the deer population over a given time period. Your report should discuss things such as the initial and final populations, the years when the deer population was relatively stable or experienced fluctuations, the maximum and minimum populations, etc. Conjecture as to the reason for the population fluctuations.

Modeling with Polynomials: Deer Population

analysis

3. Analyze the behavior of the graph over the domain of all real numbers. In the analysis list the intervals of increase/decrease, the coordinates of the local extrema, the x- and y- intercepts, and the end behavior. Write a paragraph that compares and contrasts the graph of the function relating to the scenario with its restricted domain and range and the graph of the function over the entire domain.

input

Introduction to Rational Functions

You were absent on Wednesday and asked your friend for her math notes. Her notes contained the following information:

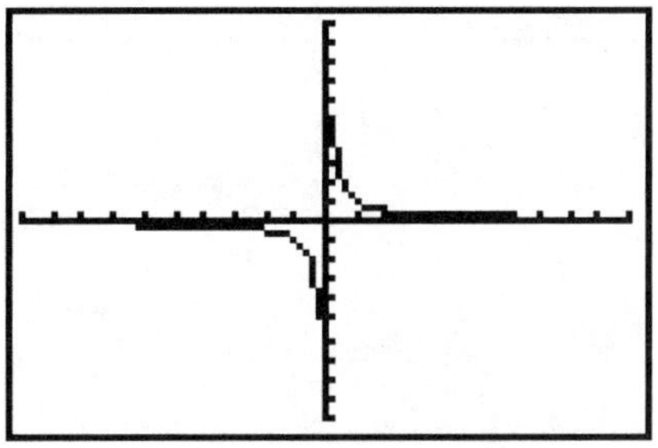

x	y
1	1
2	1/2
3	1/3
-1	-1
-2	-1/2
1/2	2

When you finally got a chance to look at the notes, you were not sure what you were looking at, so you called your friend to ask her for some help but she was not home. At that moment, you realized you would have to figure things out on your own.

analysis

Introduction to Rational Functions

1. Does the graph in your friend's notes look similar to the graphs of any of the other functions that you have studied (polynomial and exponential)?

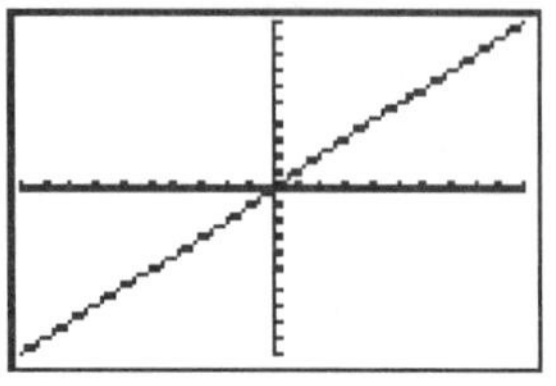
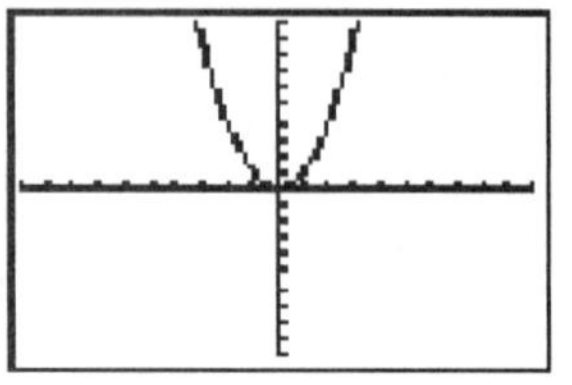
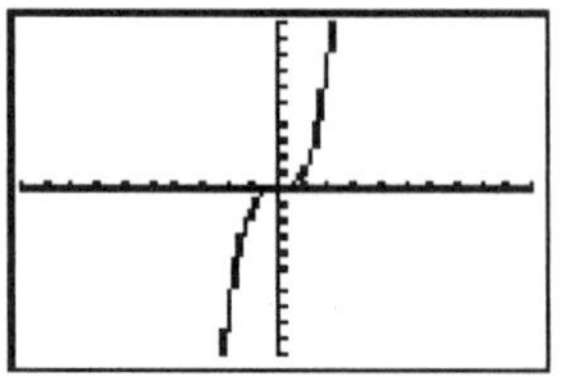
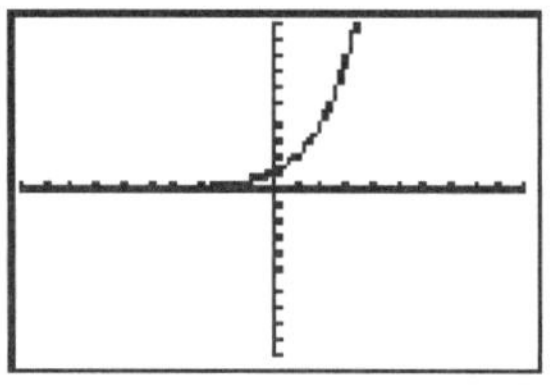

   a. What are the similarities between the graphs of the functions? What are the differences? Explain.

2. Looking at the graph of the function in the notes, what can you say about the output values when the input values are:

   a. Small positive values

   b. Large positive values

   c. Small negative values

   d. Large negative values

analysis

Introduction to Rational Functions

3. Using the data in the table, how would you write the function in symbolic form?

4. What is the:
   a. Domain of the function

   b. Range of the function

5. If there are restrictions on domain or the range, explain why this is the case.

input

Introduction to Rational Functions

The function that has been defined is called a rational function. A **rational function** is defined as the quotient of polynomials where the denominator is non-zero. If a function is defined at every point, as in the case of the polynomial function, we say the function is **continuous**. A function that is not defined for a point or set of points is **discontinuous** or that there is a **discontinuity** at the point(s).

Introduction to Rational Functions

analysis

6. Examine the functions below and determine which, given the definition, is a rational function. Explain why the function meets or does not meet the criteria. If the function is a rational function, indicate any values where the function may be discontinuous.

   a. $f(\ x) = \dfrac{x^2 - 1}{5x}$

   b. $f(x) = \dfrac{1}{x+1} - \dfrac{2}{x}$

   c. $y = \dfrac{x^3 + 3x^2 - 15x + 1}{6x^2 - 1}$

   d. $y = \dfrac{1}{x} \div \dfrac{2x \quad 3}{x^3 + x^2 + 1}$

   e. $f(x) = 16x^2 + 50x + 173$

7. Describe the characteristics that are shared by all functions in question 6.

8. Based on the information above, what do you think is a key difference between a polynomial function and a rational function with a non-constant denominator?

Analyzing Rational Functions

process: model

1. Sketch the graph of each of the following rational functions.

a. $f(x) = \dfrac{5x+1}{x+2}$

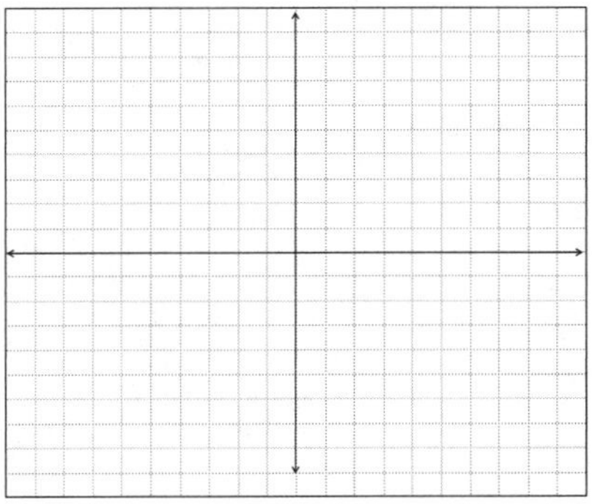

b. $y = \dfrac{1}{x^2+x+2}$

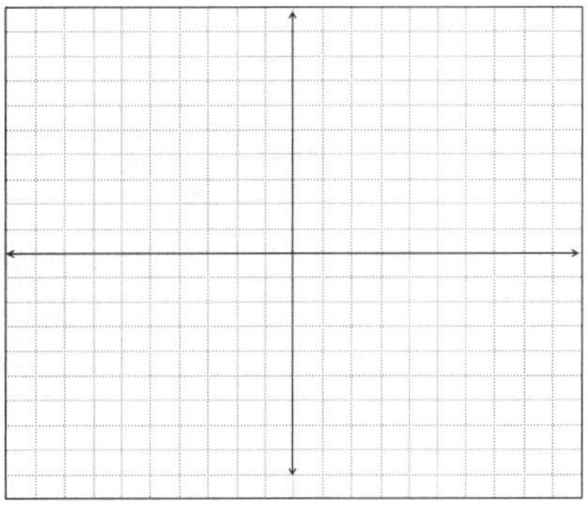

c. $y = \dfrac{x^3-4x^2+2}{x-1}$

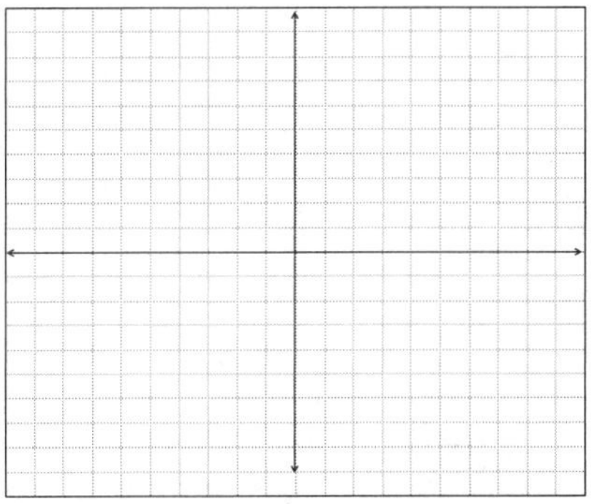

d. $f(x) = \dfrac{x^2-3x-4}{x-1}$

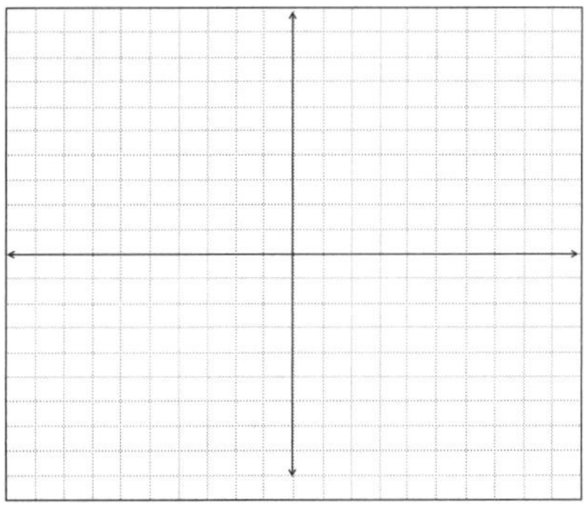

e. $f(x) = \dfrac{x+1}{x^2-2x-8}$

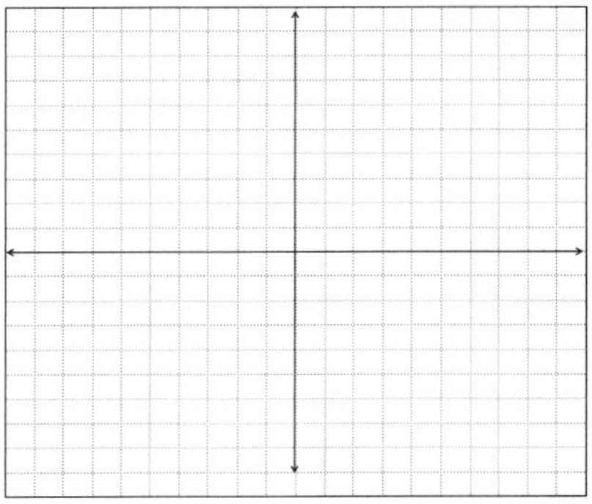

Analyzing Rational Functions

analysis

2. Describe how the graphs of rational functions are similar to the graphs of other functions.

3. Which characteristics of a rational function are unique? .

Analyzing Rational Functions

input

As you may have noted in your study of rational functions, when there is a variable in the denominator, as x approaches zero, the function approaches the y-axis but never touches it, and as x gets very large, y approaches or gets closer and closer to the x-axis.

This type of behavior by the y-axis is called **asymptotic behavior,** and the y-axis is called the **vertical asymptote** of the function. The x-axis functions as the **horizontal asymptote** for the function.

Analyzing Rational Functions

analysis

4. Indicate the vertical asymptotes for each of the functions graphed and the domain and range.

   a.

   b.

   c.

   d.

   e.

Analyzing Rational Functions

extension

5. What will the vertical asymptote be for a function f(x) if it is not defined for x = 1? What about the case where the function is not defined for x = -1? What if the function is not defined for x = 2 and x = -2?

Analyzing Rational Functions

analysis

6. Based on the procedure for determining the inverse of a function, find the inverse of $f(x) = \frac{1}{x}$ and draw the graph.

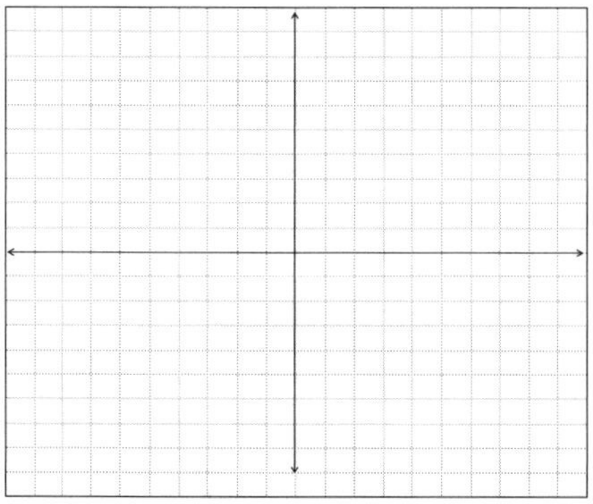

7. Given $f(x) = \frac{1}{x}$, does the inverse exist? Explain your reasoning.

process: evaluate

Analyzing Rational Functions

8. Find the inverse of the following functions:

   a. $f(x) = \dfrac{x+3}{x+1}$

   b. $f(x) = \dfrac{x+3}{x+1}$

   c. $f(x) = \dfrac{x-5}{x+2}$

9. Are all the functions f(x) in question 7 one-to-one? Explain why or why not.

process: model

Transforming Rational Functions Graphically and Algebraically

1. If $f(x) = \frac{1}{x}, x \neq 0$ is the parent function, based on what you know about transforming polynomial and exponential functions, provide an example of a rational function that has undergone a:
   a. Vertical shift upward

   b. Vertical shift downward

2. Using your graphing calculator, verify your answers by graphing the parent function and the transformed functions.

3. Using the generalizations you developed earlier in this unit, provide a general equation representing a rational function that has undergone a:
   a. Vertical shift upward

   b. Vertical shift downward

4. Provide an example of a rational function that has undergone a:
   a. Horizontal shift to the right

   b. Horizontal shift to the left

analysis

Transforming Rational Functions Graphically and Algebraically

5. Using a graphing calculator, verify your answer by graphing each of the functions and comparing it to the parent function. If your answer does not match the graph determine the correct transformation and your answer in question 1.
6. Represent each of the following transformations in general form:

   a. Vertical shift

   b. Horizontal shift

   c. Dilation

   d. Vertical shift, horizontal shift, and dilation

7. Graph using your calculator and compare the following functions. Note your observations.

   $f(x) = \frac{1}{x} + 1$ and $f(x) = \frac{1 + x}{x}$

8. Show how you could have determined the same information about the two functions algebraically. (Hint: Think about how to combine fractions.)

process: practice

Transforming Rational Functions Graphically and Algebraically

9. Use algebraic techniques to rewrite each of the following expressions as a single fraction.

   a. $\frac{3}{x} + 2$

   b. $\frac{2}{x-2} + 4$

   c. $\frac{4}{3(x+1)} - 2$

   d. $\frac{x}{x+2} - \frac{1}{2}$

   e. $\frac{5}{x+3} - 1$

input

Transforming Rational Functions Graphically and Algebraically

Following from the problems on the previous page, let's consider how to add two rational numbers with variables in the denominator. Similar to numerical fractions, when you combine two rational expressions under addition or subtraction, they must have the same denominator. Given a problem of the form:

$$\frac{x^2+x-3}{2x+3}+\frac{2x^2+6}{2x-3}$$

how will you find a common denominator? Think about the sum of simple rational expressions involving constants, such as $\frac{1}{2}+\frac{1}{3}$. What is the common denominator?

Next, consider the sum of rational expressions like $\frac{1}{2}+\frac{1}{x}$ and $\frac{3}{x}+\frac{4}{(x+1)}$. What is the common denominator for $\frac{1}{2}+\frac{1}{x}$?

What is the common denominator for $\frac{3}{x}+\frac{4}{(x+1)}$?

What is the common denominator for the $\frac{x^2+x-3}{2x+3}+\frac{2x^2+6}{2x-3}$?

Knowing the common denominator, write the equivalent expression for the last example.

Combine like terms in the numerator, holding the denominator fixed.

process: summary

Transforming Rational Functions Graphically and Algebraically

A necessary condition for two or more rational expressions to be combined is that the denominators are the same.

- If the denominators are not the same, then find a common denominator. The least common denominator (LCD) is the product of each distinct factor raised to the highest power.
- Given the common denominator or LCD, write equivalent expressions. As in the case of fractions, this requires multiplying the numerator and denominator by the "missing" factors.
- Combine numerators over the common denominator by applying the rules for combining of operations.

process: practice

Transforming Rational Functions Graphically and Algebraically

10. Combine each of the following expressions:

a. $\frac{2}{x+3}+\frac{1}{x}=$

b. $\frac{4}{2x}-\frac{3}{(x-1)}=$

c. $\frac{1}{x+2}+\frac{2}{x-2}=$

d. $\frac{3}{x^2+2x+1}+\frac{1}{x-1}=$

e. $\frac{1}{x^2+5x+4}+\frac{2x}{x^2+7x+12}=$

extension

Transforming Rational Functions Graphically and Algebraically

11. Some of the earlier examples, such as $\frac{1}{x+1} - \frac{2}{x} =$, involve simple operations on rational expressions such as addition and subtraction. If you add or subtract two rational expressions, will the sum or difference be a rational expression? Write a few sentences to justify your response.

12. The set of rational numbers is said to be **closed** with respect to the operations of addition and subtraction if the sum or difference of two rational numbers always results in another rational number. Based on question 11, do you think the set of rational expressions closed under addition and subtraction? Explain

13. Based on the previous question, given a rational expression $\frac{1}{x} \div \frac{2x-3}{x^3+x^2+1} =$, involving simple operations on rational expressions such as multiplication and division, if you were to multiply or divide two rational expressions, would the product or quotient always be a rational expression? Write a few sentences to justify your response.

14. The set of rational numbers is said to be **closed** with respect to the operations of multiplication and division if the product or quotient of two rational numbers always results in another rational number. Is the set of rational expressions closed under multiplication and division? Explain.

input

Transforming Rational Functions Graphically and Algebraically

To simplify a rational expression or simplify the product or quotient of rational expressions you must, where possible, write each polynomial as a product of polynomial factors (since only factors can be cancelled, not terms).

process: practice

Transforming Rational Functions Graphically and Algebraically

15. Simplify the following functions:

a. $\dfrac{3x^2(x+1)}{(x+1)^2} =$

b. $\dfrac{x^2+5x+6}{x^2+6x+8} =$

c. $\dfrac{4(x^3+1)}{2x^2+16x+14} =$

d. $\dfrac{2t^2-3t-2}{2t^2-10t+12} =$

e. $\dfrac{a^2-ab-2b^2}{2b^2-ab-a^2} =$

input

Modeling with Rational Functions: Buffo Nutrition

Buffo Nutrition stores have decided to produce their own line of nutrition shakes. The shakes will be sold in a cylindrical container holding 12 fluid ounces (approximately 21.6 cubic inches). The sides of the container will be made of plastic that costs 1 cent per square inch. The top and bottom of the container will be made of aluminum that costs 1.5 cents per square inch.

process: evaluate

Modeling with Rational Functions: Buffo Nutrition

1. Determine the cost of the aluminum lid.
2. Determine the cost of the plastic sides.
3. What is the total cost of the lids and side?
4. Given the V = 21.6 cubic inches, solve for h.
5. Substitute h into the equation for total cost.
6. What are the dimensions of the can that will result in the least cost to manufacture each container?
7. What will be the minimum cost to produce each container?

input

Modeling with Rational Functions: Buffo Nutrition

Buffo Nutrition stores have unveiled a new barbell-shaped promotional cup consisting of two equal-sized cylinders representing the weight connected by a thinner cylinder representing the barbell. Each cup will hold 32 fluid ounces (approximately 43.2 cubic inches). The "weights" will be made of a black-colored plastic costing 1 cent per square inch. The "barbell" will be made of clear plastic costing 2 cents per square inch. The radius of the weight should be twice as much as the radius of the barbell, and the height of the barbell should be 3 inches.

process: evaluate

Modeling with Rational Functions: Buffo Nutrition

8. Determine an equation to compute the total cost to produce each cup as a function of the radius of the barbell by completing the following steps:
   a. Determine the cost of the weights.
   b. Determine the cost of the barbell.
   c. Write the equation showing the cost of the weights and the barbells.

9. Determine the equation for the total volume of the cylinders.

10. Using the fact that V = 43.2 cubic inches, solve for h.

11. Using the value for h, write the equation representing the total cost.

process: evaluate

Modeling with Rational Functions: Buffo Nutrition

12. What is the total cost per cup if the radius of the barbell is:

    a. One-half inch

    b. One and a quarter inches

13. Determine algebraically what barbell radius will result in a total cost of:

    a. $1.50

    b. $2.75

14. What barbell radius will result in the minimum cost per cup? What is the minimum cost?

preview

Solving Rational Equations

You were always told that if you start counting when you see a bolt of lightning and stop counting when you hear the thunder, you could figure out how far the storm is from you. Each second represents a mile's distance between you and the storm.

In science class, however, you found that the time between seeing the lightning and hearing thunder was a function not only of distance, but also of temperature. You learned:

$$\text{Time between seeing the lightning and hearing the thunder} = \frac{d}{1.09T + 1050}$$

where d is the distance you are from the lightening and T is the temperature in Fahrenheit. Suppose you know that t = 10 seconds and d = 2 miles; what is the temperature in Fahrenheit?

input

Solving Rational Equations

In general, solving a rational equation is similar to solving any polynomial equation. You want to find a value for the variable such that the equation is true. In the case of a rational equation, you will need to check the solution(s) to make sure they make sense in the problem. What would cause a solution to fail?

process: evaluate

Solving Rational Equations

1. Using your knowledge of solving equations involving ratios, show how you would solve the rational equation $\frac{1}{a+7} = \frac{2}{a+2}$ for a

2. Use your knowledge of combining rational expressions and solving equations to solve the rational equation for a.

$$\frac{1}{a+7} - \frac{7}{a^2+9a+14} = \frac{-2}{a+2}$$

input

Solving Rational Equations

The two methods used to solve the problems are formalized below.

Method 1

- Find the least common denominator.
- Multiply both sides of the equation by the LCD (each and every term) to transform the rational equations to polynomial form (clearing the denominator).

Solve for the unknown using appropriate techniques

Method 2

- Combine expression as required on the left and right sides by applying the rules for combining expressions and writing equivalent expressions.
- Multiply by the reciprocal of the denominator on both sides to transform the rational equations to polynomial form (clearing the denominator).

Solve for the unknown using appropriate techniques.

process: practice

Solving Rational Equations

3. Solve the following rational equations for the variable and test to make sure the solutions do not represent discontinuities.

a. $\frac{4}{a-2}=4$

b. $\frac{1}{3}+\frac{1}{2}=\frac{1}{x}$

c. $\frac{2x}{x+2}=\frac{6}{x-1}$

d. $\frac{x^2+2x+2}{x-1}=\frac{2x+3}{x-1}$

e. $\frac{2}{y}+\frac{2y-6}{y-6}=\frac{2y-4}{y}$

process: summary

Summarizing Polynomial and Rational Functions

1. Define a polynomial function.

2. What is the domain and range of a polynomial function?

3. What does the y-intercept of a polynomial function represent?

4. What do the zeros of the polynomial function represent?

5. Based on the functions you have observed, what is the maximum number of zeros a function of degree 1 will have? Degree 2? Degree 3? Degree 4?

process: summary

Summarizing Polynomial and Rational Functions

6. What do the local maxima and minima of the polynomial function represent?

7. Describe the end behavior of a:
   a. Linear function

   b. Quadratic function

   c. Cubic function

8. What happens to a polynomial function at a local maximum or minimum value?

process: summary

Reviewing Polynomial and Rational Functions

9. Are all polynomial functions continuous? Explain.

10. Define a rational function.

11. Define an asymptote.

12. Define a discontinuity.

Reviewing Polynomial and Rational Functions

extension

13. Compare and contrast polynomial functions and exponential functions.

14. Compare and contrast rational functions and exponential functions.

15. Compare and contrast polynomial and rational functions.

Table of Contents

Assignments

Unit 1: Probability

Unit 2: Linear Functions

Unit 3: Parallel and Perpendicular Lines

Unit 4: Triangles and Quadrilaterals

Unit 5: Transformations and Symmetry

Unit 6: Similarity

Unit 7: Exponential and Logarithmic Functions

Unit 8 Quadratics

Unit 9 Higher Order and Rational Functions

Assignment for The Lady or the Tiger

Name: ______________________________

1. Rate each of the following events according to the probability that the event will take place.
   a. The school cafeteria will serve hamburgers today.

   b. You will have homework tonight.

   c. More than 5 students will be absent from math class.

   d. Your family will win the lottery.

2. Consider a family consisting of four children.
   a. Use a tree diagram to show the sample space for the children.

   b. Use an organized list to show the sample space for the children.

c. What is the size of the sample space?

d. What is the probability that there will be at least two girls in the family?

e. What is the probability that there will be exactly two girls in the family?

f. What is the probability that the oldest child will be a boy?

3. Use the chart below to list the product of two number cubes when tossed.

	1	2	3	4	5	6
1						
2						
3						
4						
5						
6						

a. Use set notation to show the sample space for the product.

b. What is the size of the sample space?

c. What is the probability that the product will be exactly 6?

d. What is the probability that the product will be less than 20?

e. What is the probability that the product will be an odd number?

f. What is the probability that the product will be a prime number?

Assignment for Relative Frequency vs Probability

Name: ______________________________

Cut the net of a square pyramid out, number the faces1 through 5, fold along the dotted segments, and tape the edges together to form the square pyramid.

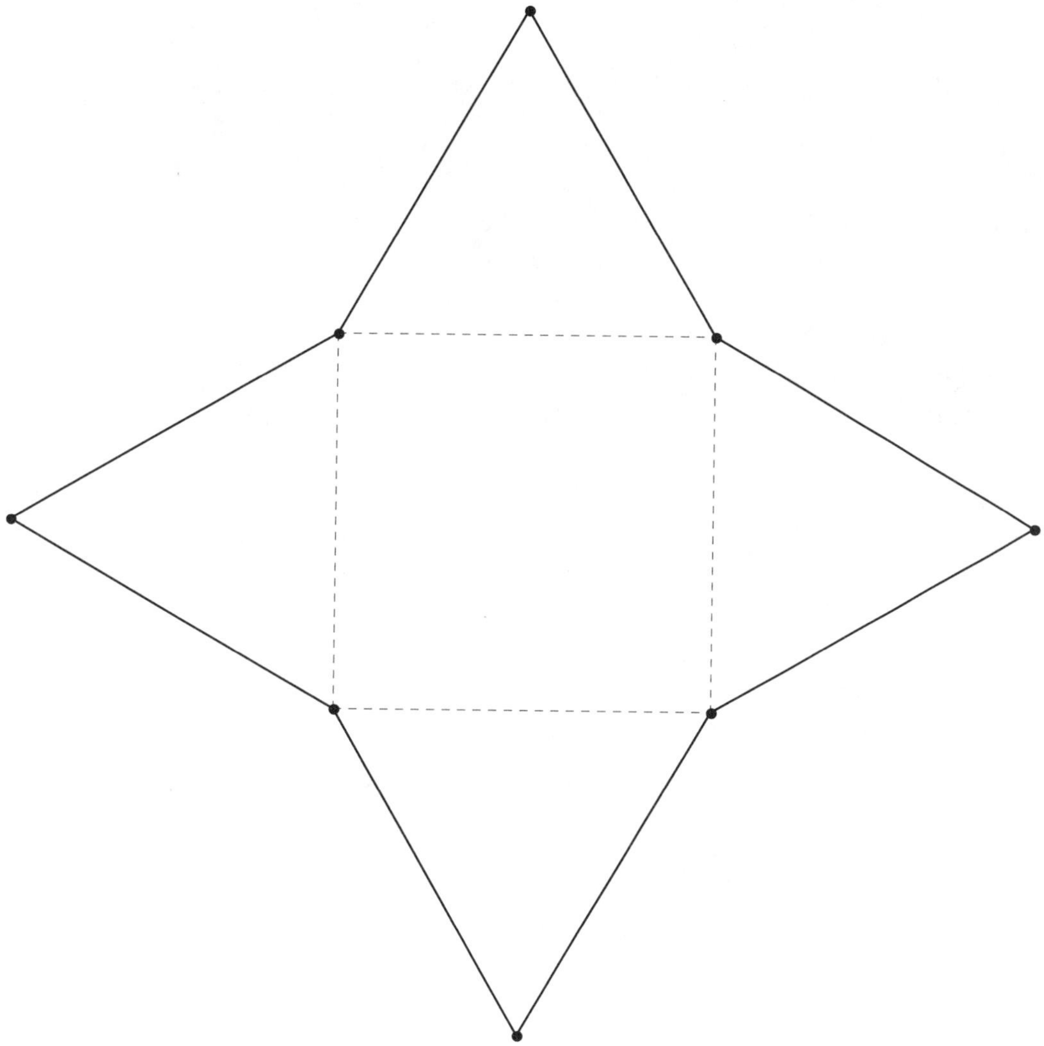

The page left intentionally blank so you can cut out the square pyramid.

1. What is the probability that the pyramid will land on its square base?

2. Are there any possibly reasons why the probability of landing on the square base would NOT be $\frac{1}{5}$?

3. Toss the pyramid a number of times. Keep track of the frequency with which it lands on each of its faces in a tally chart.

1	2	3	4	5

4. Use the data from your tally chart to create a relative frequency histogram at intervals of 10 tosses.

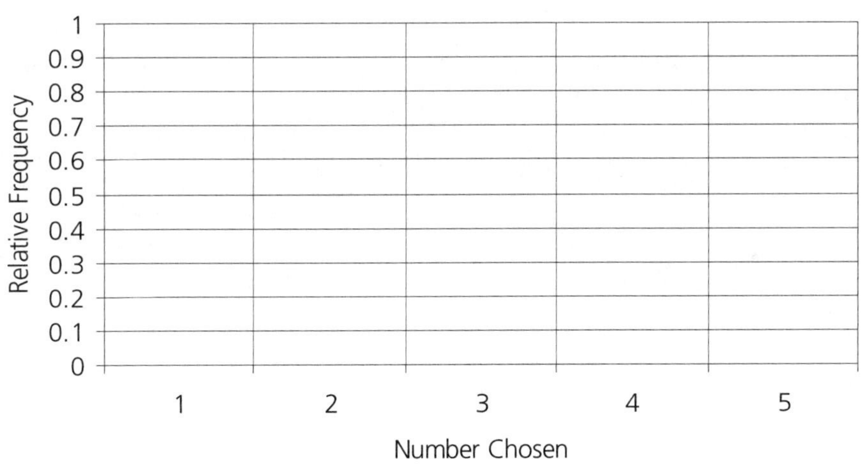

5. Does the relative frequency histogram change as the number of tosses increases? If so, how does is change? If not, why not?

6. Stop after about 50 tosses of the pyramid and create your final relative frequency histogram.

7. Each of the edges of the square pyramid is 5 cm. Calculate the area of each of the five faces of the pyramid.

8. Based upon these calculations, why does the pyramid have a greater possibility of landing on the square base?

Assignment for The Queen of Hearts

Name: ___________________________

1. Determine if each pair of events are mutually exclusive.

   a. Having pizza as your favorite food and having cheeseburgers as your favorite food.

   b. Having homework in math class and having homework in English class.

   c. Having blue as your favorite color and having 2 brothers.

   d. Drawing a spade from a deck of cards and drawing a diamond from a deck of cards.

2. A number of high school students were asked about their future plans and the amount of sleep they get each night. The results of the two question survey are shown in the chart below. Use the chart to answer the following questions.

What do you plan to do after high school?

How much sleep do you normally get each night?	Attend a 4 year college	Attend a 2 year college or tech school	Go to work or the military
0-4 hours	15	17	8
4-6 hours	27	31	15
6-8 hours	39	21	12
Over 8 hours	13	17	11

   a. What is the total number of students in the survey?

b. How many students in the survey plan to attend a four year college?

c. How many students in the survey get between six and eight hours of sleep a night?

d. How many of the students in the survey get less than four hours of sleep a night *and* plan to attend a two year college or tech school?

e. Create a Venn Diagram to illustrate the relationship between the students who get more than eight hours of sleep a night and the students who plan to go to work or the military.

f. Use the Venn Diagram to determine how many students in the survey get more than eight hours of sleep a night *or* plan to go to work or the military. Use set notation in your answer.

g. What is the probability that a student plans to attend a four year college?

h. What is the probability that a student gets between four and six hours of sleep a night?

i. What is the probability that a student who gets less than four hours of sleep also plans to attend a two year college or tech school?

j. Create a Venn Diagram to illustrate the relationship between the students who get between four and six of sleep a night and the students who plan to attend a four year college.

k. Use the Venn Diagram to determine how many students in the survey get between four and six hours of sleep a night *or* plan to attend a four year college. Use set notation in your answer.

Assignment for To Be or Not To Be

Name: ____________________________

1. Describe the complement of each event.

   a Drawing a diamond from a standard deck of cards.

   b. Today is your birthday.

   c. You get an A on your next math test.

2. "If events are complementary then they are also mutually exclusive." Do you agree or disagree with this statement? Explain. If the statement is false, provide a counter-example.

3. "If events are mutually exclusive then they are also complementary." Do you agree or disagree with this statement? Explain. If the statement is false, provide a counter-example.

4. All students in grades 9-12 are being asked to sell candy bars to help raise money for their class. The student who sells the most candy bars will win a new bicycle. Use the table below to answer the following questions.

	9th grade	10th grade	11th grade	12th grade
Boys	112	134	141	101
Girls	131	150	128	113

a. What is the probability that the winning student will be a 10th grader?

b. What is the probability that the winning student will be a girl?

c. Create a Venn Diagram displaying the set of 10th grade students and the set of girls.

d. What is the probability that the winning student will be a 10th grade girl?

e. What is the probability will neither be a 10th grader or a girl?

Assignment for Pizza and a Movie (Part 1: The Movie)

Name: ____________________________

1. Calculate the following.

   a. $8!$

   b. $\dfrac{7!}{4!}$

   c. $\dfrac{5!4!}{3!}$

   d. $\dfrac{9!4!}{8!}$

2. A school lunch consists of an entrée, a vegetable, a dessert, and a drink. Entrees include pizza, hamburger, or chicken. Side dishes include green beans or corn. Desserts include cookie, brownie, ice cream, or pudding. Drinks include juice, milk, or soda pop. How many different lunch combinations are there?

3. How many different batting orders are possible on a nine person baseball roster?

4. How many different lineups are possible on a six person hockey roster?

5. Custom Technology sells customized desktop computers available with several options. Monitors sized include 17 inch, 19 inch, or 23 inch. Drives include CD drive or DVD drive. Hard drive space includes 10GB, 20GB, 40GB, or 80GB. RAM includes 256 MB, 512 MB, or 1.0 GB. Software bundles include games, office, music, or video. How many different computers can be ordered?

6. You are in charge of putting together the program for this year's school talent show. The show consists of 13 acts. How many different ways are there to order the acts?

7. As part of student government, students can run for president, secretary, and treasurer. The student who comes in second place in the presidential election will be appointed vice president. This year 5 students ran for president, 3 for secretary and 4 for treasurer. How many different studies bodies are possible?

Assignment for Pop Quiz

Name: ____________________________

1. Each division in the NBA consists of 7 teams. How many different division standings are possible at the end of the season?

2. The registration code for a new software game you purchased consists of a string of 3 letters followed by 5 digits.

   a. How many different registration codes are possible?

   b. If you lost your registration code and had to guess, what is the probability that you will enter the correct registration code?

3. Your next quiz will consist of 10 multiple choice questions. Each question will have choices A through D. How many possible ways are there to answer the questions on the quiz?

4. Fruiteez candies come in five fruity flavors: grape, orange, cherry, lemon, and sour apple. The candies come in a package of five.

   a. How many different ways are there to package the candies if each flavor only appears once in each package?

   b. How many different ways are there to package the candies if each flavor can appear more than once in each package?

5. Your school is considering modifying class schedules for next year so that all students will have Math, Science, History, English and Foreign Language in the morning. How many different morning schedules based on subject are possible?

6. Your favorite CD consists of 12 songs. If you play the CD randomly on your CD player, how many possible play lists are possible?

Assignment for Pizza and a Movie (Part 2: The Pizza)

Name: ______________________________

1. Compute the following

   a. ${}_5P_2$

   b. ${}_5C_2$

   c. ${}_7P_4$

   d. ${}_7C_4$

   e. ${}_{10}P_6$

   f. ${}_{10}C_6$

2. The Pick 5 lottery involves correctly picking 5 numbered balls from a tub of 40 numbered balls.
   a. How many different combinations of numbered balls are possible?

   b. If you randomly pick 5 numbers, what is the probability of picking all five numbers correctly?

3. The track team consists of 12 members.
   a. How many ways are there to put together a team for the four member relay?

   b. What is the probability that you will be in the first position on the relay team?

4. Pick four cards from a standard deck of cards
   a. How many possible combinations of four cards are possible?

   b. What is the probability of getting four of a kind?

5. After practicing almost every day, your band knows how to play 40 different songs. You have been invited to be the opening act at an upcoming concert. As the opening act, you are only able to play five songs.
   a. How many possible set lists are possible for your set?

   b. If you randomly generate the set list, what is the probability that you will end your set with your favorite song?

   c. If you randomly generate the set list, what is the probability that your favorite song will be included?

Assignment for Socks and Free Throws

Name: ______________________________

1. During the 2002-2003 NHL season, Alexander Mogilny of the Toronto Maple Leafs had a shooting percentage of 20%. This means that he scored a goal 20% of the time he shot the puck.

   a. Create a pipe diagram showing the possible outcomes if he had three shots during a game.

   b. What is the probability that he had a hat trick (three goals) in a game in which he had three shots on goal?

   c. What is the probability that he had two goals in a game in which he had three shots on goal?

d. What is the probability that he had one goal in a game in which he had three shots on goal?

e. What is the probability that he did not score a goal in a game in which he had three shots on goal?

2. On your drive home from school, you pass four separate traffic lights. In the last 100 days of school, you have had to stop at the first light 25 times, the second 20 times, the third light 50 times, and the fourth light 33 times. Only once did you have to stop at all four lights. Based on the data you collected, do you think the traffic lights are operating independently? Explain.

3. If you pick 5 cards from a standard deck of cards, what is the probability of getting a royal straight flush? A royal straight flush is made up of the 10, Jack, Queen, King, and Ace of the same suit.

Assignment for No Fair!

Name: ______________________________

Heads Up is a four player game. Each round of play involves each player flipping a quarter.

If all four quarters are heads or if all four quarters are tails, player 1 gets one point.

If three heads are showing, player 2 gets one point.

If two heads are showing, player 3 gets one point.

If one head is showing, player 4 gets one point.

The player with the most points at the end of all rounds wins.

1. Does each of the four players in the game have an equal chance of winning the game? Explain.

2. Modify the points awarded to each player to make the game fair.

Assignment for Linear Equations and Inequalities

Name: ______________________________

You decide to mow lawns over the summer to make money. You have to spend $547 initially to purchase equipment such as lawn mowers, rakes, and garbage bags. After asking several adults what they would be willing to pay, you decide to charge $20 per lawn.

1. Write an algebraic equation to represent your profit.

2. Generate a graph of the problem situation

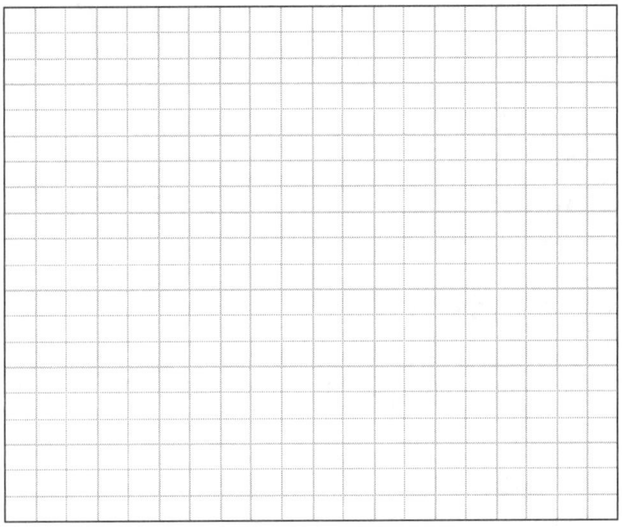

3. Identify the slope of the line. What does the slope mean in this problem situation?

4. Identify the y-intercept. What does the y-intercept mean in this problem?

5. Identify the x-intercept. What does the x-intercept mean in this problem?

6. Write the domain for this problem as an inequality.

7. Write the range for this problem as an inequality.

8. Calculate your profit if you mow
   a. 10 lawns

   b. 70 lawns

9. The newest video game system costs $250. How many lawns would you need to mow to be able to buy it?

10. Your neighbor is selling their computer for $500. How many lawns would you need to mow to be able to buy it?

Assignment for Linear Transformations

Name: ______________________________

1. Write the resulting function if the following transformations are applied to the function y = x and graph the parent function and transformed function.

   a. reflect about the x-axis and shift up 4 units

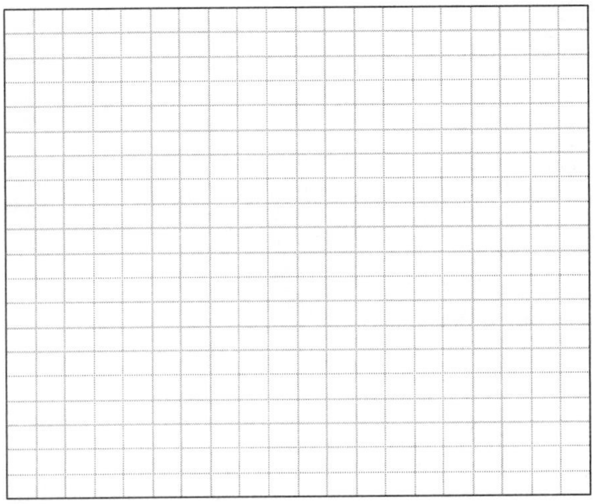

   b. Dilate by a factor of 5 and shift down 3 units

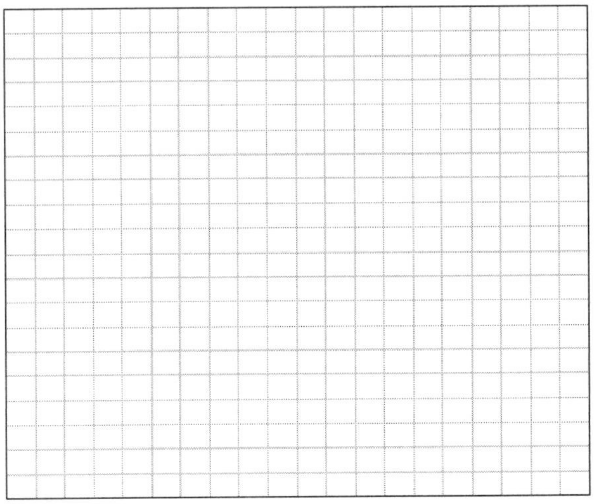

c. Dilate by a factor of 2, reflect about the x-axis and shift down 3 units

d. Dilate by a factor of 0.75 and shift up 3

2. Describe the transformations applied to the function y = x to result in each function and graph the parent function and transformed function.

   a. $y = -x - 5$

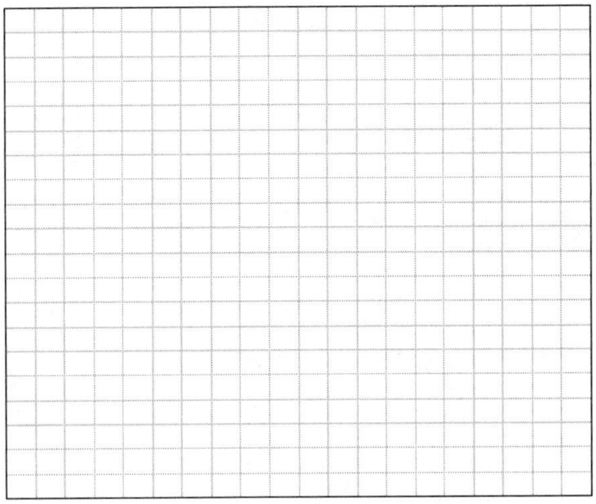

   b. $y = -3x$

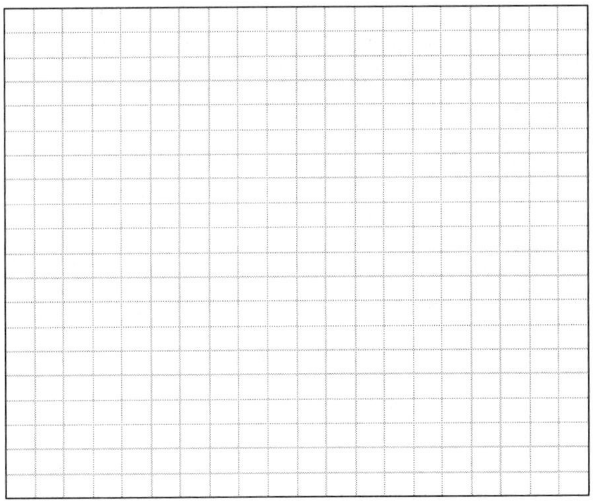

c. $y = 1.5x + 1$

d. $y = -10x + 2$

Assignment for Systems of Linear Equations: Producing and Selling Markers

Name: ______________________________

While working for the marker company you are contracted by another company that agrees to make markers for \$3.00 each but with a startup cost of only \$50. You determine you will be able to sell the new markers for \$3.50 each.

1. Write algebraic equations for

   a. Cost

   b. Income

2. Create a table of values displaying cost and income

Labels	Number of markers	Cost	Income
Units			
Expression			

3. Create a graph displaying cost and income

4. What is the point of intersection of the cost and income graphs? What does the intersection point mean in this problem?

5. Determine the range of values for the number of markers sold to make a profit.

Assignment for Making and Selling Shirts

Name: ______________________________

The local fire department is sponsoring a raffle in order to raise money for a new fire truck. Tickets will be sold for $5.00 each and cost 10 cents per ticket to print. The raffle winner will receive a cash prize of $25,000.

1. Write algebraic equations for

   a. Cost

   b. Income

2. Create a table of values displaying cost and income

Labels	**Number of tickets made and sold**	**Cost**	**Income**
Units			
Expression			

3. Create a graph displaying cost and income

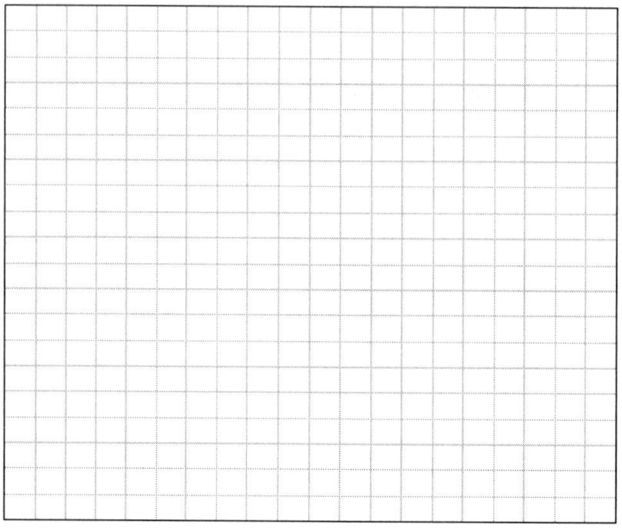

4. What is the point of intersection of the cost and income graphs? What does the intersection point mean in this problem?

5. Determine the range of values for the number of markers sold to make a profit.

Assignment for Connecting Algebraic to Graphical Representations

Name: ______________________________

Based on the algebraic representation for each system of equations, determine the number of solutions to the system. Verify graphically.

1. $y = 2x + 5$

   $3x + y = -4$

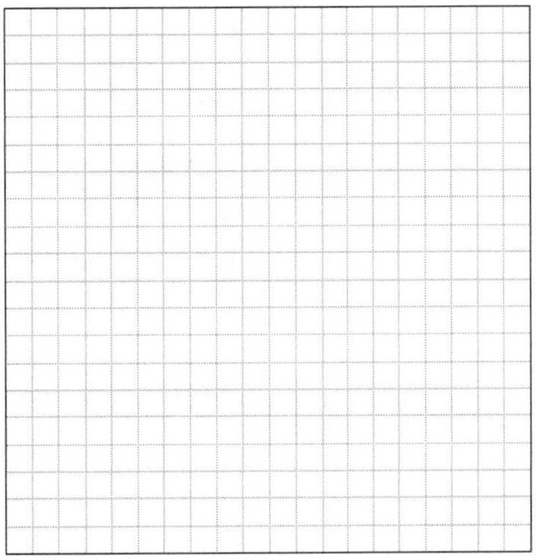

2. $-x + 2y = -6$

   $-2x + 4y = 4$

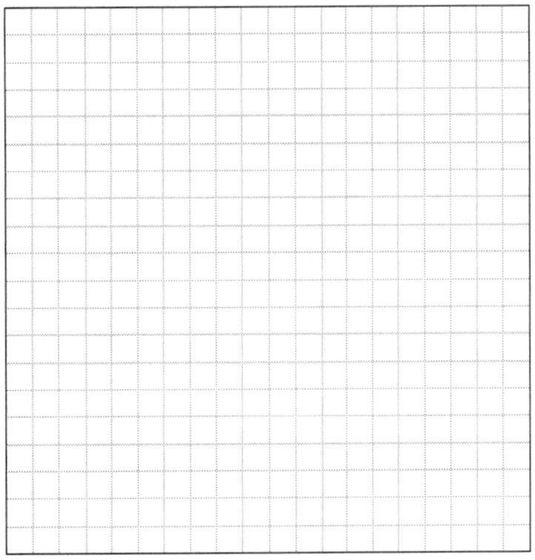

3. $3x + 5y = 1$

x – 2y = 6

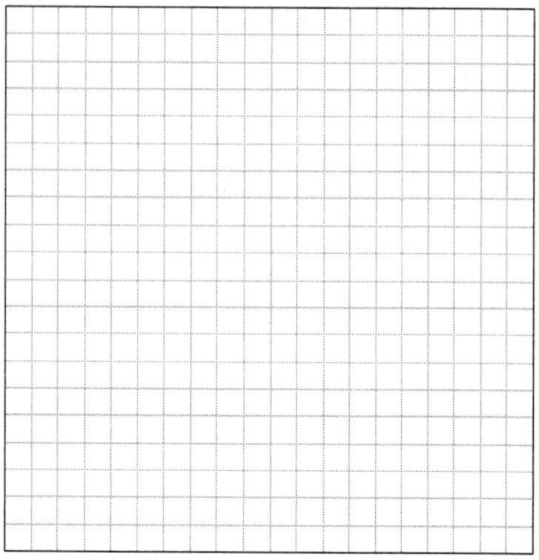

4. 4x + 2y = 2

   y = -2x + 1

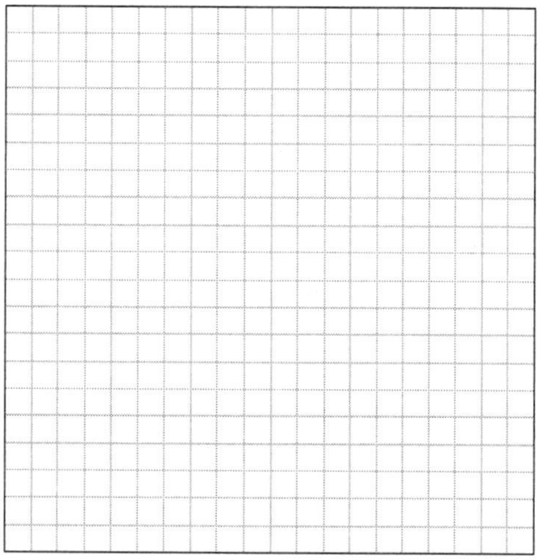

Assignment for Finding the Better Paying Job

Name: ______________________________

You just purchased a new CD burner for your computer and are interested in burning your own music CDs. You gather information about some music download services. Tunes.com charges a monthly fee of $20 and 50 cents per song. MusicNet has no monthly fee but charges $1 per song.

1. Write equations representing the monthly charge for Tunes.com and MusicNet.

2. Create a table of values summarizing the monthly cost for each service.

3. Create a graph summarizing the monthly cost for each service.

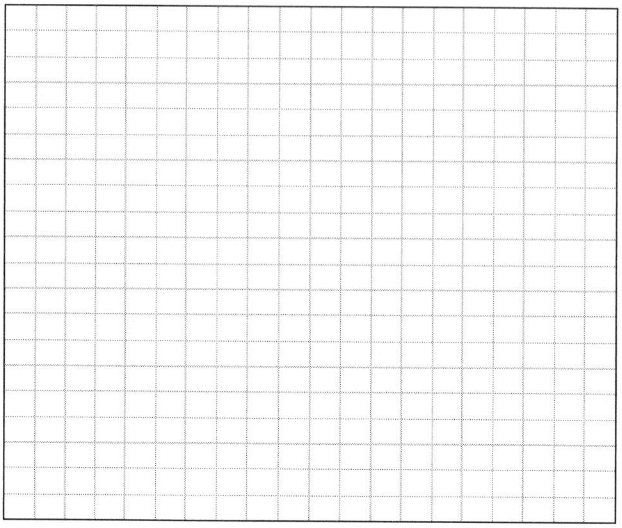

4. Will the monthly charge for the two services ever be the same? If so, how many songs can you download and what is the monthly charge?

5. Based on your findings, what service would you choose and why?

Assignment for Solving Systems of Two Equations Algebraically

Name: ______________________________

1. Solve the following systems of equations

   a. $3x + 2y = -11$

      $x + y = -4$

   b. $x + y = 4$

      $3x - 4y = -23$

c. $y = 5x - 3$
   $y = 2x + 12$

d. $3x + 2y = 12$
   $y = x + 1$

e. $y = 6x - 5$
   $7x - y = -7$

f. $-4x + 2y = 12$
   $3x + y = -4$

2. Your local theater group sold 568 tickets to its last performance. Tickets sold for $5.00 for adults and $2.00 for children. Total ticket sales were #2345. The theatre manager expects to sell the same number of tickets to the next play, but wants to increase the ticket sales to $3554 by increasing the price of adult tickets. Find the new price of an adult ticket.

Assignment for World Oil: Supply and Demand

Name: ______________________________

You and your friends have decided to form a band. After practicing for several months, you record a number of songs and create a demo CD. You plan on selling the CD at shows, around school, and on your band's web page. In preparation, you have already made 200 CDs and plan on creating 10 more every day. After one week, you find that you are selling an average of 20 CDs every day.

1. Write an equation to model the supply of CDs over time.

2. Write an equation to model the demand for CDs over time.

3. Create a table of values showing the supply and demand for the CD over the first month.

4. Create a graph of the supply and demand for the CD.

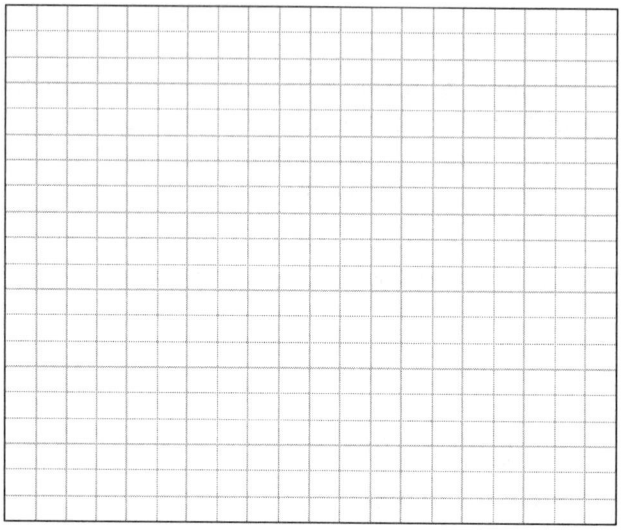

5. Determine the point of intersection algebraically. What does the intersection point mean in terms of the problem situation?

6. Interpret the portion of the graph to the left and right of the intersection point.

Assignment for Solving Inequalities

Name: ______________________________

Solve the following systems of inequalities graphically and shade the solution set. For each, identify two point contained within the solution set and two points not contained in the solution set.

1. $\begin{cases} y < 3x + 2 \\ y \geq -x \end{cases}$

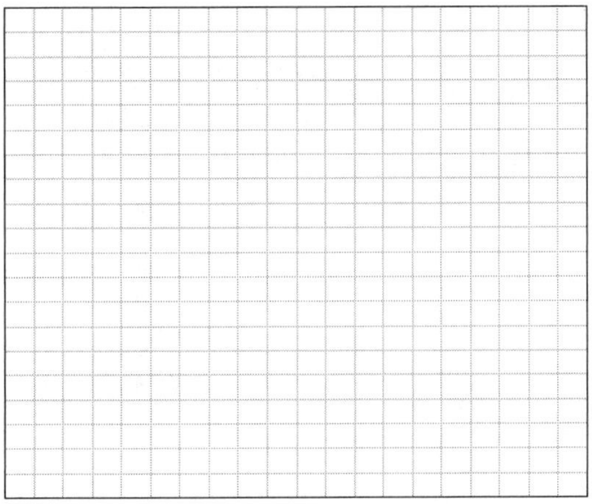

2. $\begin{cases} y \leq 2x + 1 \\ y \geq 2x - 6 \end{cases}$

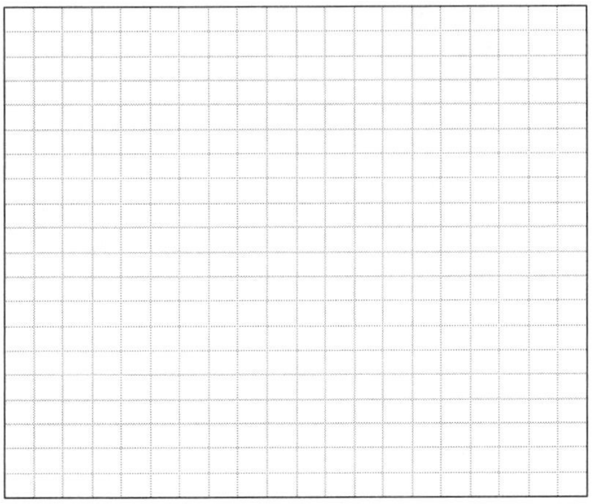

3. $\begin{cases} y < x + 1 \\ y > 2x + 1 \end{cases}$

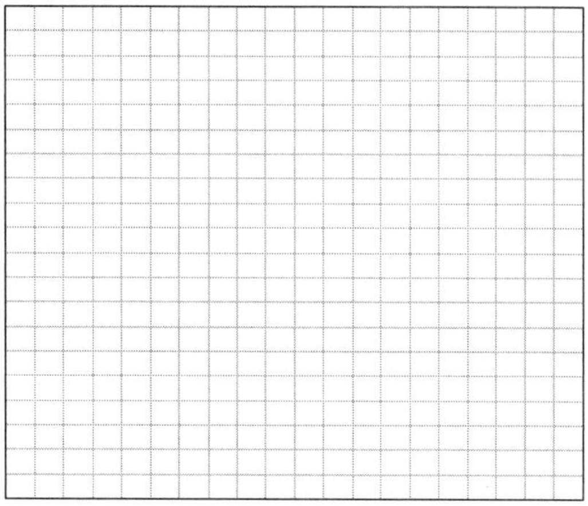

4. $\begin{cases} y < 6 \\ x \geq -3 \end{cases}$

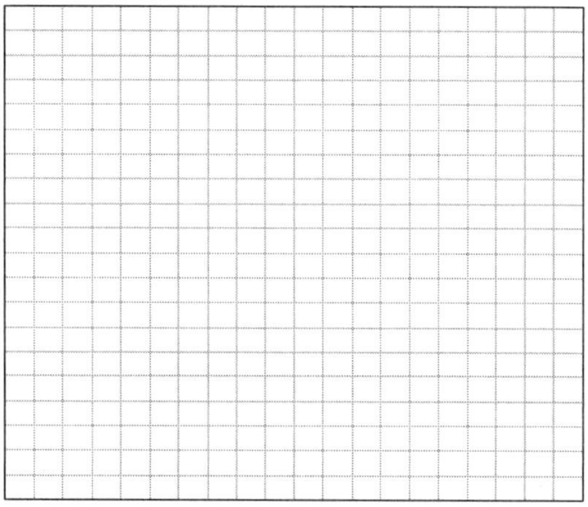

Assignment 1 for Systems of More Than Two Linear Equations and Inequalities

Name: ______________________________

You really enjoy ice cream and decide to buy an ice cream making machine. The cost of the machine is $175 and it will make up to a half gallon of ice cream. Supplies to make a half gallon of ice cream such as eggs, milk, and sugar cost $2.75. Your friends wonder why you want to make your own ice cream since good ice cream can be bought at the store for $5.50 for a half gallon.

1. Write an equation showing the cost of making your own ice cream.

2. Write an equation showing the cost of buying ice cream from the store.

3. A new local ice cream manufacturer advertises a special; offering all you can eat ice cream for $500 per year. Write an equation showing the cost of buying ice cream from the manufacturer.

4. Graph the three equations.

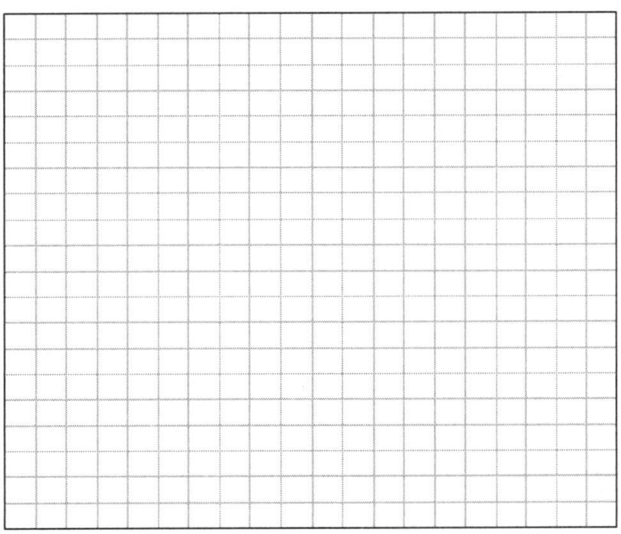

5. When is it better to make your own ice cream?

6. When is it better to buy from the store?

7. When is it better to buy from the manufacturer?

8. Is there a solution to the system of three equations representing the cost of ice cream? If so, calculate the solution. If not, why not?

Assignment 2 for Systems of More Than Two Linear Equations and Inequalities

Name: ______________________________

1. The following data tables show the number of bank accounts for Civilians Bank by type and branch location.

   Table A: Total number of accounts at Civilians Bank as of January 1

	Checking	Savings	Certificate of Deposit	Money Markets
Main Branch	254	1087	931	836
North Side Branch	1705	645	345	241
West End Branch	191	583	356	307

   Table B: Number of new accounts opened during the first quarter

	Checking	Savings	Certificate of Deposit	Money Markets
Main Branch	149	73	64	57
North Side Branch	89	49	53	30
West End Branch	93	35	62	49

   Table C: Number of accounts closed during the first quarter

	Checking	Savings	Certificate of Deposit	Money Markets
Main Branch	85	34	51	72
North Side Branch	115	29	60	19
West End Branch	77	42	41	23

a. Represent each table using a matrix.

b. Compute D = B – C. Describe what matrix D represents including the meaning of positive and negative values.

c. Compute E = A + D. Describe what matrix E represents.

d. A manufacturing plant in the area is closing. The bankers at Civilians Bank are expecting a 5% decrease in all accounts at all branches during the second quarter. Based on the number of accounts at the end of the first quarter, write a matrix F to represent the number of accounts the bank will have at the end of the second quarter.

2. The president of Civilians Bank has decided to offer each branch a bonus (to be used for the summer picnic) for each new checking, savings, or certificate of deposit account opened during the first quarter. The main branch was awarded a total of \$1944, the north side branch was awarded a total of \$1289, and the west end branch was awarded a total of \$1218. Determine the amount awarded for each type of account.

   a. Write a matrix equation to model the problem scenario.

   b. Compute the inverse matrix needed to solve the matrix equation using the formula below.

$$\begin{bmatrix} a & b & c \\ d & e & f \\ g & h & i \end{bmatrix}^{-1} = \frac{1}{\text{DET}}\begin{bmatrix} ie-hf & -(ib-hc) & fb-ec \\ -(id-gf) & ia-gc & -(fa-dc) \\ hd-ge & -(ha-gb) & ea-db \end{bmatrix}$$

Where $\text{DET} = a(ie-hf)-d(ib-hc)+g(fb-ec)$

c. Use the inverse matrix to solve the matrix equation.

d. How much did the bank president award for each type of new account?

Assignment for Maximizing and Minimizing Functions

Name: ______________________________

A company produces two models of portable stereos, one with a CD player and one without. The stereo with a CD player requires 4 man-hours to assemble and the stereo without a CD player requires 3 man-hours to assemble. The stereo with a CD player requires $75 worth of materials for production and the stereo without a CD player requires $60 worth of materials for production. There are a total of 480 man-hours of labor available every day. The budget for materials each day is $9300.

The profit for the stereo with a CD player is $35 and the profit for the stereo without a CD player is $27.50. Determine how many stereos of each model should be produced to maximize profit.

1. Write each constraint of the problem as a linear inequality.

2, Write an expression to calculate the profit.

3. Graph the system of inequalities representing the constraints.

4. Find the coordinates of each vertex of the solution region.

5. Determine the profit for each vertex.

6. How many of each model stereo should the company produce to maximize profit?

7. Another company who makes stereos just lowered the price of their stereos without CD players. In order to remain competitive, you lower the price of your non-CD stereos so the profit on each is $25. Does this affect the number of stereos the company should produce to maximize profit?

Assignment for Arithmetic Sequences

Name: ________________________________

1. For each sequence below, identify the next two terms. Write an explicit or recursive rule to generate the sequence and write a sentence describing the rule.

   a. 1, 3, 7, 15, 31, ...

   b. 1, 1, 2, 3, 5, 8, 13, ...

   c. 30, 24, 18, 12, 6, ...

   d. -1, 3, -9, 27, ...

2. Determine if the following sequences are arithmetic. If the sequence is arithmetic, write a recursive and explicit formula for each.

   a. 9, 5, 1, -3, …

   b. 3, -6, 12, -24, …

   c. 1, 4, 7, 10, …

   d. $1, \frac{4}{3}, \frac{5}{3}, 2, \frac{7}{3}, \ldots$

Assignment for Arithmetic Series

Name: ______________________________

For each arithmetic sequence:

a. Write an explicit formula to compute the nth term.

b. Calculate the specified sum.

1. 3, 7, 11, 15, …; sum of the first 25 terms

2. 20, 17, 14, 11, …; sum of the first 10 terms

3. -7, -5, -3, -1, …; sum of the first 15 terms

4. 21, 15, 9, 3, ...; sum of the first 30 terms

5. -37, -32, -27, -22, ...; sum of the first 18 terms

6. 52, 44, 36, 28, ...; sum of the first 14 terms

Assignment for Functions and Inverse Functions

Name: ______________________________

For each function:

a. Find the inverse.

b. Graph the function and its inverse.

1. $f(x) = 4x - 2$

2. $f(x) = 3 - 2x$

3. $f(x) = \dfrac{x - 2}{2}$

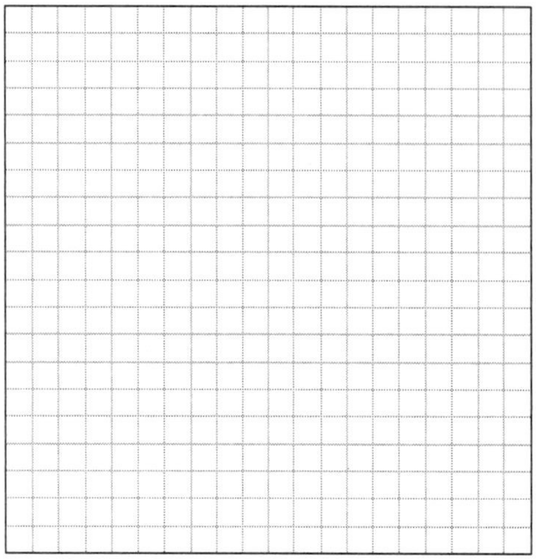

4. $f(x) = \dfrac{2}{3}x + \dfrac{1}{2}$

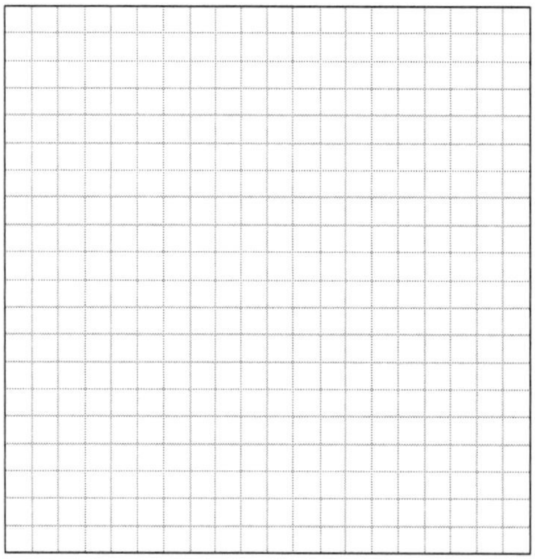

Assignment for Symbols and Geometric Objects

Name: ______________________________

1. List the three sides of ΔYOU. Use the correct symbolic notation.

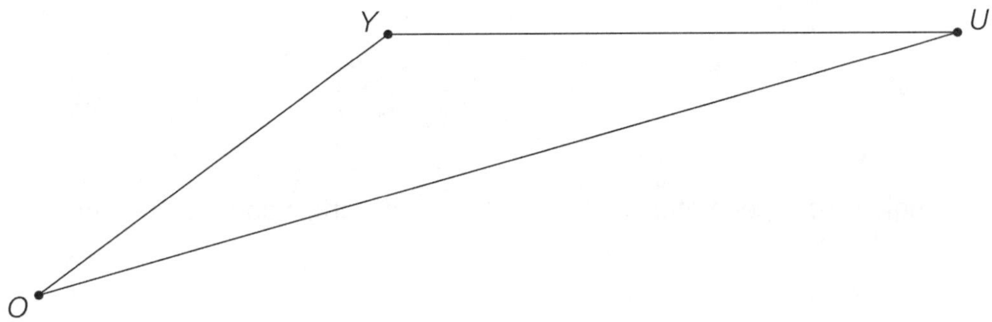

   a. Measure the three sides of ΔYOU, and list their measurements below, using the correct symbolic notation.

2. Draw ΔSHE that has these characteristics: $HE = 5$ cm, $SH = 6$ cm.

   a. List the three sides of this triangle, using the correct symbolic notation.

b. What is the length of the third side of the triangle?

c. Will your classmates' triangles all have a third side with the same length as yours? Why or why not?

3. Draw a triangle whose perimeter is 13 cm. Label your triangle and give its name using the correct symbolic notation.

a. List the names of the sides of your triangle using the correct symbolic notation.

b. Measure each of the three sides of your triangle, and list their measures using the correct symbolic notation.

c. Do you think other students drew the same triangle you did? Why or why not?

Assignment for Relationships between Lines

Name: ______________________________

1. What are the three ways that lines can relate to one another?

2. What are the two ways in which coplanar lines can relate to one another?

3. Define the term *parallel* in your own words.

4. Define the term *skew* in your own words.

5. Use mathematically correct symbols to write about the relationship between the following two lines.

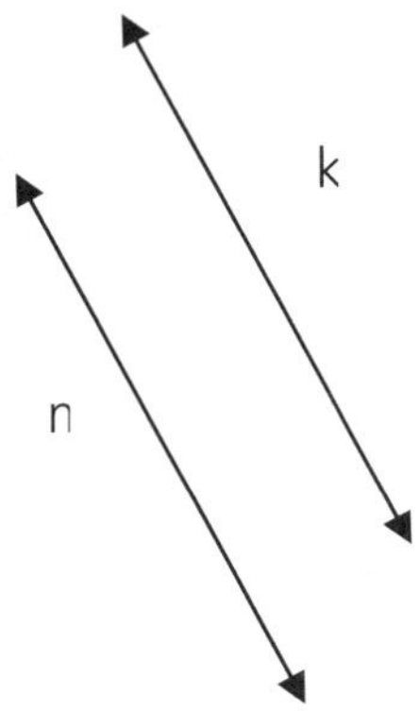

4. Cats-R-Us finally started selling a lot of the squealing mouse toys. As the toys became more popular, the company decided to raise the price of the toy to consumers to \$13. Recall that the cost to produce the squealing mouse toys was \$10 per toy with a set-up fee of \$200.

   a. Model the cost to produce the toys with an algebraic equation.

   b. Model the income from the toys with an algebraic equation.

   c. What is the relationship between these two lines? If the lines are parallel, how do you know? If the lines intersect, what is their point of intersection?

5. As soon as the company that produces the cat toys for Cats-R-Us heard about the price increase on the squealing mouse, it changed its production price to \$13 per mouse.

   a. Model the cost to produce the toys with an algebraic equation.

   b. Model the income from the toys with an algebraic equation.

   c. What is the relationship between these two lines? If the lines are parallel, how do you know? If the lines intersect, what is their point of intersection?

Assignment for Angles and Measuring Angles

Name: ____________________________________

For the four triangles in this assignment, follow these five steps:

1. Name each of the angles in the following triangles using the correct symbolic notation with three letters.
2. Shade what looks to be the largest angle in each of the triangles.
3. Estimate the measures of each of the angles in the triangles. List the estimates.
4. Use your protractor to measure each angle in these triangles. List measurements of each and compare these measurements with your estimates.
5. Label each of the angles in the triangles as *acute, right, obtuse, or straight.*

 a.

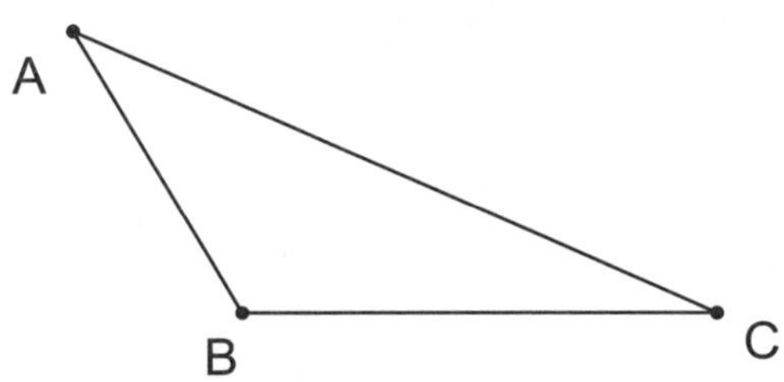

 b.

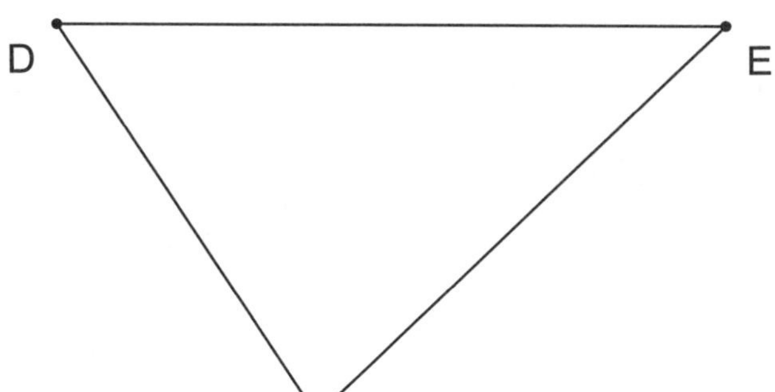

c.

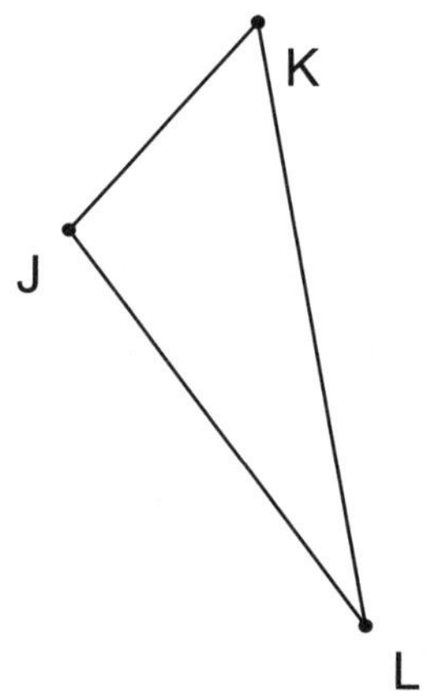

d.

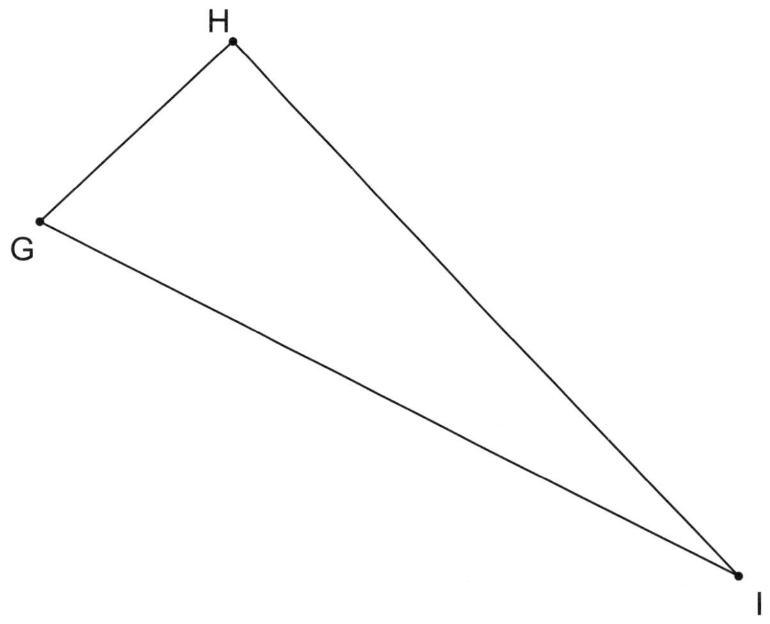

Assignment for Relationships between Parallel Lines and Relationships between Perpendicular Lines

Name: ______________________________

State whether each pair of lines is parallel, perpendicular, or neither. Explain why.

1. $y = 4x - 18$

   $y = 4x$

2. $y = 2x - 6$

   $y = -\frac{1}{2}x + 9$

3. $y = 6x$

   $y = \frac{1}{6}x - 5$

4. $2y = 2x + 10$

   $y = -\frac{1}{2}x$

5. $7x - 2 = y$

   $y = 7x$

6. $y = x$

   $y = -x$

7. $x = 5$

   $y = 2$

8. $4x + 6 = y - 2x$

   $y = -\frac{1}{6}x + 22$

9. State the relationship between each of the following pairs of lines.

   Line *A*: $y = 4x + 2$

   Line *B*: $y = 4x - 3$

   Line *C*: $y = -0.25x$

   Line *D*: $y = -\frac{1}{4}x + 1$

   Line *A* and Line *B* are ________________ to each other.

   Line *A* and Line *C* are ________________ to each other.

   Line *A* and Line *D* are ________________ to each other.

   Line *B* and Line *C* are ________________ to each other.

   Line *B* and Line *D* are ________________ to each other.

   Line *C* and Line *D* are ________________ to each other.

10. Write each of the following sentences symbolically.

    a. Line *NV* is perpendicular to line *AR*.

    b. Line segment *KY* is parallel to line *TN*.

    c. Line *OR* is parallel to line *WA*.

    d. Line segment *OH* is parallel to line *IN*.

    e. Line *TX* is perpendicular to line segment *CO*.

Assignment for Perpendicular Lines and Segments

Name: ______________________________

1. Use a ruler and protractor to create all three altitudes in $\triangle ABC$. Label the points where the altitudes intersect the sides as D, E, and F respectively.

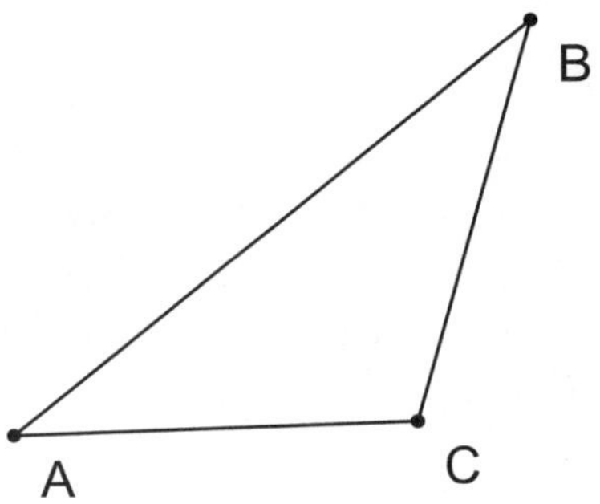

a. Measure the length of $\overline{AB}$ and the length of the altitude drawn to it. Use these measurements to calculate the area of $\triangle ABC$.

b. Measure the length of $\overline{AC}$ and the length of the altitude drawn to it. Use these measurements to calculate the area of $\triangle ABC$.

c. Are the results from parts a and b the same. Explain.

2. The following figure is not drawn to scale. Rely only on the measurements given in each problem. Create the altitude of ΔXYZ from point Y to side XZ. Label this altitude as $\overline{YK}$.

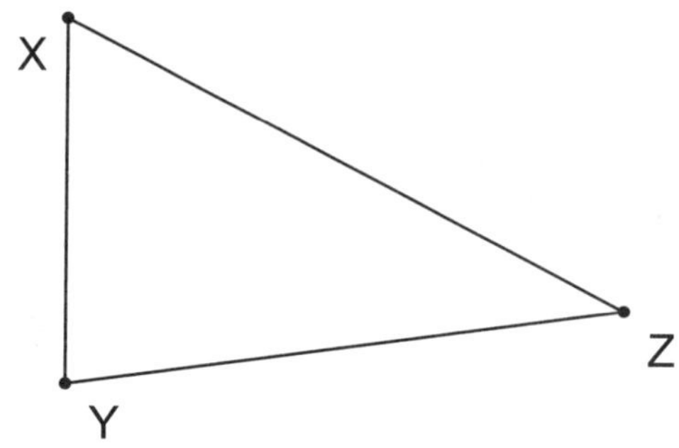

   a. If the area of ΔXYZ is 52 cm^2, and side XZ is 18 cm in length, determine the length of $\overline{YK}$.

Assignment for Vertical Angles

Name: ______________________________

Do not use a protractor to measure any given angles in this assignment. The figures may not be drawn to scale.

1. If $m\angle 1 = 38°$, find the measure of any other angles that you can and justify your conclusions.

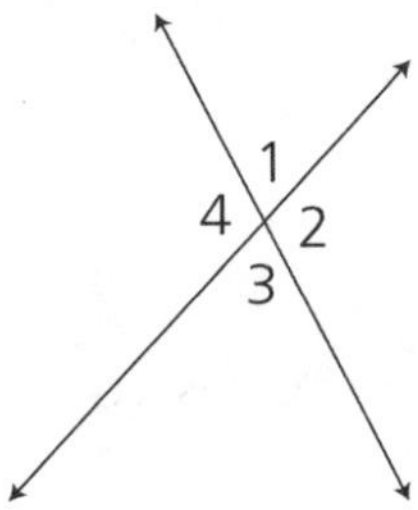

2. Use the given information to find any other angle measures that you can in the following figure. Write the measure of each angle you find in the figure.

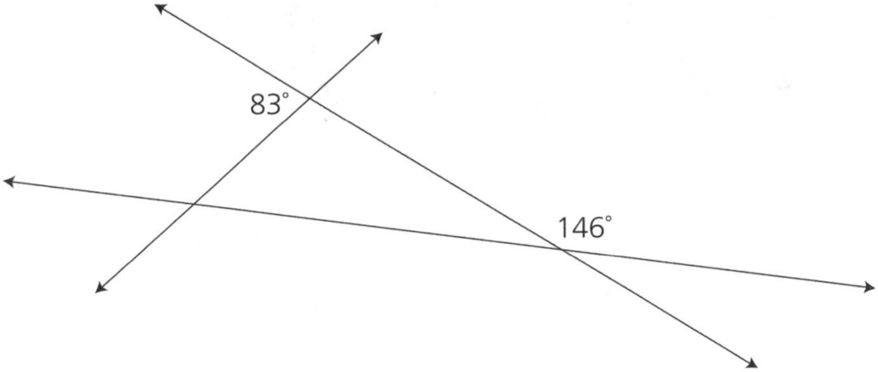

3. Find the measures of any other angles that you can. Write the measure of each angle you find in this figure.

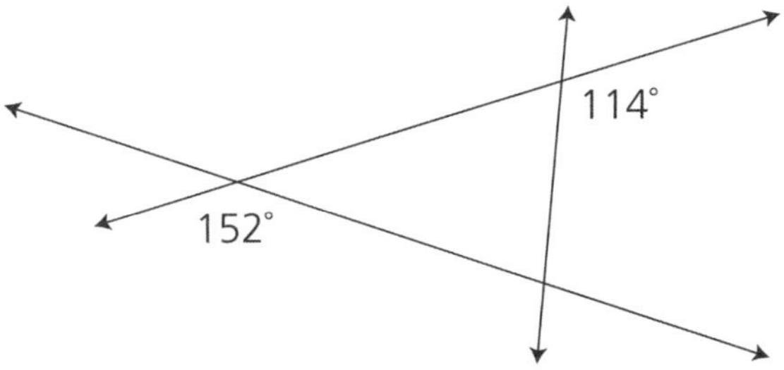

4. Find the measures of each of the missing angles in this figure. Write the measure of each angle in the figure.

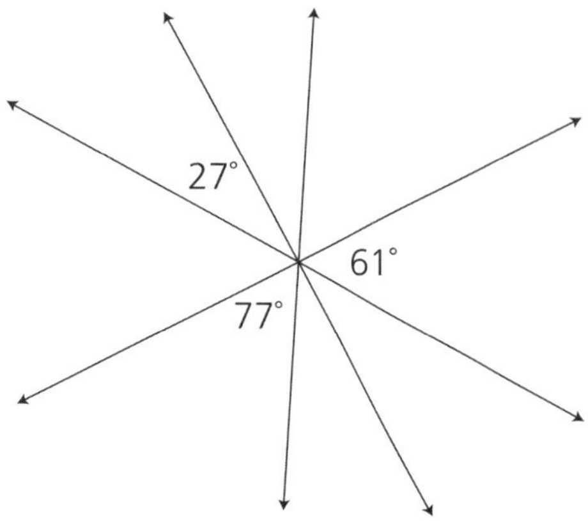

5. Create two intersecting lines to form four angles where the measure of one angle is 104°. Find the measure of any other angles that you can. Draw a picture of this situation.

6. How many different pairs of vertical angles are formed when five lines intersect at the same point? Draw a picture to support your answer.

Assignment for More Angle Pairs

Name: ____

Invent and describe a problem situation from everyday life that fits the diagram below. (Use a different idea than the gas tank situation described in your text.) Take measurements from within the diagram, using tools of your choice, and write several conclusions about your problem situation.

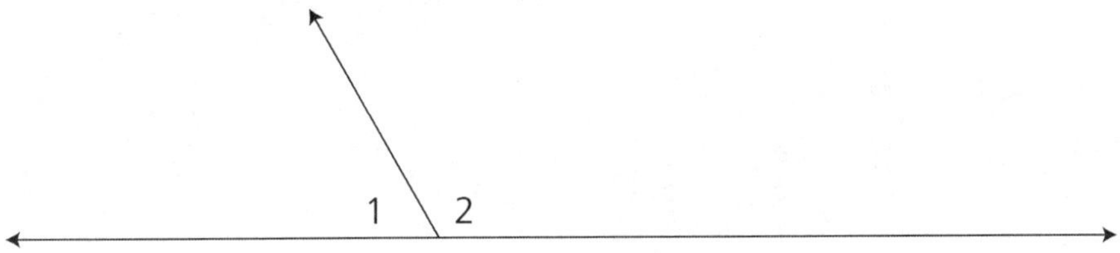

Assignment for Angles Formed by Transversals

Name: ____

Do not use a protractor in this assignment. Rely only on the measurements given in each problem.

1. Find the measures of the missing angles in the following figure. List the angle measures and the reasons you used for finding them in the table.

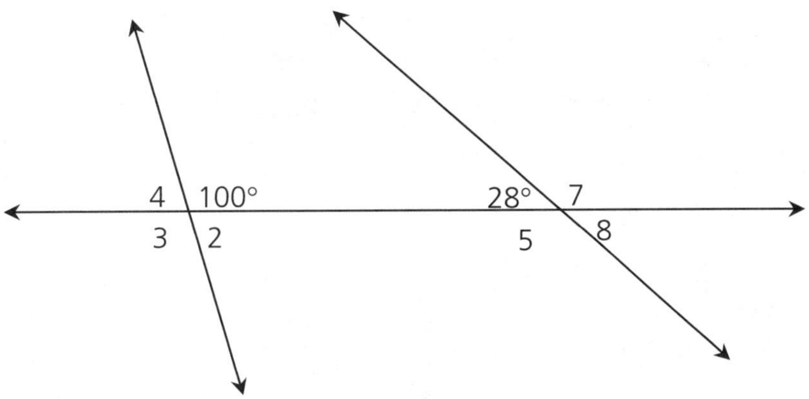

	Measure	**Reason or Justification**
∠8		
∠2		
∠3		
∠4		
∠5		
∠7		

2. How many transversals are pictured above? Explain.

3. Find the measures of the missing angles in the following figure. List the angle measures and the reasons you used for finding them in the table.

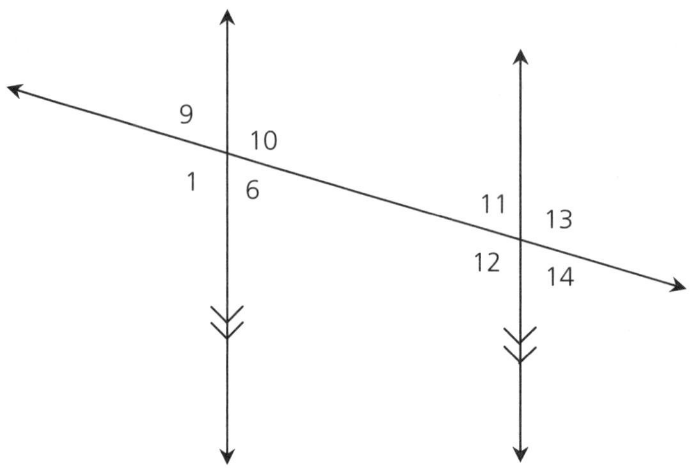

Given: m∠11 = 68°

	Measure	Reason or Justification
∠9		
∠6		
∠1		
∠10		
∠12		
∠13		
∠14		

4. Using the figures in this assignment, name at least two pairs of the following types of angles.

a. Same side exterior angles

b. Corresponding angles

c. Alternate interior angles

d. Alternate exterior angles

Assignment for Parallel Lines and Transversals

Name: ____

1. In parallelogram *GEOM*, segment *GE* is parallel to segment *OM* and segment *GM* is parallel to segment *EO*. Find the measure of each interior angle. Show your work.

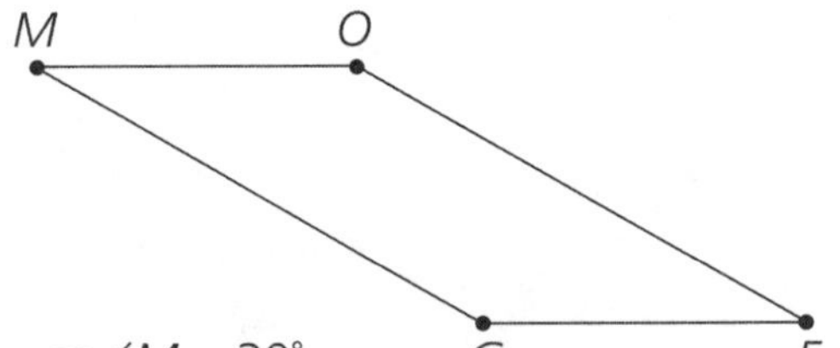

$m\angle M = 30^\circ$

2. In this figure, *AB* is parallel to *CD*. Find the value of *x*. Show your work.

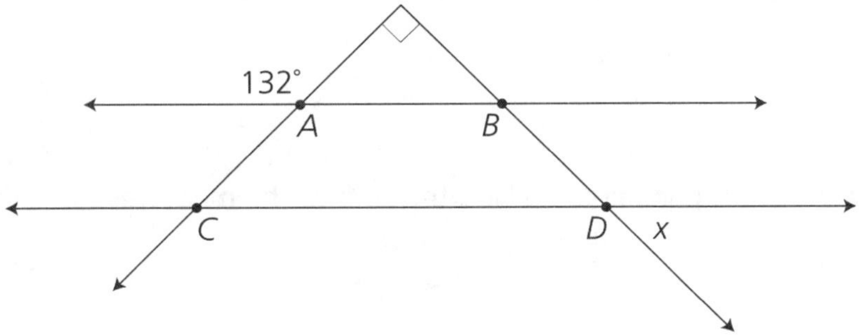

3. In this figure, *AB* is parallel to *CD*. Find the value of *x*. Show your work.

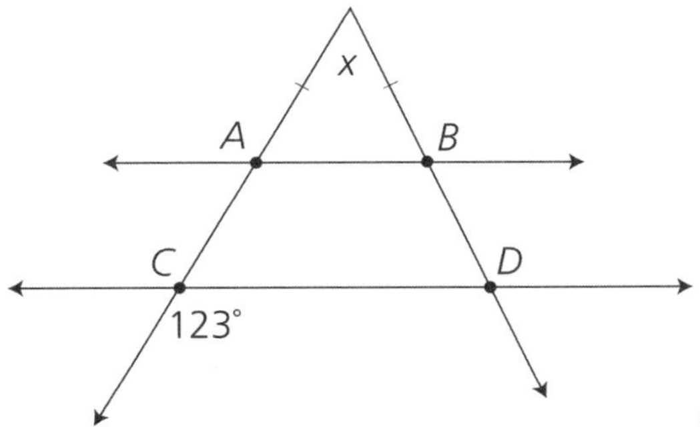

4. Find the measure of each angle in this figure. The arrow markers identify parallel lines.

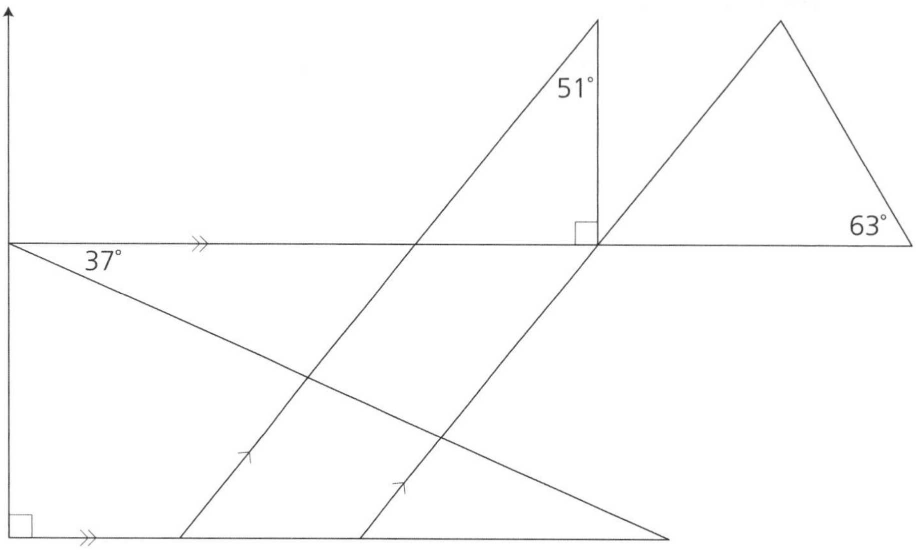

5. In this figure, ℓ_1 is parallel to ℓ_2 and ℓ_3 is parallel to ℓ_4. Find the sum of the measures of all the angles marked •. Explain how you found your answer.

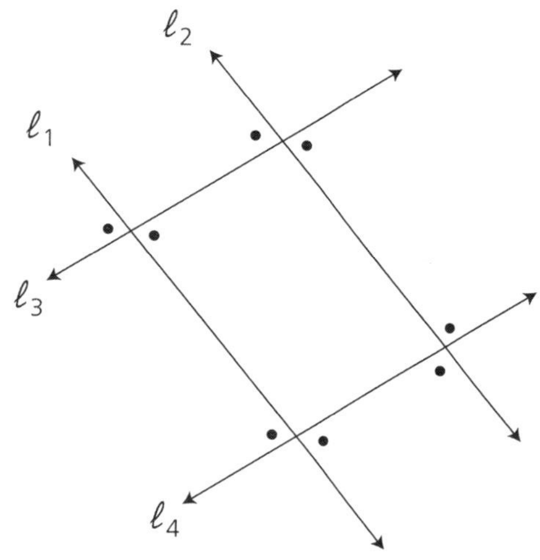

Assignment for Names and Symbols for Triangles

Name: ______________________________

1.

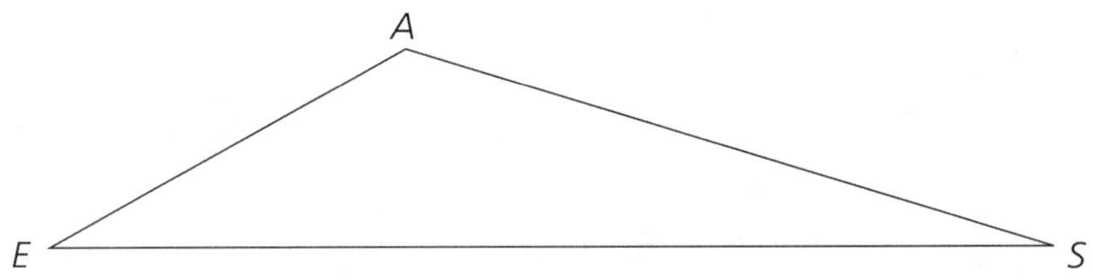

a. How many vertices does this triangle have?

b. What are their names?

c. Name this triangle in six different ways. Use the correct symbolic notation when naming the triangle.

2.

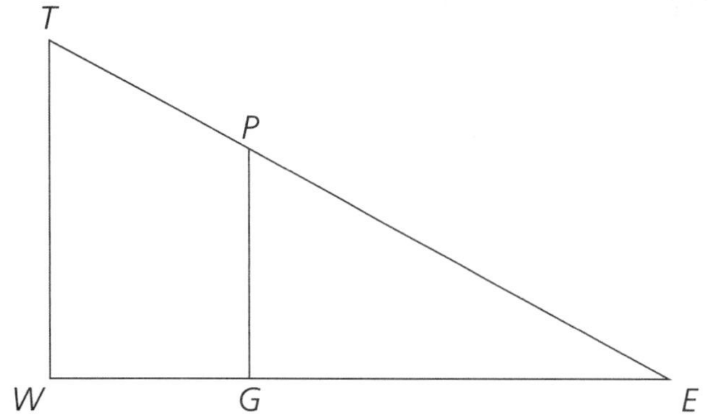

a. How many triangles do you see in this figure?

b. Name each triangle in the figure using the correct symbolic notation.

Assignment for Interior Angles of a Triangle

Name: ______________________________

1. Draw a triangle that has these characteristics: One side is 5 cm and another side is 7 cm.

   a. What is the length of the third side?

   b. Should everyone doing this problem get the same measure for the third side? Explain why or why not.

2. Draw a triangle that has these characteristics: One angle is 20° and another angle is 120°.

   a. What is the measurement of the third angle?

   b. Should everyone doing this problem get the same measure for the third angle? Explain why or why not.

3. Draw a triangle that has these characteristics: One angle is 35° and another angle is 80°.

   a. What are the lengths of the three sides of this triangle?

   b. What is the measurement of the third angle?

   c. Should everyone doing this problem get the same lengths of the sides? Explain why or why not.

   d. Should everyone doing this problem get the same measure for the third angle? Explain why or why not.

4. Draw a triangle that has these characteristics: One side is 4 cm and another side is 5 cm.

   a. What are the measures of the three angles of this triangle?

   b. Should everyone doing this problem get the same angle measures? Explain why or why not.

   c. Should everyone doing this problem get the same sum of the measures of the three interior angles? Explain.

   d. What is the measure of the third side of this triangle? Should everyone get the same measurement of the third side?

   e. Should everyone doing this problem get the same sum of the measures of the three sides? Explain.

Assignment for Classifying Triangles by the Measures of Their Angles

Name: ______________________________

1. Look for triangles around your house. Where did you find them? Name the type of each triangle: acute, obtuse, right, or equiangular.

Where you found the triangle Name of the item	Type of Triangle Classified according to its angles

Assignment for Classifying Triangles by the Lengths of Their Sides

Name: ______________________________

Use a ruler and a protractor to classify each triangle by the measure of its sides *and* by the measure of its angles. For example, a triangle that is *isosceles* (classified by its sides) and *right* (classified by its angles) is an *isosceles right* triangle. Explain each of your answers, using measurements from the triangle.

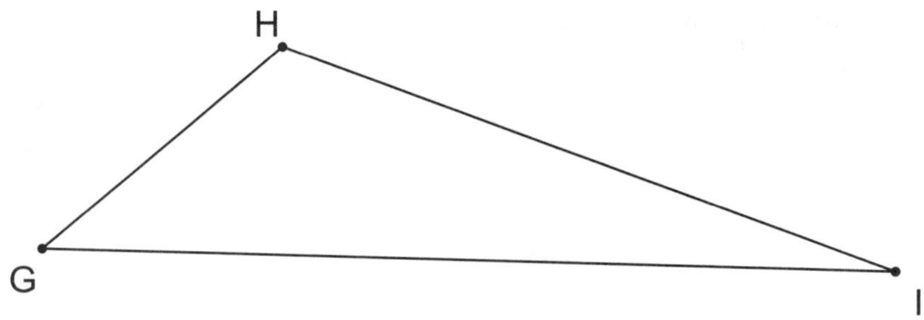

1. This is a ______________________ triangle because:

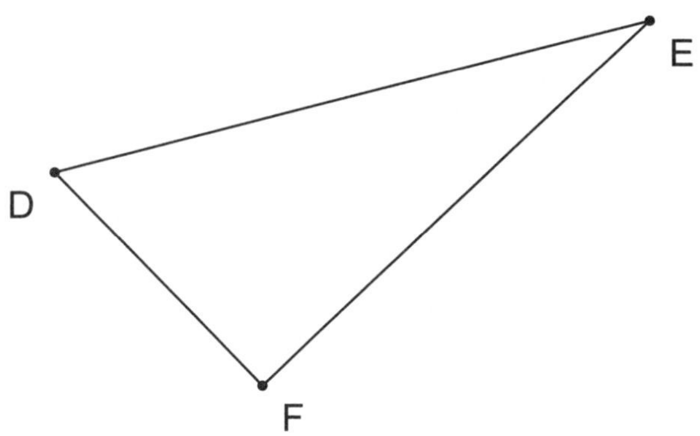

2. This is a ______________________ triangle because:

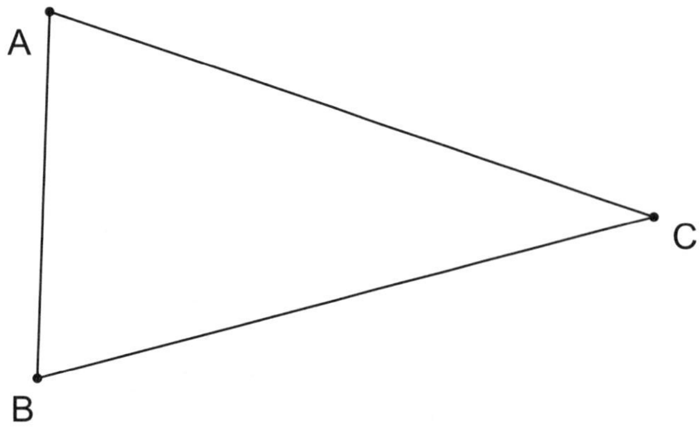

3. This is a _____________________ triangle because:

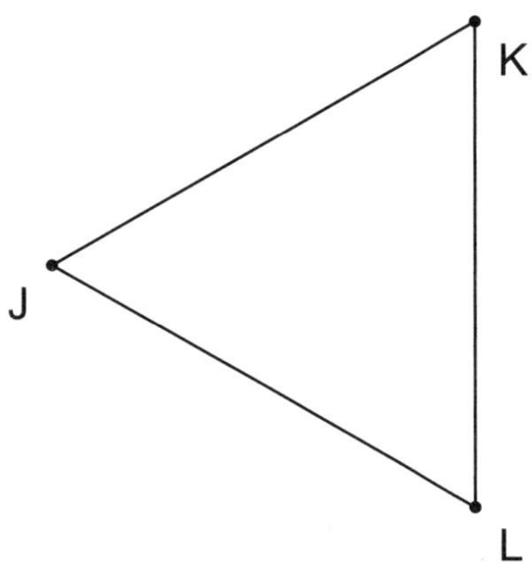

4. This is a ____________________ triangle because:

Assignment for the Triangle Inequality

Name: ______________________________

1. Draw ΔSHE that has these characteristics: $HE = 5$ cm, $SH = 6$ cm, and $SE = 11$ cm.

   a. Explain what you just did.

b. Why were you unable to create the triangle on the previous page?

2. Examine each set of three measurements below. Is it possible to create a triangle from the measurements?

   a. 26 units, 41 units, 4 units

   b. 15 cm, 76 cm, 86 cm

   c. 9.7 inches, 10 inches, 3.6 inches

   d. 7 m, 3 m, 24 m

Assignment for Exterior Angles

Name: ______________________________

Do not use a protractor to measure any angles in this assignment. The figures may not be drawn to scale. Rely only on the measurements given in each problem.

1. Find the measure of the exterior angle in this triangle. Explain how you found your answer.

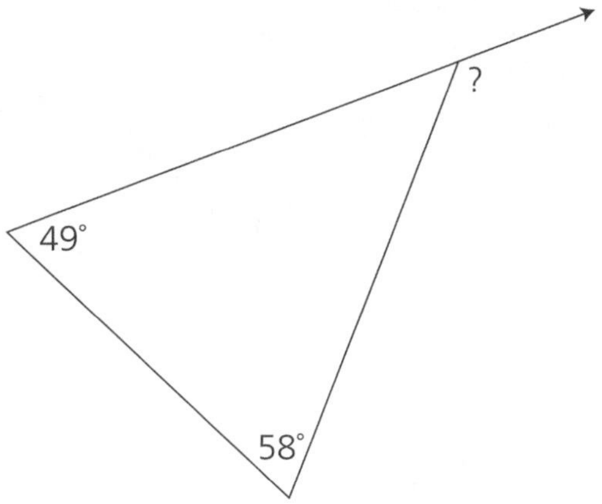

2. Find the measure of the exterior angle in this triangle. Explain how you found your answer.

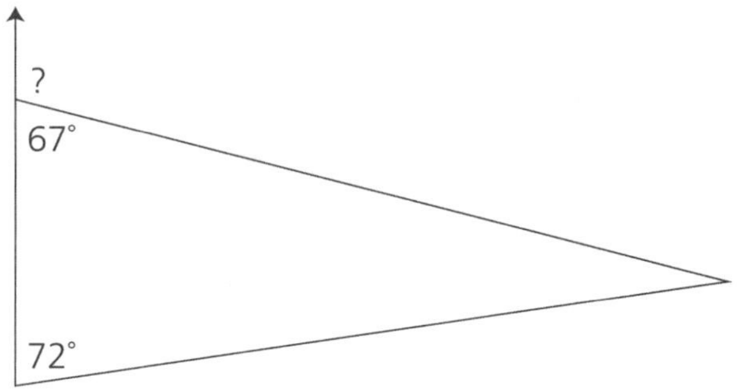

3. In ΔCUT, with exterior angle CTE, $m\angle C = 36°$, and $m\angle CTE = 71°$. Find $m\angle U$. Explain how you found your answer.

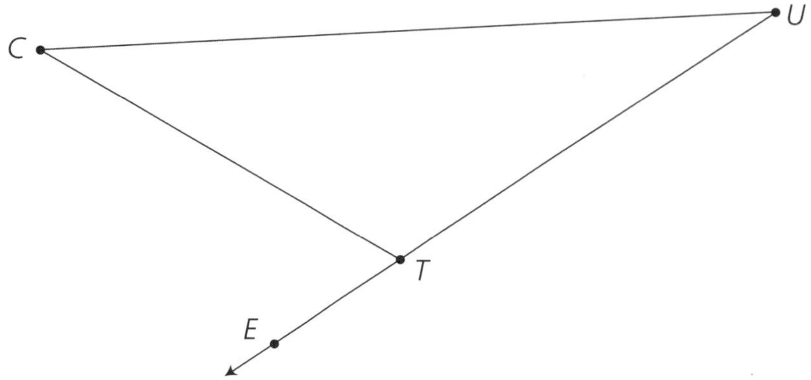

4. In the figure, $m\angle PBM = 35°$, $m\angle BPM = 21°$, and $m\angle PUM = 62°$. Find $m\angle PMU$. Explain how you found your answer.

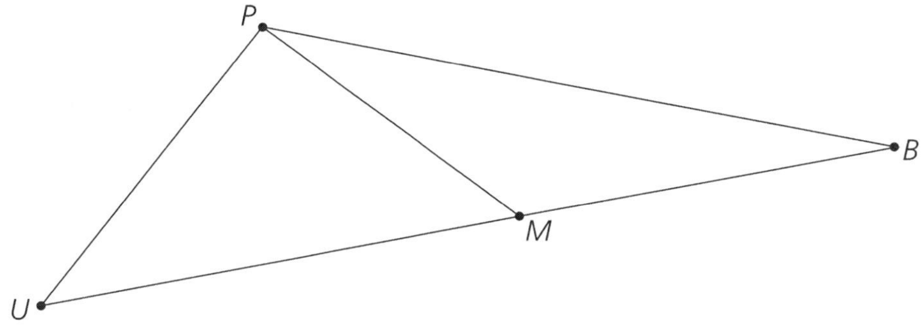

5. Find the sum of the measures of all the angles marked •. Explain how you found your answer.

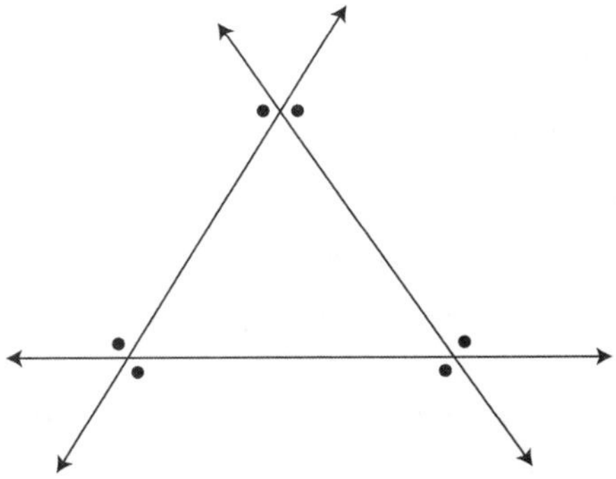

Assignment for Bisectors

Name: ______________________________

1. If you bisect an angle that measures 137°, what are the measures of the two newly formed angles? Explain how you found your answer.

2. Measure each angle in this triangle and write your results in the figure. Then draw three angle bisectors in the triangle, measure each newly formed angle, and write your results in the figure.

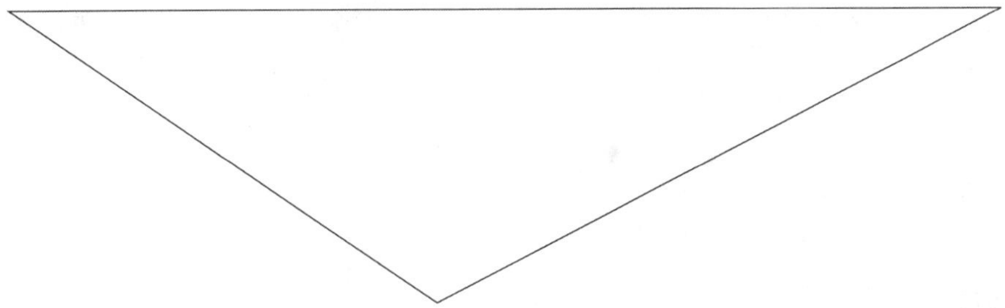

3. Can *all* angles be bisected? Why or why not?

4. If you bisect an angle that measures 56°, and then bisect each newly formed angle, how many angles do you create? What is the measure of each angle?

Now practice bisecting angles.

5. Draw an acute angle on the starter ray in the figure, and draw the angle bisector. Justify your results by measuring the angles with a protractor and recording the measures in the figure.

6. Draw an obtuse angle on the starter ray in the figure, and draw the angle bisector. Justify your results by measuring the angles with a protractor and recording the measures.

7. If you bisect a 15-cm line segment, what are the measures of the two newly formed segments? Explain how you found your answer.

8. Measure each side of the triangle below and write your results in the figure. Then draw three segment bisectors in the triangle, measure each newly formed segment, and write your results in the figure.

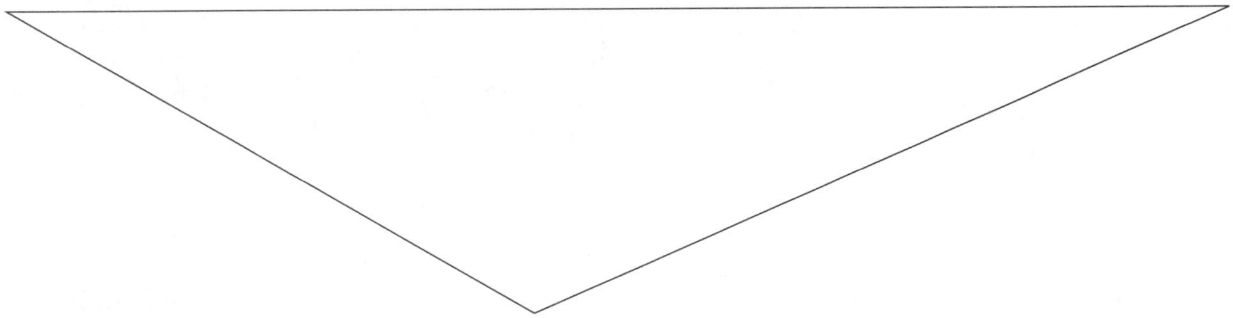

9. Can *all* segments be bisected? Why or why not?

10. If you bisect a 20-cm segment and then bisect each newly formed segment, how many segments do you create? What is the measure of each segment?

Now practice bisecting segments.

11. Draw an 8-cm segment on this starter ray and bisect it. Then bisect the resulting segments. How many segments do you create? What is the measure of each segment?

6. Draw a 6-inch segment on this starter ray and draw the segment bisector. Repeat this process two more times: Bisect the results of the first bisection, and then bisect the results of the second bisections. (If you are not constructing with a straightedge and compass, then just use your ruler to bisect the 6-inch segment three times.) How many segments do you create? What is the measure of each segment?

Assignment for Parts of a Quadrilateral

Name: ______________________________

1. List all of the things you know to be true about this quadrilateral. You may use a protractor and ruler.

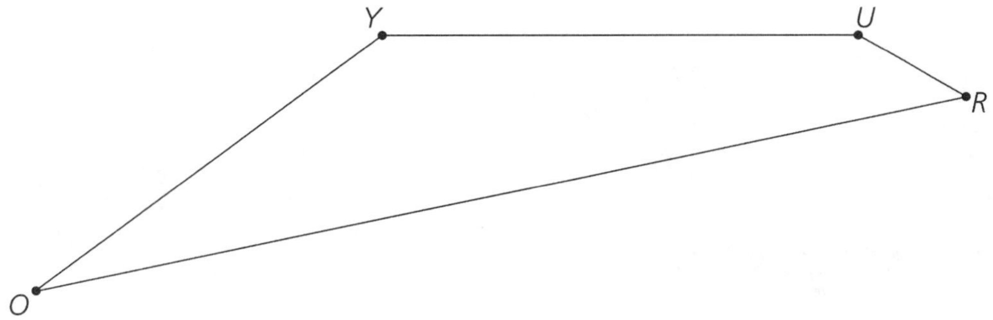

2. Draw quadrilateral *SHIM* that has these characteristics: $m\angle S = 122°$, $m\angle H = 79°$, and $m\angle I = 51°$.

   a. Find m∠M, SH, HI, IM, MS.

   b. List all of the things you know to be true about this quadrilateral.

   c. Do you think other students drew the same quadrilateral? Explain.

3. How many parts does a quadrilateral have?

Assignment for Special Quadrilaterals

Name: ______________________________

1. Draw and label a parallelogram that has these characteristics: Its area is 35 cm^2 and its base is 7 cm.

   a. Use your protractor to find the measure of each interior angle in your parallelogram.

   b. Do you think other students drew the same parallelogram? Explain.

2. Draw and label a rhombus that has these characteristics: Its area is 56 cm^2 and its height is 7 cm. (Remember that a rhombus is a parallelogram whose sides have equal measure.)

   a. Could the interior angles of this rhombus be right angles? Why or why not?

   b. Use your protractor to find the measure of each interior angle in your rhombus.

   c. Do you think other students drew the same rhombus? Explain.

3. Draw and label a quadrilateral whose area is 16 cm^2.

   a. Use your protractor to find the measure of each interior angle in this quadrilateral.

   b. Do you think other students drew the same quadrilateral? Explain.

4. A side of a rhombus measures 8 cm and one angle in the rhombus measures 110°. Draw the rhombus. (Remember that a rhombus is a parallelogram whose sides have equal measure.)

a. Do you think everyone drew the same rhombus? Explain.

Assignment for Parallelograms

Name: ______________________________

1. In parallelogram *GRAM*, $GR \parallel AM$ and $GM \parallel RA$.

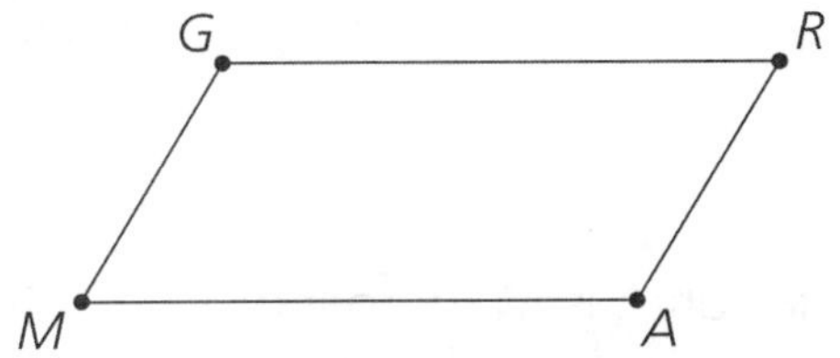

   a. If $m\angle G = 107°$, what is $m\angle A$? Justify your conclusion.

   b. If $m\angle R = 77°$, what is $m\angle G$? Justify your conclusion.

   c. If $GR = 14$ yd, what is the length of *MA*? Justify your conclusion.

2. If the measure of one angle of a parallelogram is 57°, find the measures of the other angles of the parallelogram.

3. The measures of two consecutive angles of a parallelogram are given by the expressions $(m + 46)$ and $(3m - 90)$. Find the measure of each angle of the parallelogram in degrees. Show your work.

4. The measures of two opposite angles of a parallelogram are given by the expressions $(2k + 24)$ and $(8k - 30)$. Find the measure of each angle of that parallelogram in degrees. Show your work.

5. In parallelogram *YARD*, *YA* || *RD*, *YD* || *AR*, and *YR* is a diagonal. Is ΔYAR congruent to ΔRDY? Justify your conclusion.

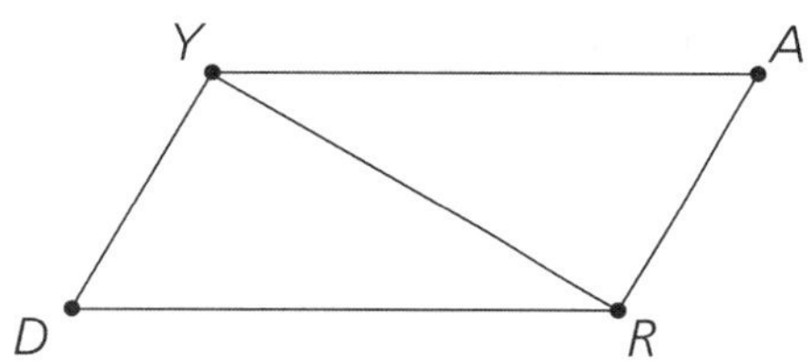

6. In parallelogram *YARD*, *YA* || *RD*, *YD* || *AR*, and the diagonals *YR* and *AD* intersect at point *S*.

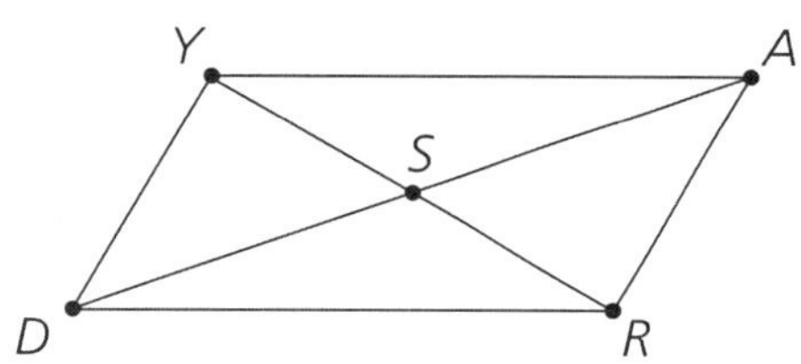

a. If *YS* = 32 mm, what other segment measures do you know in this diagram?

b. If *DS* = 43 mm, what other segment measures do you know in this diagram?

c. If segment *YR* bisects segment *AD*, what does that mean?

d. If segment *AD* bisects segment *YR*, what does that mean?

Assignment for Rhombi

Name: ____________________

1. In rhombus $MBUS$, $MB \parallel US$ and $MS \parallel UB$.

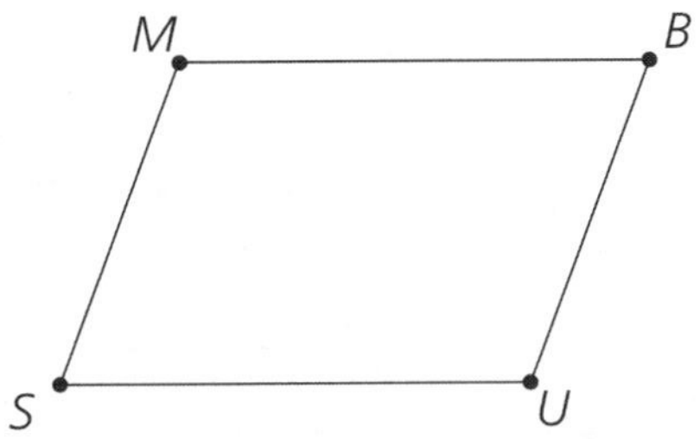

   a. If $m\angle B = 33°$, what is $m\angle U$? Justify your conclusion.

   b. If $m\angle U = 117°$, what is $m\angle M$? Justify your conclusion.

   c. If $MS = 121$ mm, what is the length of US? Justify your conclusion.

2. If the measure of one angle of a rhombus is 41°, find the measures of the other angles of the rhombus.

3. The measures of two consecutive angles of a rhombus are given by the expressions $(2m + 22)$ and $(4m - 60)$. Find the measure of each angle of the rhombus in degrees. Show your work.

4. The measures of two opposite angles of a rhombus are given by the expressions $(5k + 35)$ and $(25k - 65)$. Find the measure of each angle of the rhombus in degrees. Show your work.

5. In rhombus *MBUS*, $MB \parallel US$, $MS \parallel UB$, and *SB* is a diagonal. Is ΔBUS congruent to ΔSMB? Justify your conclusion.

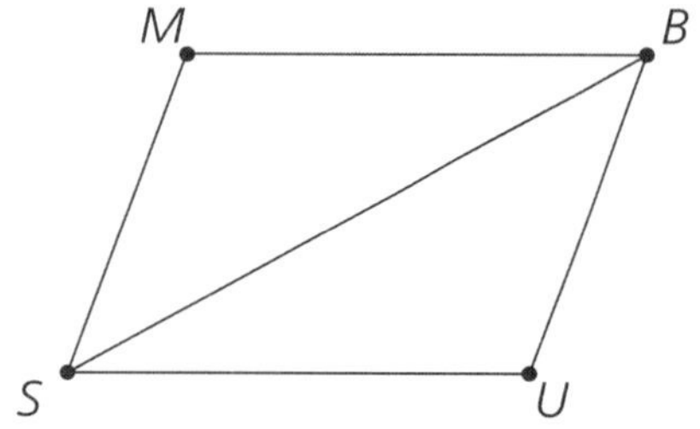

6. In rhombus *MBUS*, $MB \parallel US$, $MS \parallel UB$, and diagonals *MU* and *BS* intersect at point *O*.

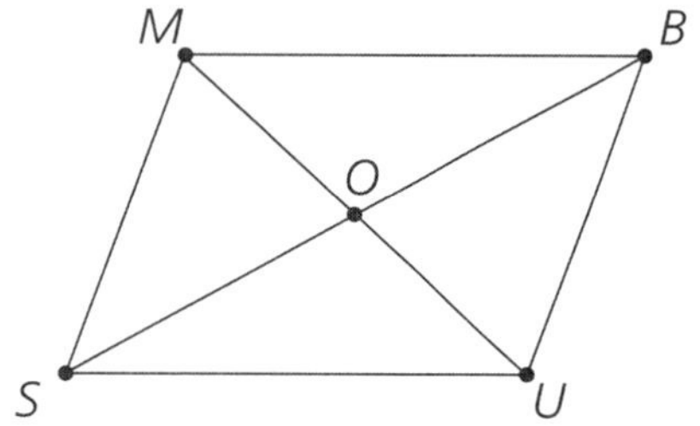

a. If $MU = 55$ mm, what other segment measures do you know in this diagram?

b. If $BO = 28$ mm, what other segment measures do you know in this diagram?

Use this figure for problems 7–9.

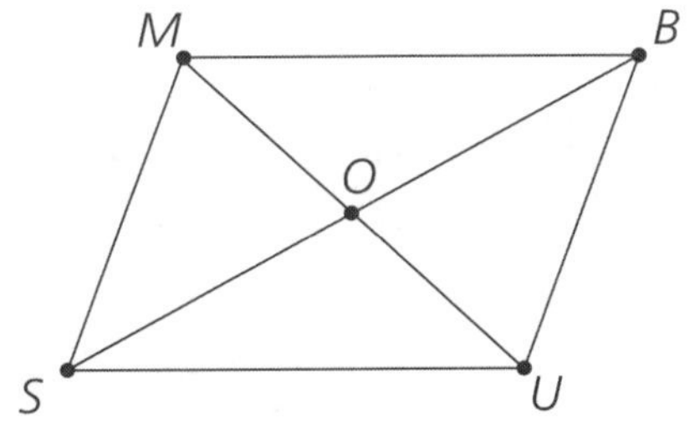

7. *The diagonals of any rhombus are perpendicular to each other.* Using *only* this fact, what can you conclude about rhombus *MBUS*?

8. *The diagonals of any rhombus bisect the angles of the rhombus.* Using *only* this fact, what can you conclude about rhombus *MBUS*?

9. If $m\angle BUM = 29°$ and $UB = 17$ mm, list everything you can conclude about rhombus *MBUS*.

Assignment for Rectangles

Name: ______________________________

1. In rectangle *RECT*, *RE* || *CT*, *RT* || *EC*, and *RC* is a diagonal.

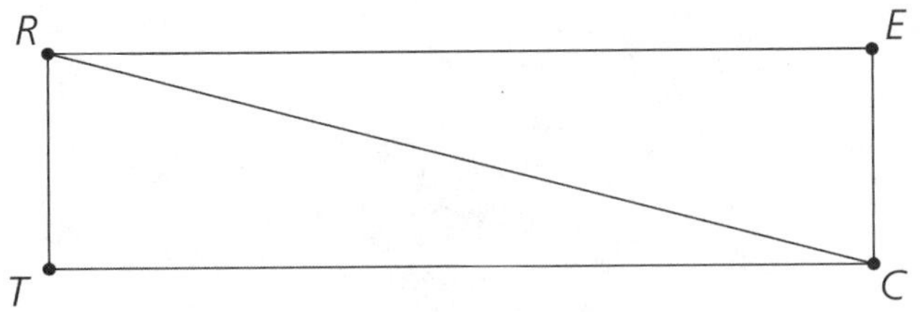

   a. Is ΔREC congruent to ΔCTR? Justify your conclusion.

   b. Is $\angle ERC$ congruent to $\angle TCR$? Justify your conclusion.

   c. Is $\angle ECR$ congruent to $\angle TRC$? Justify your conclusion.

2. In rectangle *RECT*, *RE* || *CT*, *RT* || *EC*, *RC* and *TE* are diagonals, and point *A* is the intersection of the diagonals.

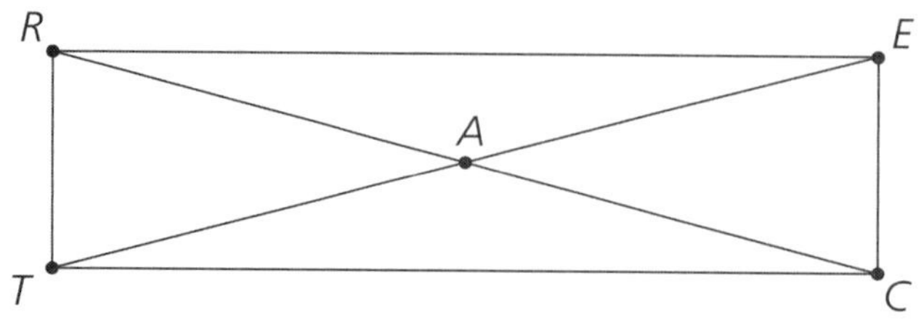

a. Does *RC* = *TE*? Justify your conclusion.

b. If *TR* = 6 ft and *TC* = 8 ft, find the length of *TE*. Show your work.

c. Name all pairs of congruent triangles you can find in rectangle *RECT*. Sketch each pair separately.

3. *WXYZ* is a rectangle with two diagonals intersecting at point *A*. If *WA* = $(5x - 14)$ ft and *AY* = $(3x + 22)$ ft, find the length of *XZ*.

Assignment for Squares

Name: ________________

1. Segment SU and segment QR bisect each other and are congruent to each other. Must quadrilateral SQUR be a square? Justify your conclusion.

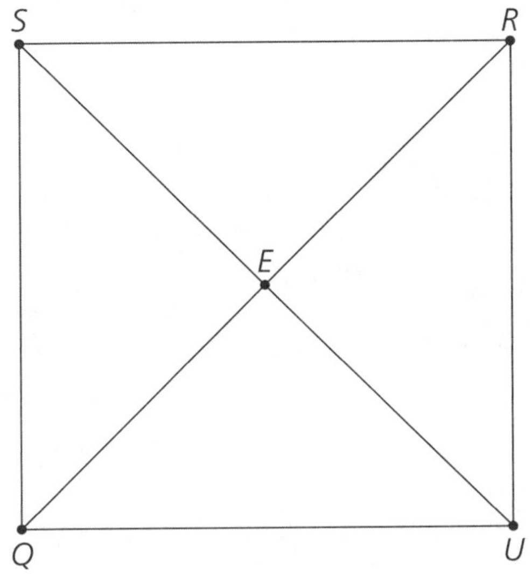

2. Segment SU bisects segment QR and segment SU is perpendicular to segment QR. Must quadrilateral SQUR be a square? Justify your conclusion.

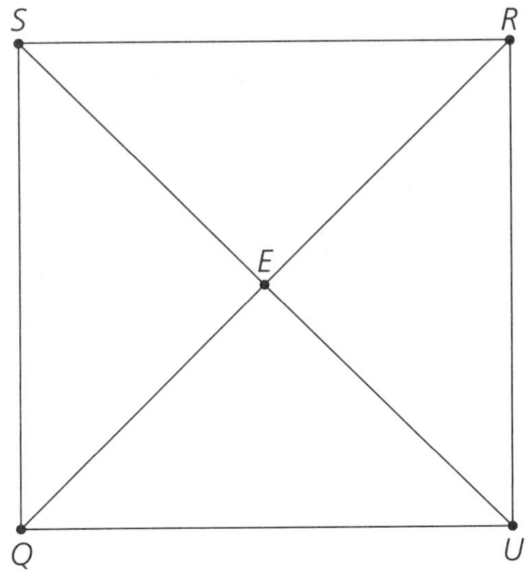

3. Segment SU bisects segment QR, SU bisects angles QSR and QUR, and QR bisects angles SQR and SRU. Must quadrilateral SQUR be a square? Justify your conclusion.

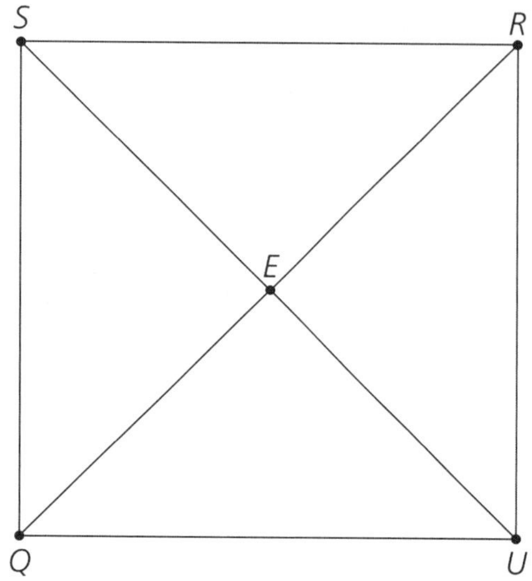

Assignment for Quadrilateral Relationships

Name: ______________________________

1. A square is an equiangular rhombus. Prove or disprove this statement. Make a sketch to illustrate your conclusion.

2. A square is an equilateral rectangle. Prove or disprove this statement. Make a sketch to illustrate your conclusion.

3. List all of the properties of a quadrilateral.

4. List all of the properties of a trapezoid.

5. List all of the properties of a parallelogram.

6. List all of the properties of a rectangle.

7. List all of the properties of a rhombus.

8. List all of the properties of a square.

9. Draw a concept map, flow chart, or picture of the relationships between these geometric figures: quadrilateral, parallelogram, trapezoid, rectangle, rhombus, and square. (Hint: You could think of this as a family tree whose members share the last name "quadrilateral.")

Assignment for Sum of the Interior Angle Measures in a Polygon

Name: ______________________________

1. Find the sum of the interior angles in a nonagon. Show your work.

2. Assume that this nonagon is a *regular* nonagon (all sides equal; all angles equal). Find the measure of *each* interior angle in the nonagon. Show your work.

3. Find the sum of the interior angles of a 50-gon. Show your work.

4. Find the measure of each interior angle in a regular pentagon.

5. Find the measure of each interior angle in an equiangular hexagon.

6. In your own words, explain how to find the sum of the interior angles in any polygon.

7. In your own words, explain how to find the measure of each interior angle in a regular polygon.

8. Given a regular polygon with n sides, create a formula for determining the measure of each interior angle.

Assignment for Sum of the Exterior Angle Measures in a Polygon

Name: ______________________________

In class, you used one method to find the sum of the exterior angles in any polygon. In this assignment, you will use a slightly different method. Show any and all work you do.

1. What is the sum of the *interior* angles in a triangle?

2. If the triangle is equiangular, what is the measure of each *interior* angle?

3. Still assuming the triangle is equiangular, what is the measure of each *exterior* angle?

4. How many exterior angles are associated with a triangle, if one is created at each vertex?

5. What is the sum of the measures of the exterior angles, if one is created at each vertex?

6. What is the sum of the *interior* angles in a quadrilateral?

7. If the quadrilateral is a square, what is the measure of each *interior* angle?

8. Still assuming the quadrilateral is a square, what is the measure of each *exterior* angle?

9. How many exterior angles are associated with a quadrilateral, if one is created at each vertex?

10. What is the sum of the measures of the exterior angles, if one is created at each vertex?

11. What is the sum of the *interior* angles in a pentagon?

12. If the pentagon is regular, what is the measure of each *interior* angle?

13. Still assuming the pentagon is regular, what is the measure of each *exterior* angle?

14. How many exterior angles are associated with a pentagon, if one is created at each vertex?

15. What is the sum of the measures of the exterior angles, if one is created at each vertex?

16. Generalize your findings about the sum of the exterior angles in any polygon.

17. If the polygon is a regular polygon, explain *two* different ways to find the measure of each exterior angle.

Assignment for Introduction to Transformations

Name: ______________________________

Take an 8 ½ by 11 piece of paper. Fold the paper along the diagonal. Cut along the fold. (Note that the two parts of the paper may not coincide at this point.)

1. Use your two triangles to create one larger triangle. Sketch your figure below, showing each of the smaller individual triangles. What transformation could have been performed to one triangle to create the larger triangle?

    a. Can you use your two smaller triangles to create a different large triangle? If so, sketch it below. If not, explain why.

2. Use your triangles to create a parallelogram that is not a rectangle. Sketch the figure showing the two triangles you used. What transformation could have been performed to one triangle to create the parallelogram?

    a. Can you use your two smaller triangles to create a different parallelogram? If so, sketch it below. If not, explain why.

3. Use your triangles to create a kite. Sketch the figure showing the two triangles you used. What transformation could have been performed to one triangle to create the kite?

   a. Recreate your original rectangle. Sketch it below, showing the two triangles. What transformation could have been performed to one triangle to create the rectangle?

Assignment for Reflections

Name: ______________________________

In this activity, you will determine where a mirror has been placed in the starting shape in order to give a particular reflection. Draw the mirror in the starting shape with a dark line segment.

Starting Shape	Preimage+Image
1. 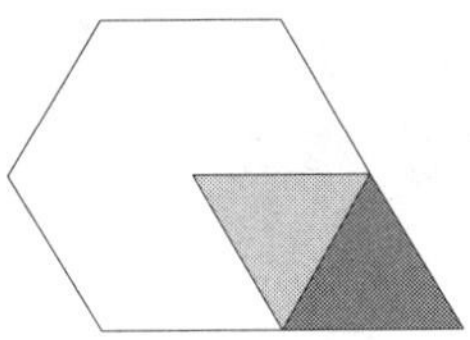	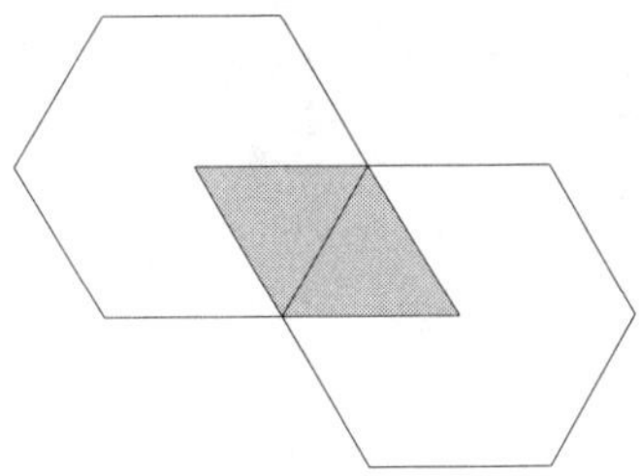
2. 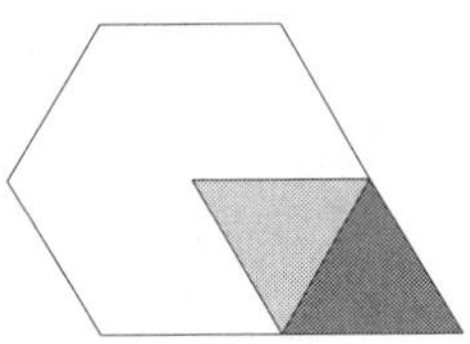	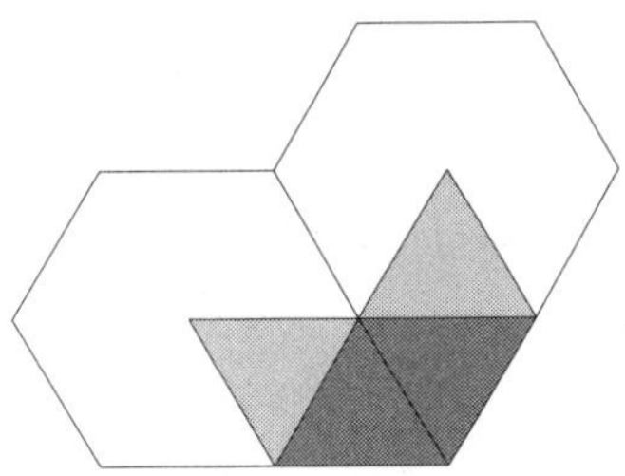
3. 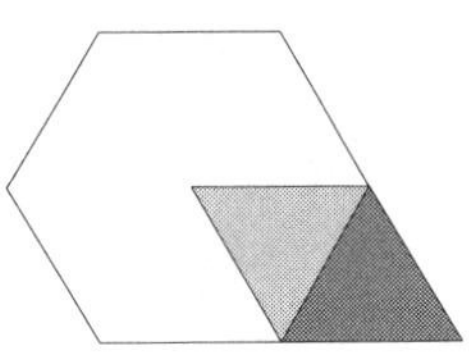	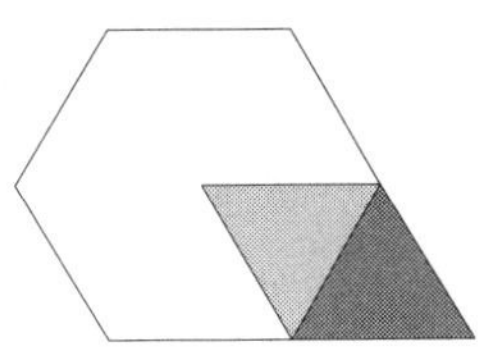
4. 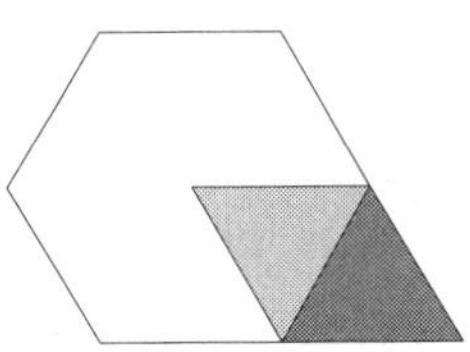	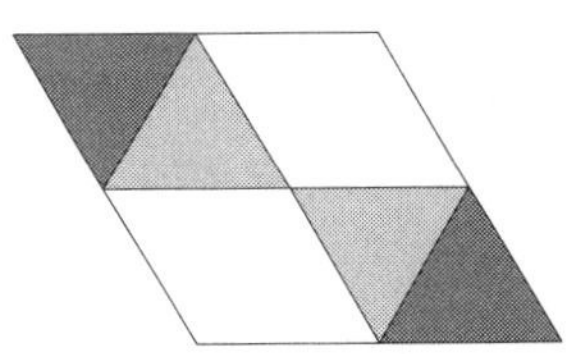

5.

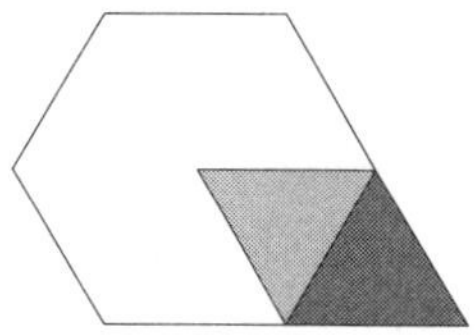

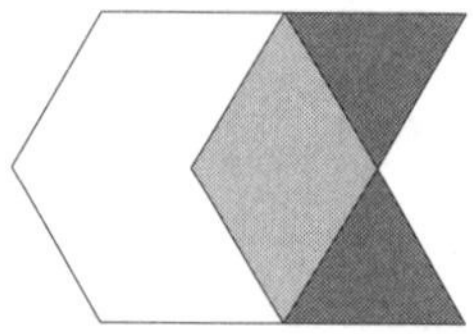

6.

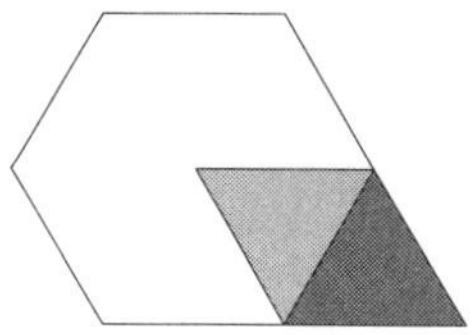

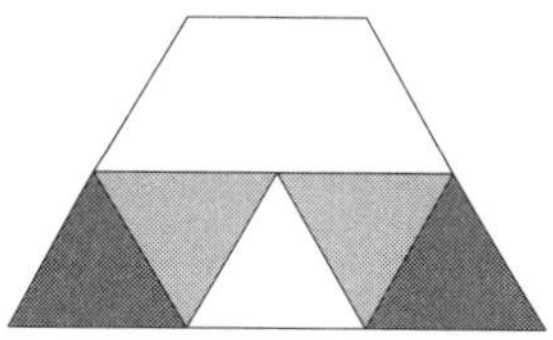

Assignment for Reflections Using a Ruler and Protractor

Name: ______________________________

Stationery and Web Pages

Back at the Acme technology company, the head of the marketing department is now experimenting with reflections to create some interesting images for the company's new brochure. She is using the company stationery as a starting place. Below, you see copies of the stationery and the placement of the mirror or line of reflection. In the space to the right, draw the figure made by the pre-image plus image. (Look into the mirror in the direction of the arrow.)

1.

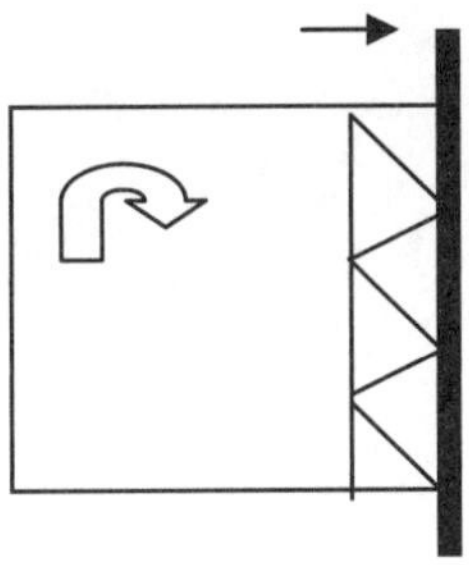

2.

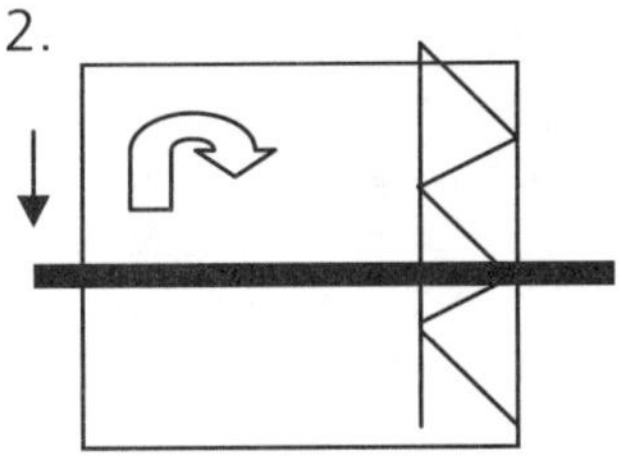

3.

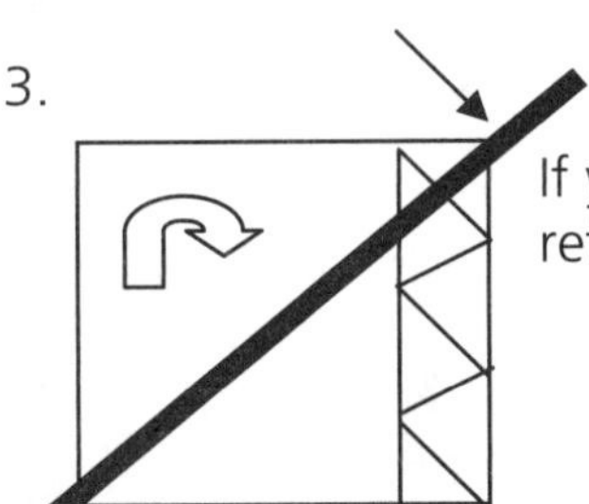

If you have some trouble with 3 and 4, use some of the properties of reflections that you discovered in the problem with your name reflection.

4.

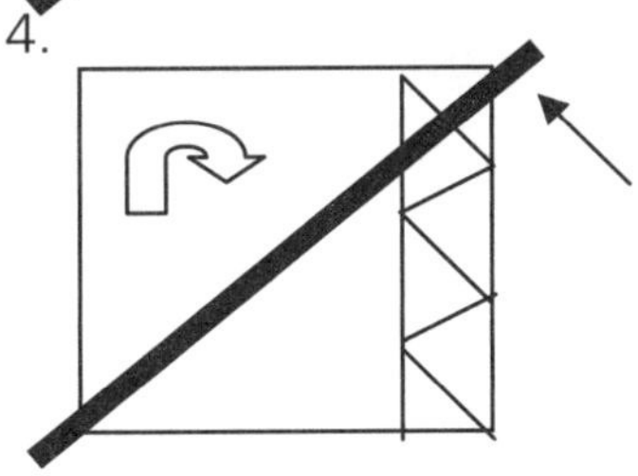

5. When can reflections be useful in the world outside your mathematics classroom?

Assignment for Reflections in the Coordinate Plane: Part 1

Name: ______________________________

1. In the following diagrams, label the vertices and give the coordinates for the vertices of the figures. Then, reflect the polygon over the y-axis. Lastly, label the new vertices and provide the coordinates of the new vertices

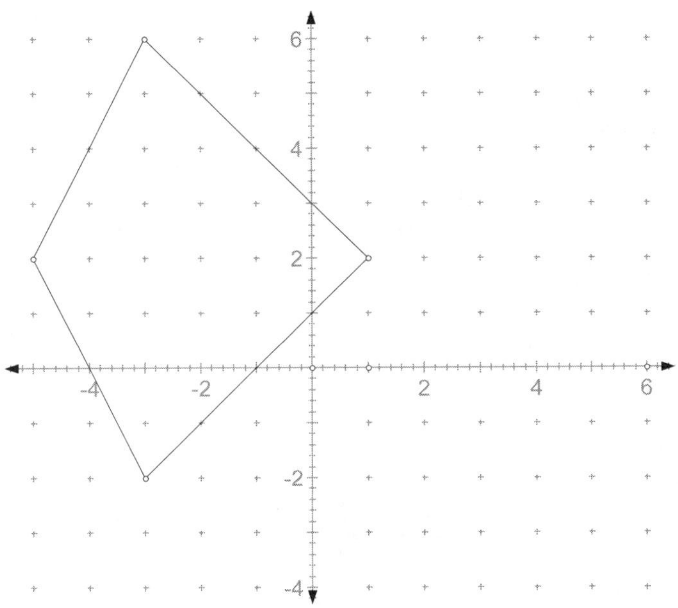

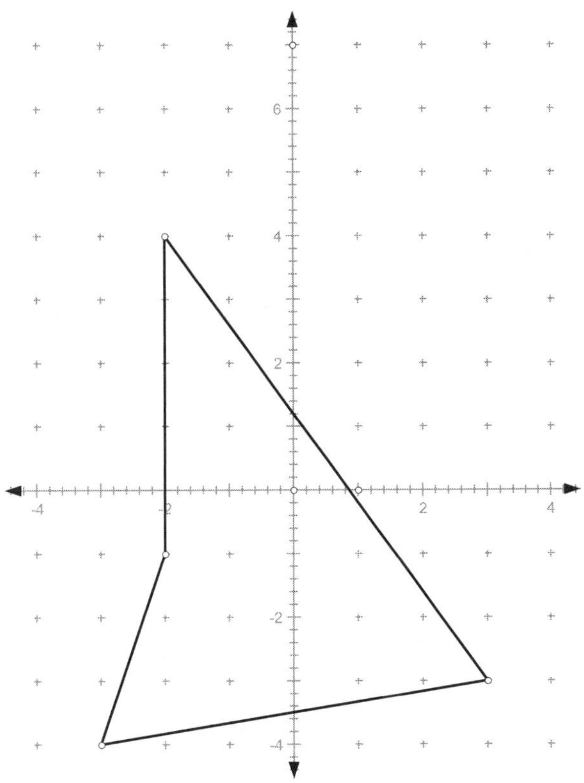

2. Using the coordinate grid provided below:
   - Plot the set of points
   - Connect the points to form a quadrilateral
   - Reflect the points over the y-axis

(-3,-3), (1,5), (7,1.5), (4.5, -6)

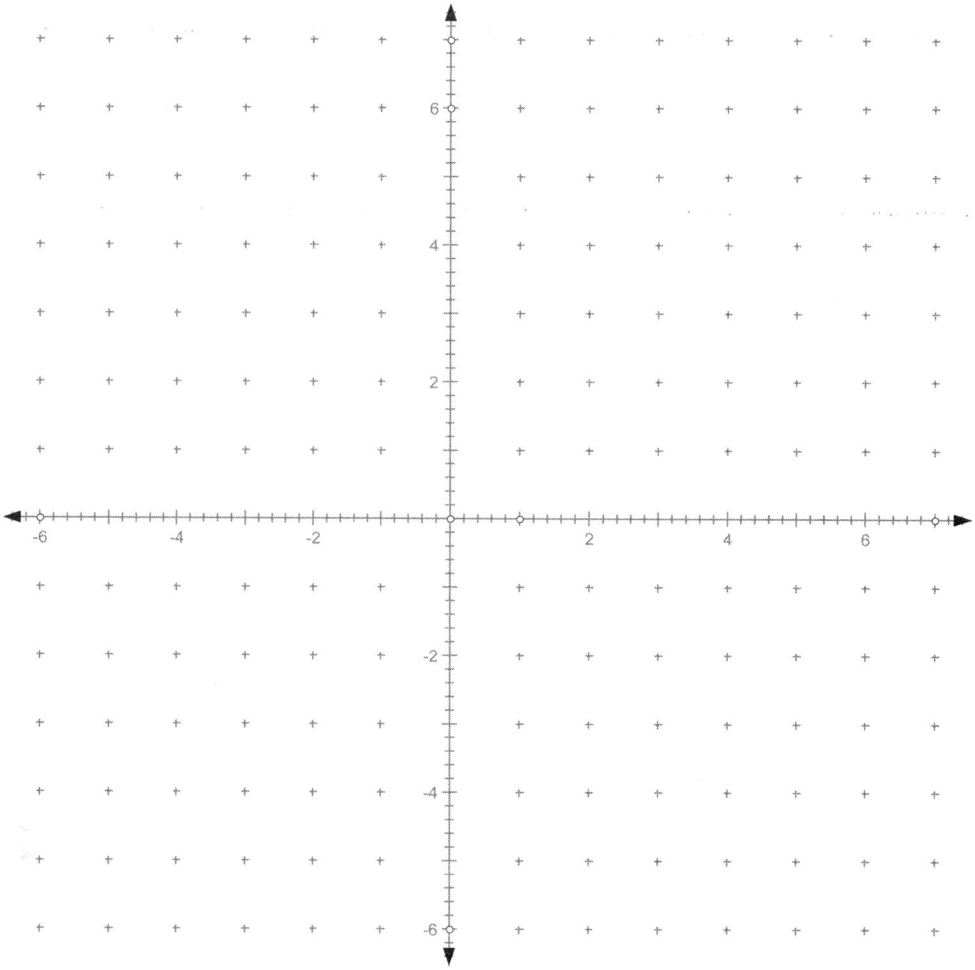

Assignment for Reflections in the Coordinate Plane: Part 2

Name: ____

1. Image points that are reflected over the x-axis have special relationships to the pre-image points. In the following diagrams, label the vertices and give the coordinates for the vertices of the figures. Then, reflect the polygon over the x-axis. Lastly, label the new vertices and provide the coordinates of the new vertices.

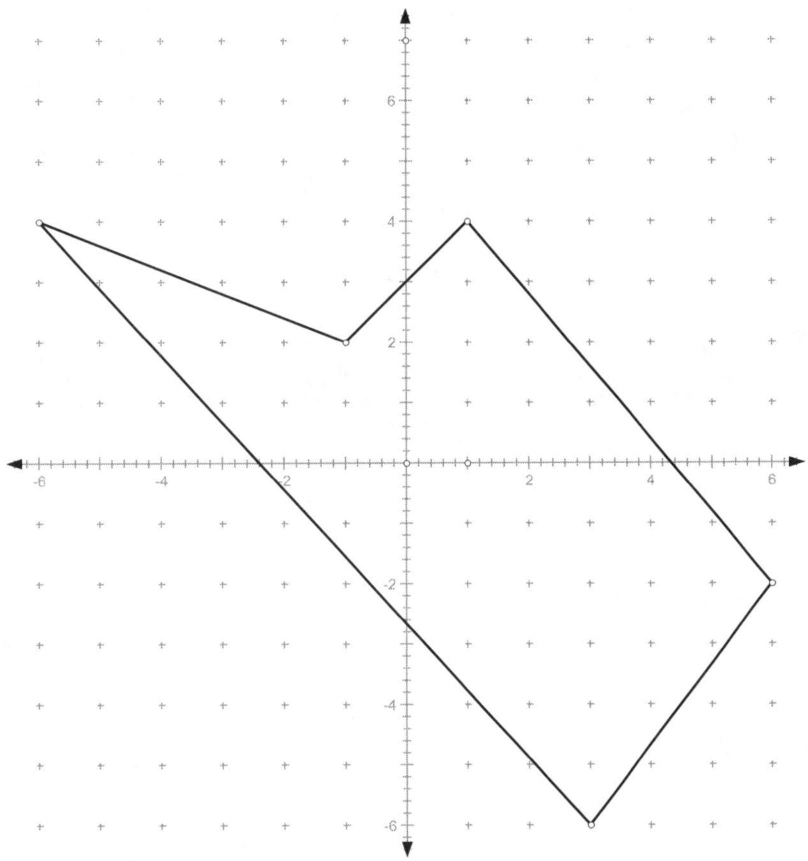

2. Using the coordinate grid provided below:
    - Plot each set of points
    - Connect the points to form a quadrilateral
    - Reflect the points over the x-axis

(5,2.5), (1.5,2), (-2,5), (-1,-6)

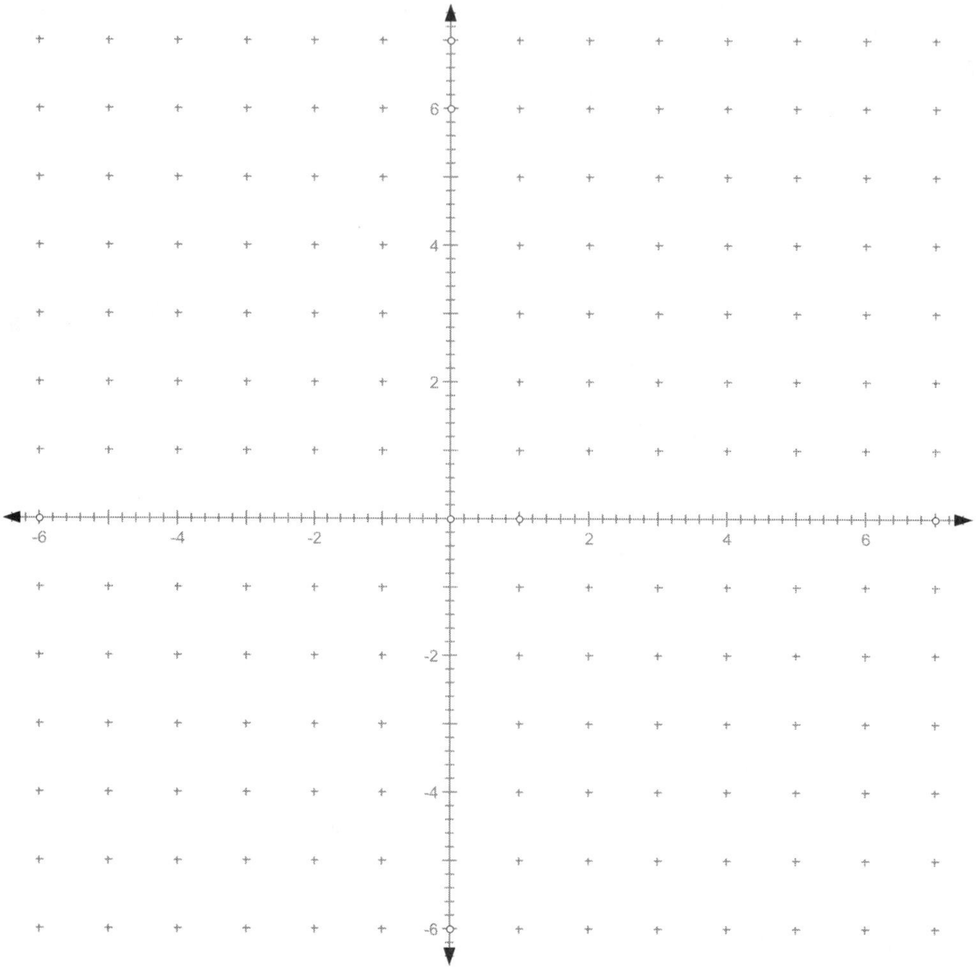

3. You have seen a lot of examples of reflections over the axes and over the line y = x. You have even formed some generalizations for these reflections. Write a convincing argument for why each of the generalizations or rules you discovered so far in class is true.

Assignment for Reflections in the Coordinate Plane: Part 3

Name: ____

1. Image points that are reflected over the line y = x have special relationships to the pre-image points. In the following diagrams, label the vertices and give the coordinates for the vertices of the figures. Then, reflect the polygon over the line y = x. Lastly, label the new vertices and provide the coordinates of the new vertices

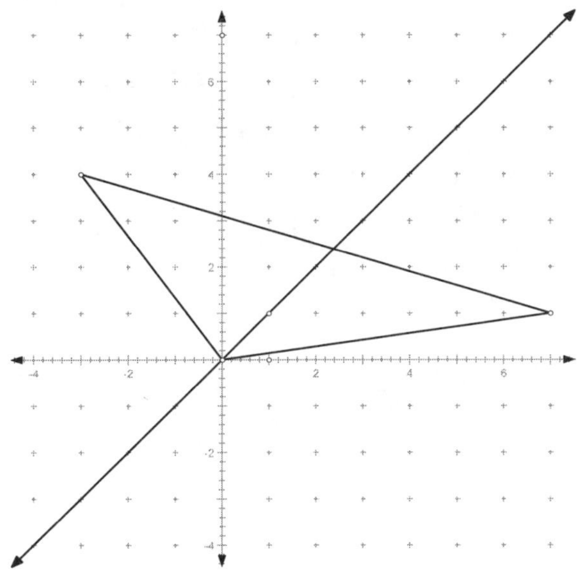

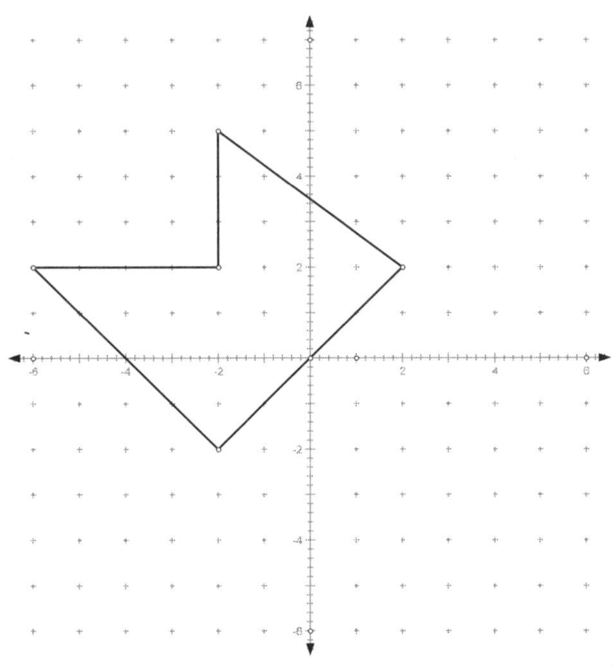

2. Using the coordinate grid provided below:
   - Plot each set of points
   - Connect the points to form a triangle
   - Reflect the points over y = x

(4,4), (0,0.5), (-4,5)

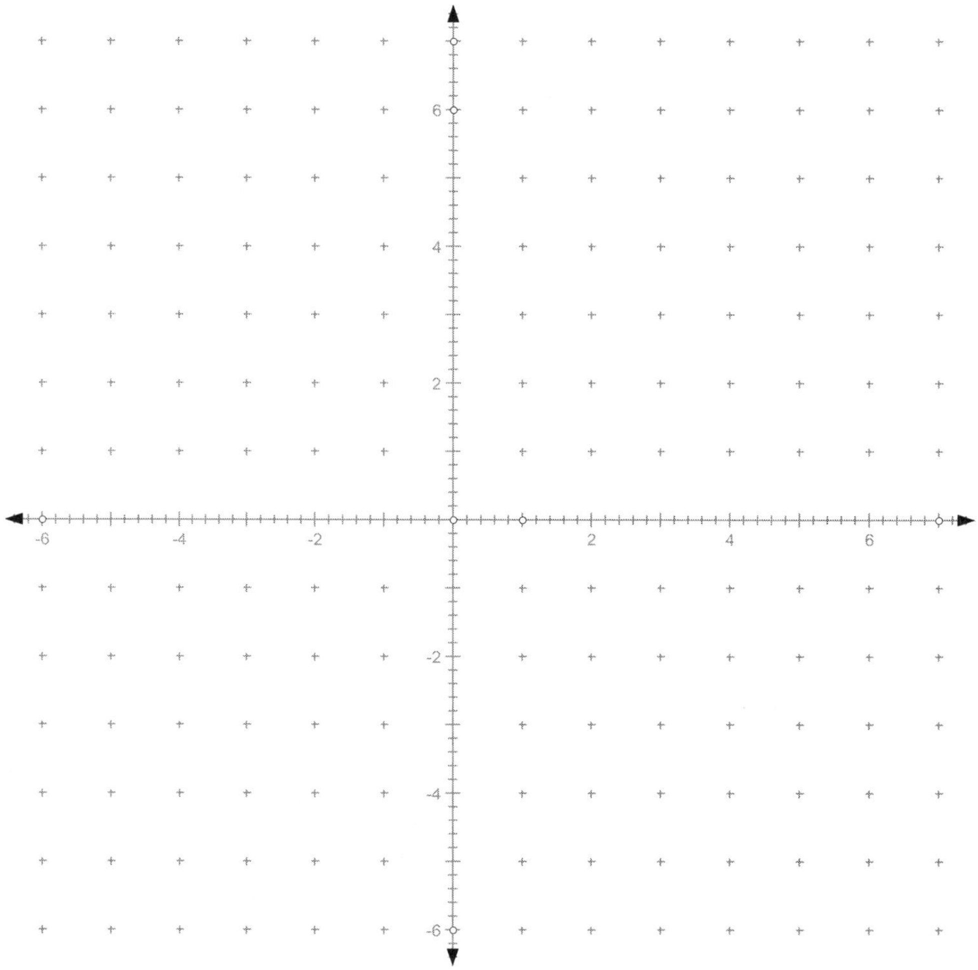

3. Design an activity like the ones you did in class in order to discover the patterns in reflections over the line **y = -x.**

    a. Use the coordinate plane provided below.

    b. Create a polygon and give the coordinates of its vertices.

    c. Graph the line, y = -x.

    d. Reflect your polygon.

    e. Provide the mapping of your pre-image points to your image points, and look for patterns.

    f. Repeat this process with a polygon that has points in other quadrants.

    g. Repeat again with a polygon that intersects the line y = -x.

    h. Create a general rule that you think will always work for reflections over the line y = -x, and write it in complete sentences. (Room on back)

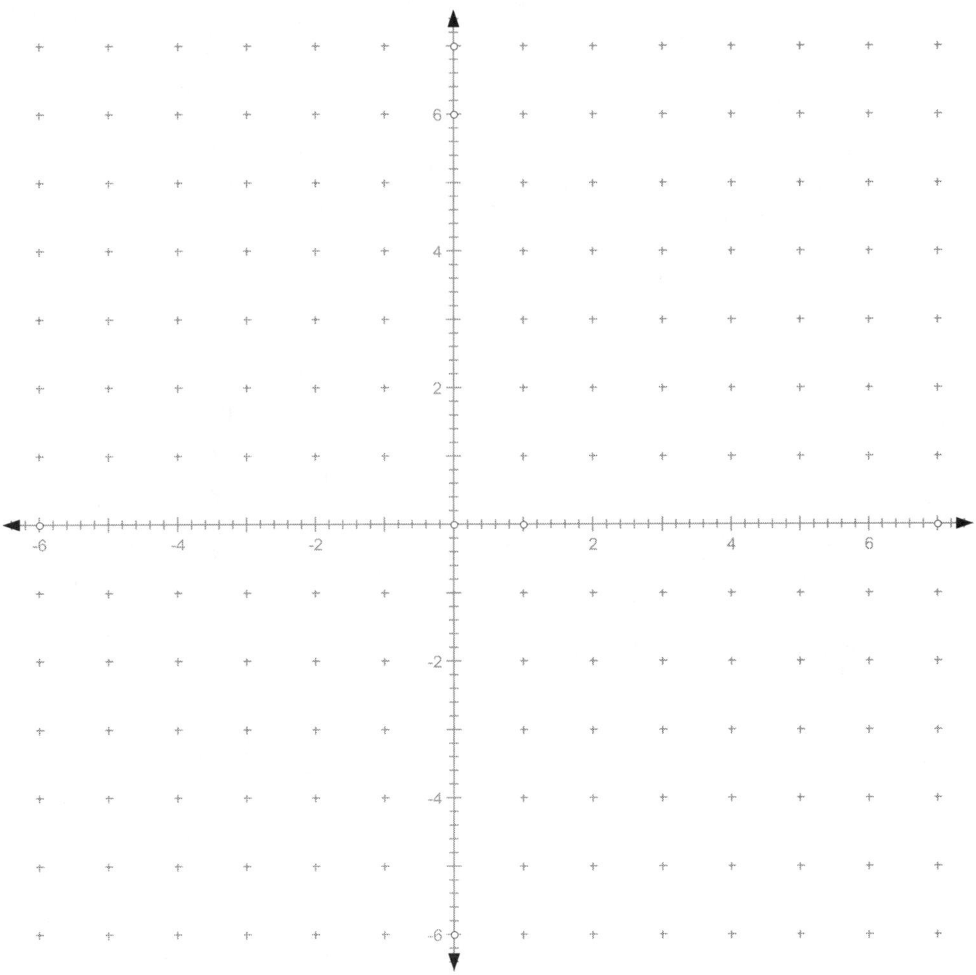

4. What are the coordinates of a point (x,y) under a reflection over the line y = -2?

5. What are the coordinates of a point (x,y) under a reflection over the line x = 6?

Assignment for Translations

Name: ____

Earlier, you learned that translations could always be viewed as double reflections. Translations can also be done directly, if you know the direction and the distance to move an object. In this assignment, you will attempt to relate the two.

The translation of the parallelogram on the coordinate grid is defined by: **T**(x,y) = (x - 6, y + 12).

1. Perform this translation.
2. Find the two parallel lines over which the parallelogram must be reflected in order to move the parallelogram to the same position. *Hint: These lines are **not** horizontal or vertical.*

You may need to use some additional graph paper for your first couple tries at this exercise.

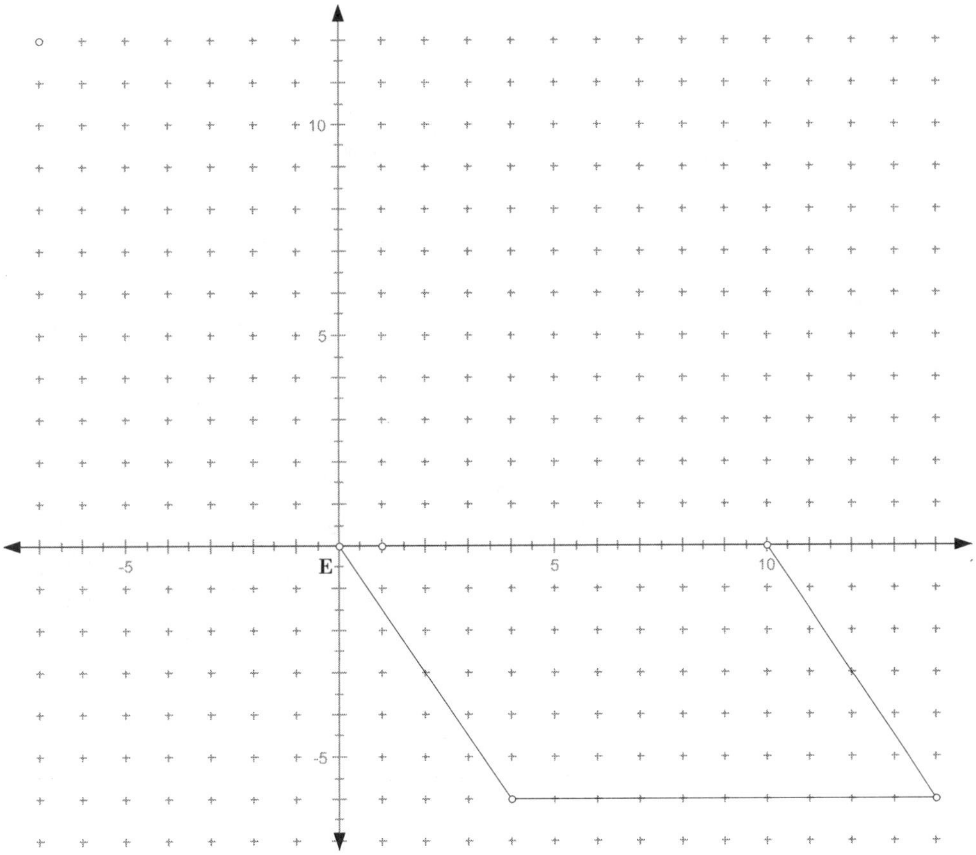

Assignment for Rotations

Name: ______________________________

Apply the procedure you used in class today to rotate the following figures using a compass and protractor. The center of rotation and the magnitude of the rotation is specified for you.

1. Rotate by 75° around point H.

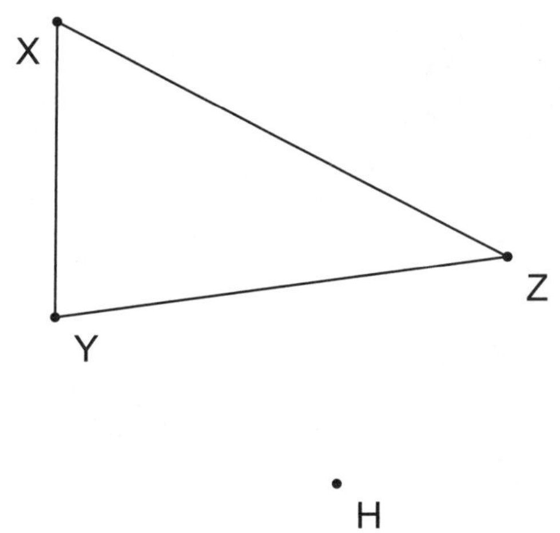

2. Rotate by -135° around point D.

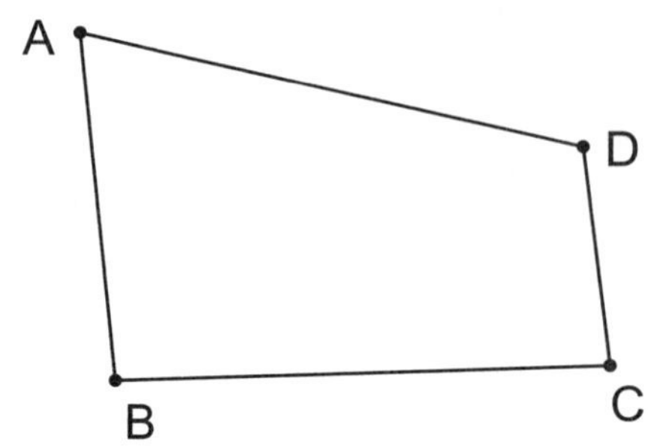

Assignment for Rotations in the Coordinate Plane

Name: ______________________________

In class today, you performed an activity similar to the one you will do for homework. However, you used different figures and rotations of different degrees.

Part 1:

1. Label the vertices of the triangle and give the coordinates of each of the vertices on the coordinate grid.
2. Rotate the figure by 90° around the origin.

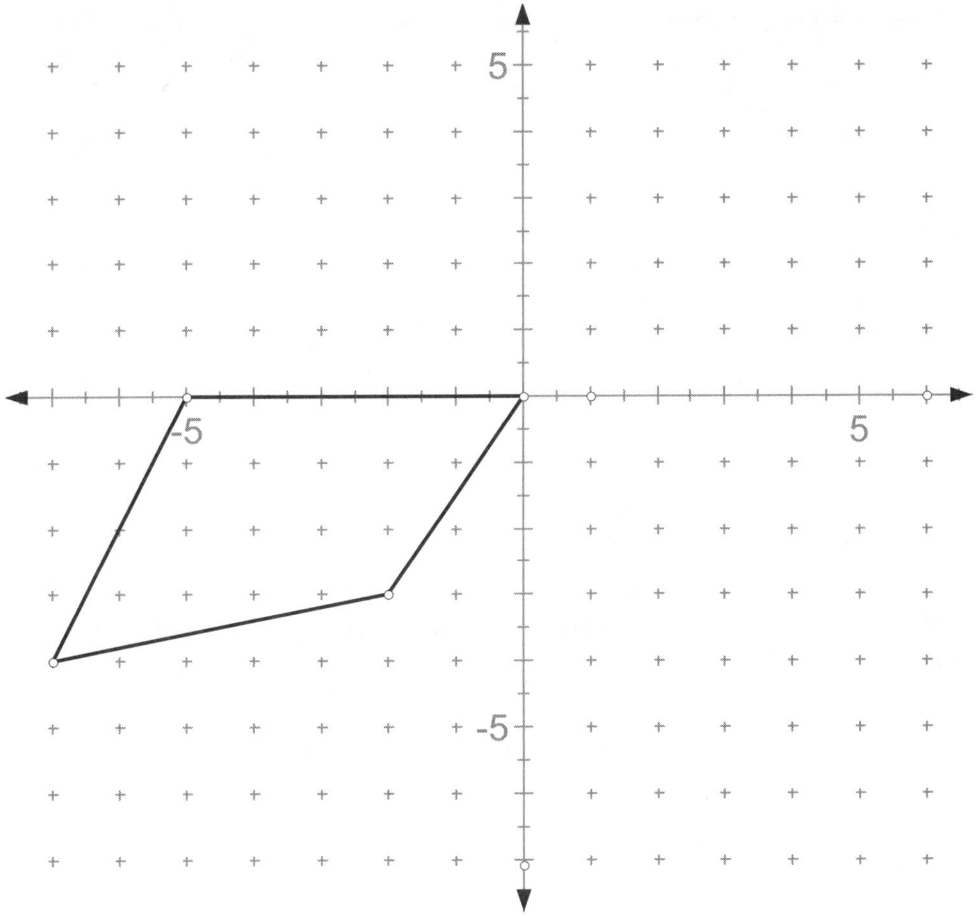

3. Label the vertices of the rotated figure and give the coordinates of the vertices on the coordinate grid.
4. Use the mapping notation [(x,y)→] to describe where each individual original point goes under the rotation.

5. Use the same figure, but this time rotate the figure by -180° around the origin. If you use only positive numbers, what is this measure?

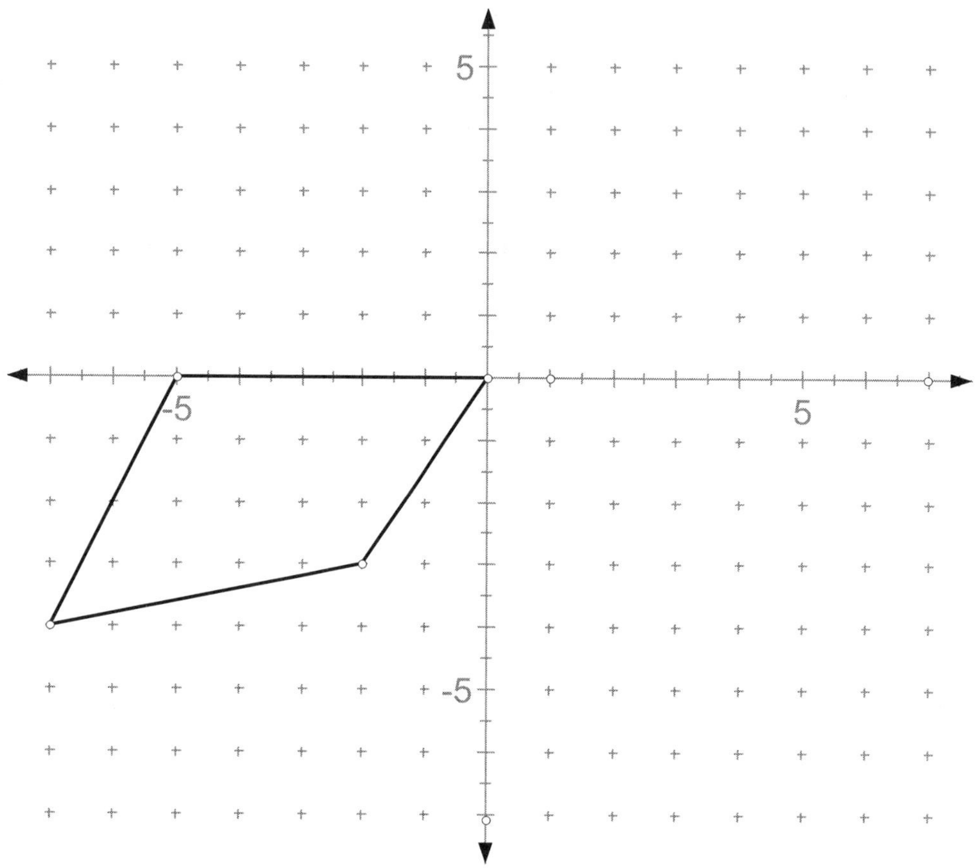

6. Label the vertices of the rotated figure and give the coordinates of the vertices on the coordinate grid.
7. Use the mapping notation [(x,y)→] to describe where each of the individual original points goes under the rotation.

8. Use the same figure one more time, but this time rotate the figure by 270° around the origin.

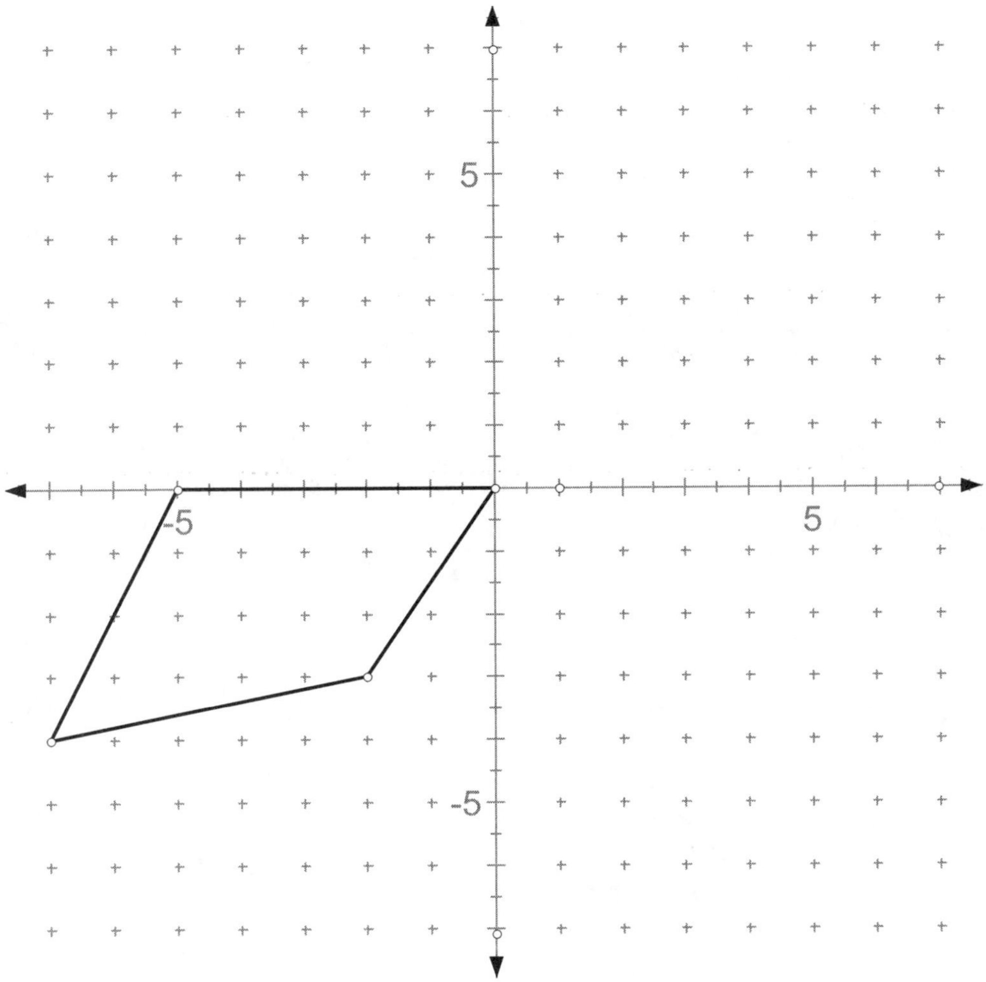

9. Label the vertices of the rotated figure and give the coordinates of the vertices on the coordinate grid.
10. Use the mapping notation [(x,y)→] to describe where each of the individual original points goes under the rotation.

Part 2:

1. Give the coordinates of the vertices of the triangle on the coordinate grid.
2. Rotate the triangle by 90° around the origin.

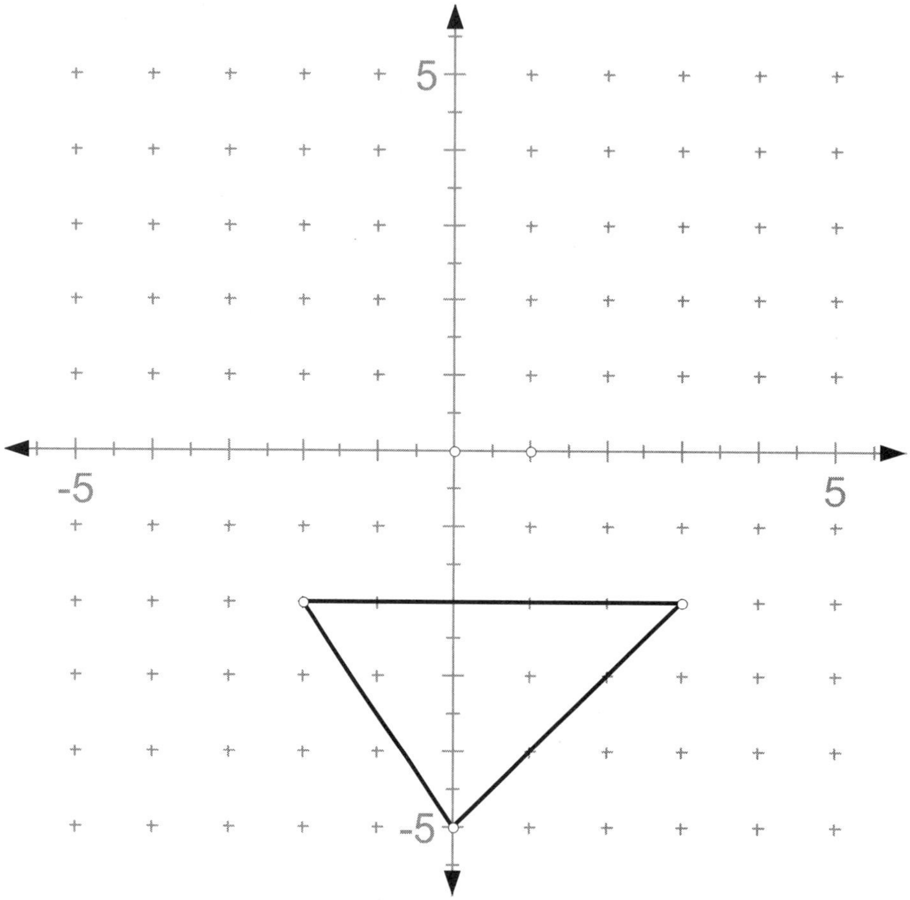

3. Label the vertices and give the coordinates of the rotated triangle on the coordinate grid.
4. Use the mapping notation [(x,y)→] to describe where each of the individual original points goes under the rotation.

5. Use the same triangle, but this time rotate by 180° around the origin.

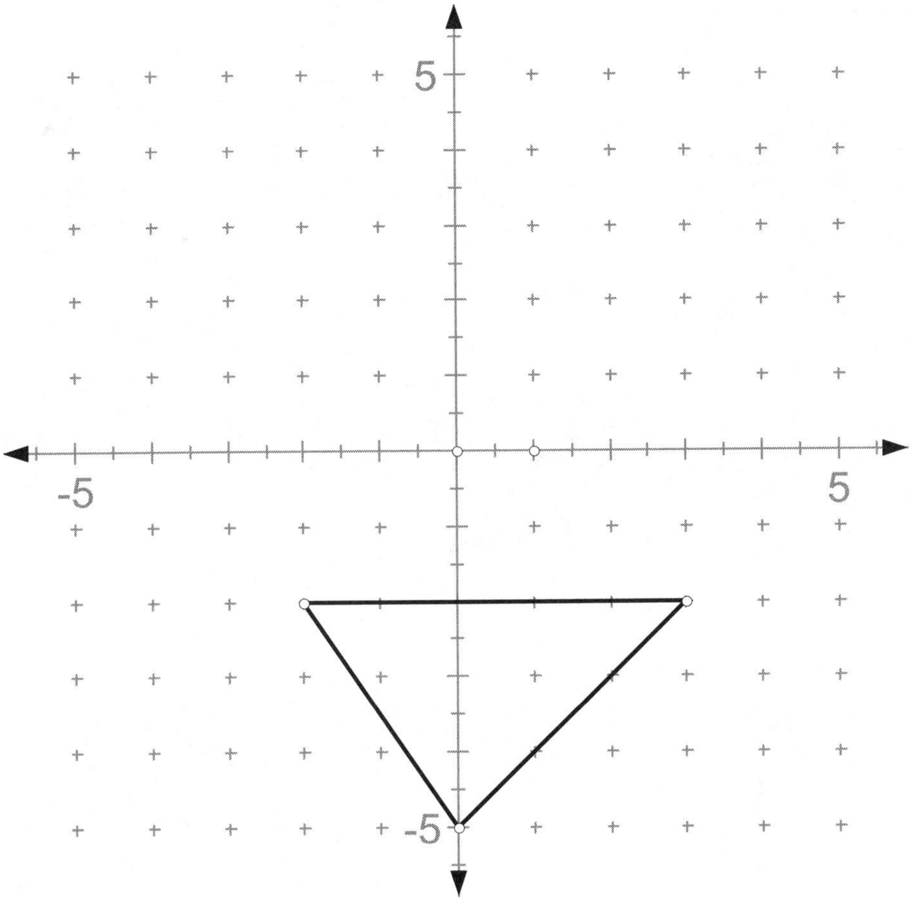

6. Give the coordinates of the transformed triangle on the coordinate grid.
7. Use the mapping notation [(x,y)→] to describe where each of the individual original points goes under the rotation.

8. Use the same triangle, but this time rotate by -90° around the origin. If you use only positive numbers, what is this measure?

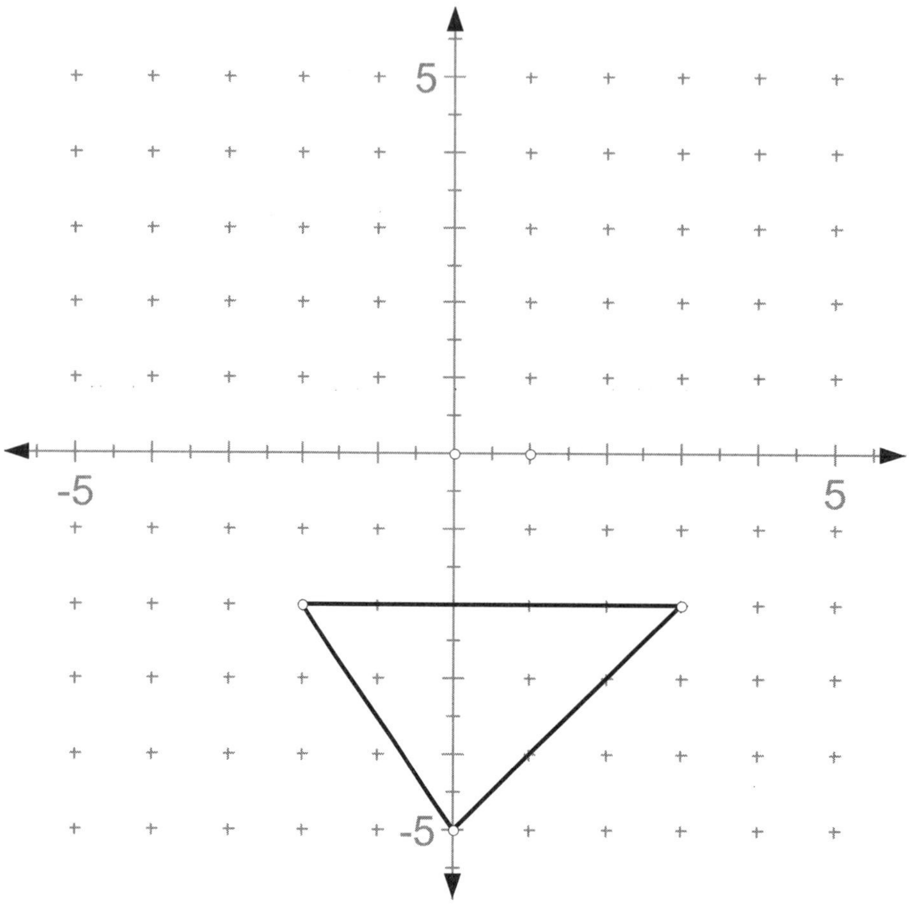

9. Give the coordinates of the transformed triangle on the coordinate grid.
10. Use the mapping notation [(x,y)→] to describe where each of the individual original points goes under the rotation.

Assignment for Dilations

Name: ______________________________

In the previous two dilations you performed in class, you made the new polygon larger than the original. In this dilation, your image should be smaller than your pre-image.

In this activity, make your center of dilation the origin, marked as point O on the coordinate grid.

Make the new octagon ¼ as big as the original octagon GHIJKLMN.

1. Describe the steps you perform as you construct this dilation of the octagon.

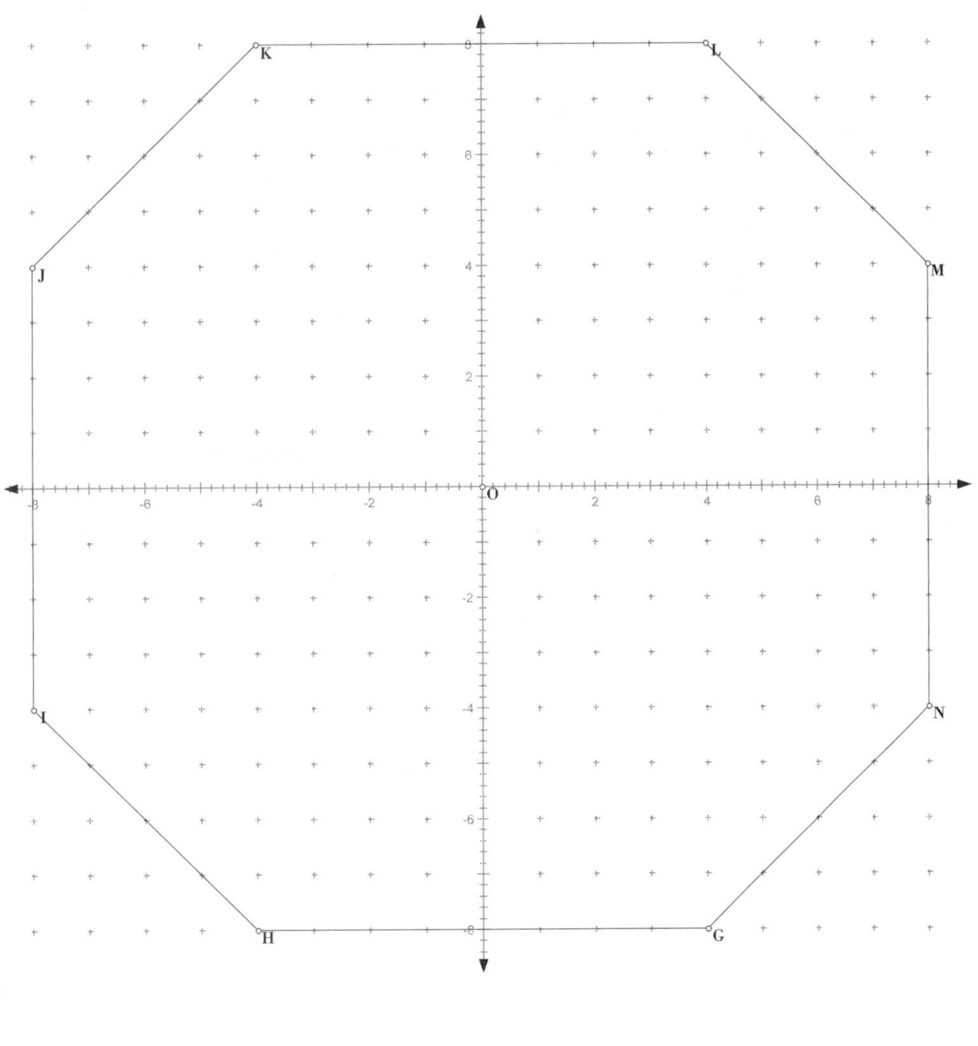

2. Measure the angles of octagon GHIJKLMN and record them on the figure.
3. Measure the angles of your new octagon G′H′I′J′K′L′M′N′ and record them on the figure.
4. How do the measures of these angles compare?

5. Does comparing these angle measures confirm or dispute your findings from the in-class activities?

6. Measure the sides of original octagon GHIJKLMN. Record the measurements on the figure.
7. Measure the sides of the image, and record these measurements on the dilation.
8. How do these measures compare?

9. What is the ratio of H′I′ to HI?

10. What is the ratio of MN to M′N′?

11. Do these measurements and ratios confirm or dispute your findings from the in-class activities? Why? Revise your conjecture if necessary.

12. Use a combination of the mapping notation and the shorthand notation for dilations to write about four of these eight points (e.g., **S**(8,4) = (,)).

13. Generalize your findings using (x,y) as the pre-image point.

Assignment for Symmetry #1

Name: ____

1. In each figure below, sketch any and all the lines of symmetry

   a.

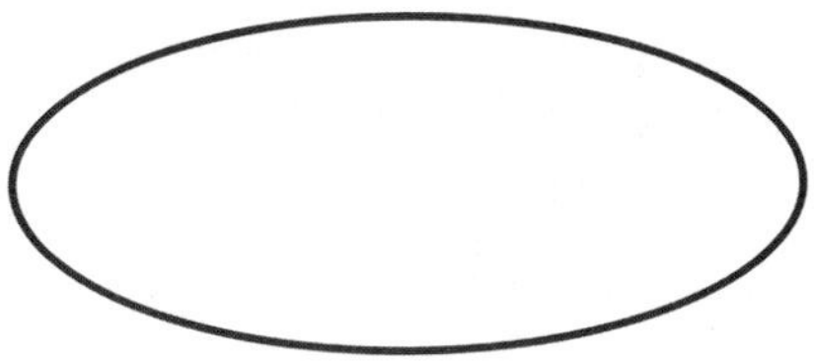

   b.

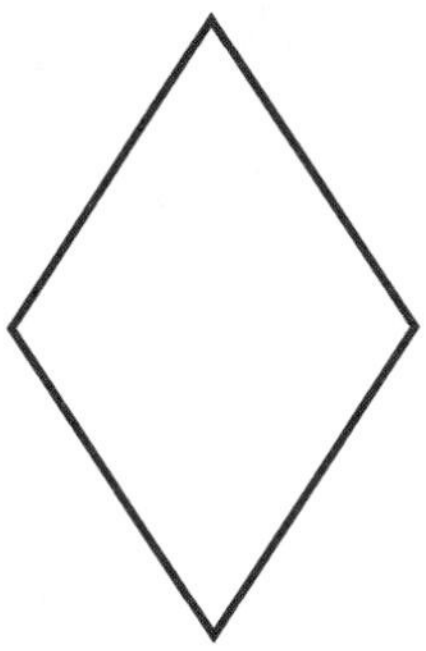

   c.

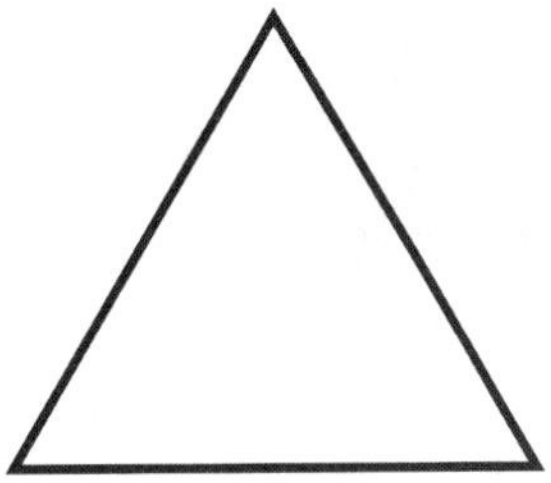

   d.

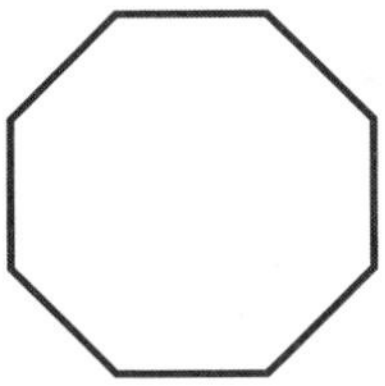

2. Complete each figure below, given its line of symmetry.

   a. The line of symmetry is the *y*-axis.

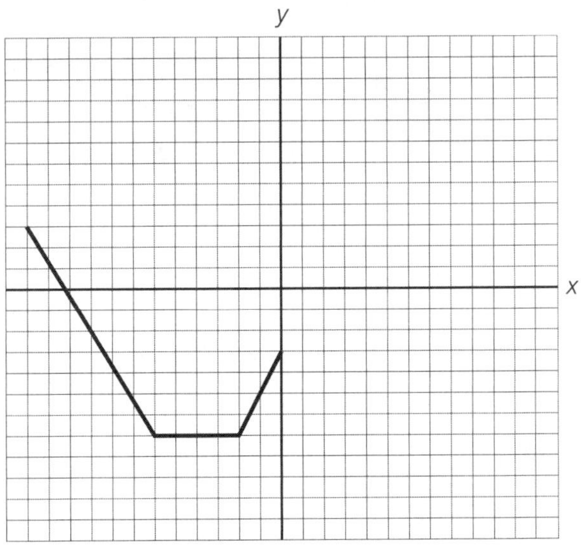

b. The lines of symmetry are the x-axis and the y-axis.

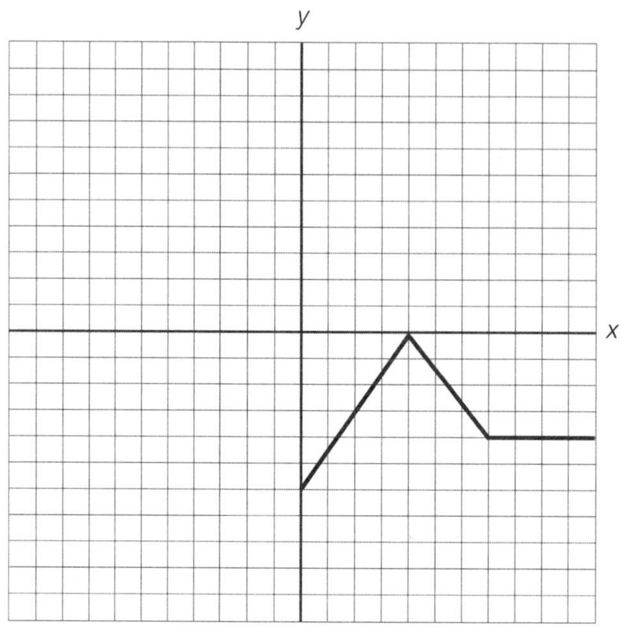

c. The line of symmetry is the line $y = x$, shown as a dotted line.

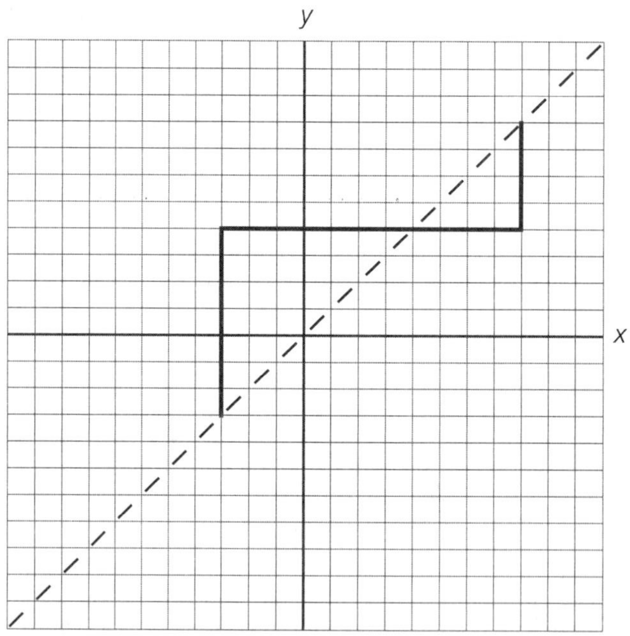

Assignment for Symmetry #2

Name: ______________________________

1. For each figure below, list any and all rotational symmetries.

   a. 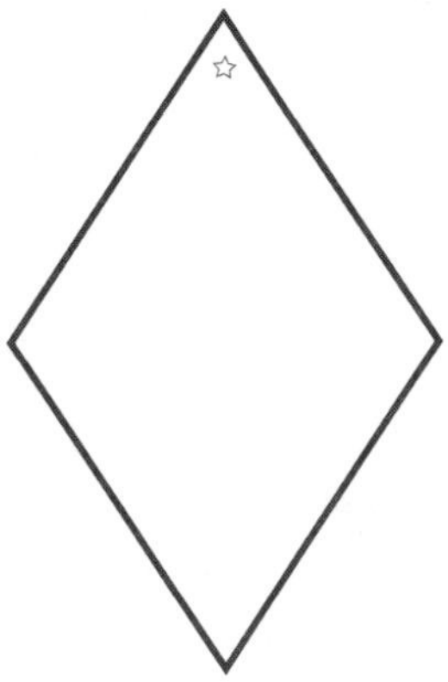

   Rotational Symmetries:

   b. 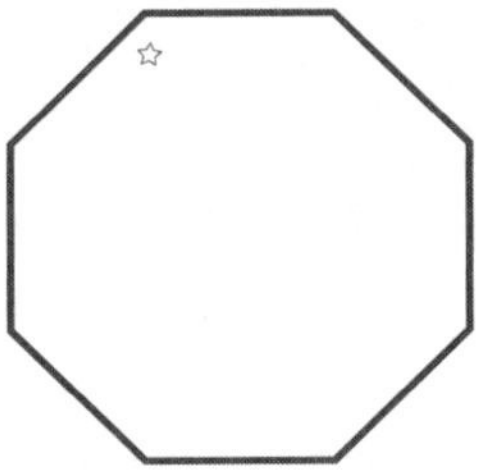

   Rotational Symmetries:

2. Complete each figure below, based on the information given about its rotational symmetry.

   a. One rotational symmetry: 180° about the origin.

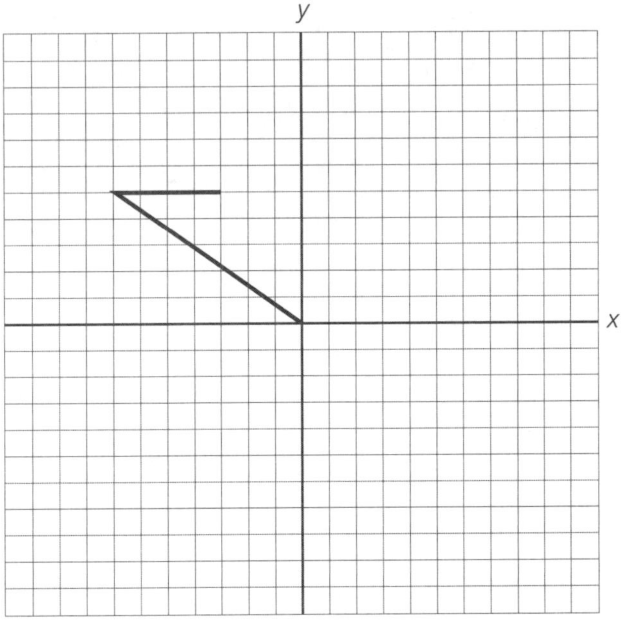

Assignment for Similar and Congruent Figures

Name: ______________________________

1. Summarize the relationship between similar figures and dilations.

2. Summarize the relationship between congruent figures and isometries.

3. A new company called Vita-Aqua is designing the bottles for its line of popular vitamin-infused drinking water. The company wants to make the small bottle in exactly the same shape as the large bottle. Data from some market surveys the company has done indicate that a small bottle approximately 6 cm in diameter and 21 cm tall is the easiest for people to use while exercising. (The small bottle holds about ½ L.) What are some possible dimensions for a large bottle? (How much will the large bottle hold? Recall that volume involves three dimensions.)

4. Are the following figures similar? Are the following figures congruent? If so, use the correct notation to describe the relationship. If not, explain why not.

   a.

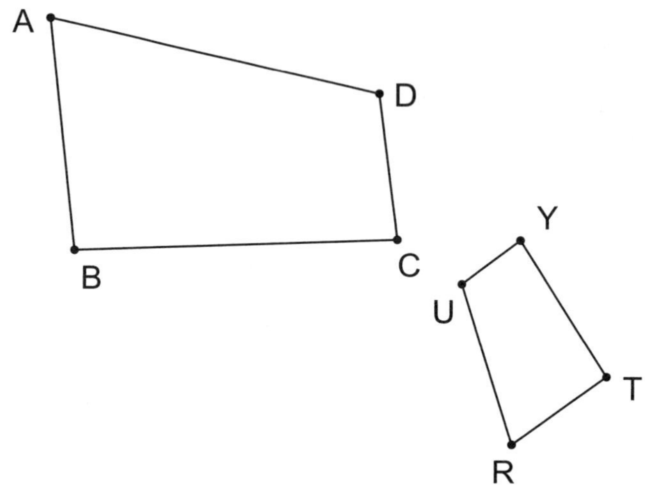

b.

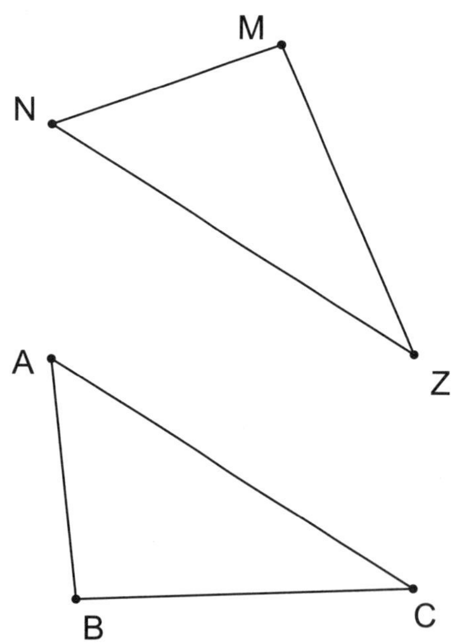

5. If Δ*EMA* is congruent to Δ*OMP*, draw a diagram of this situation and list all *congruent* parts.

   Diagram:

   Congruent and Proportional Parts:

6. If Δ*EMA* is similar but not congruent to Δ*OMP*, draw a diagram of this situation and list all *congruent* parts and all *proportional* parts.

   Diagram:

   Congruent and Proportional Parts:

7. Draw two congruent scalene triangles that share only a common vertex. Label the triangles and list all their corresponding parts. Write a congruence statement for your triangles.

8. Draw two congruent right triangles that share only a common side. Label the triangles and list all their corresponding parts. Write a congruence statement for your triangles.

Assignment for Finding Unknown Parts of Similar and Congruent Polygons

Name: ______________________________

1. Today is incredibly sunny and I just noticed my shadow on the ground. I am 5 feet 2 inches tall and my shadow is 6 feet in length.

   a. If the tree across the street casts a shadow whose length is 18 feet, find the height of the tree. Show your work. Drawing a sketch will help you solve each problem.

   b. If the building across the street is 30 feet high, find the length of its shadow.

2. Use this picture to *write* and *solve* a problem that involves similar triangles.

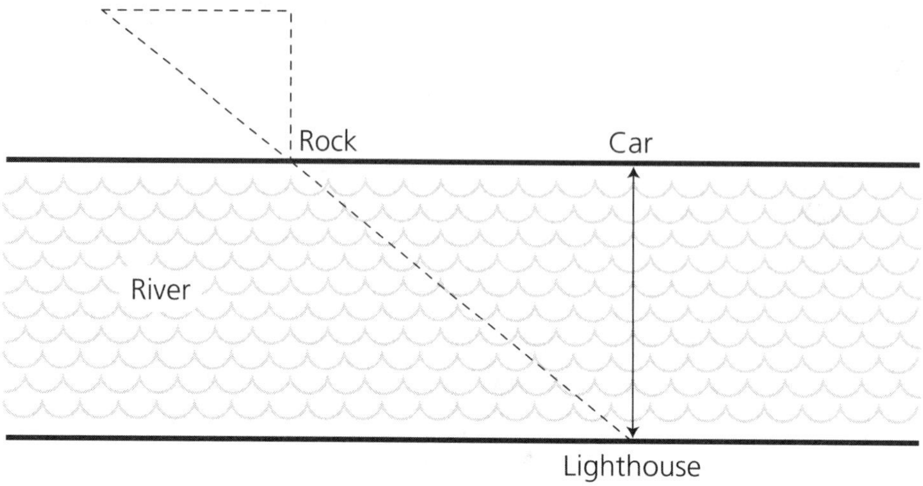

- Make up some given information using measurement units of your choice, and add these measurements to the picture. It may help to add several landmarks to the picture.
- Describe the problem situation, including the given information and the information that is to be found.
- Solve your problem.

Description of Problem Situation:

Problem Solution:

Assignment for Congruent Triangles

Name: ______________________________

A student in another class created the following set of directions based upon the in-class activity for creating congruent triangles.

- Measure all three sides of the triangle.
- Record these measurements.
- Pick one of them, and create a segment on your paper that is exactly that length.
- Now, get two pieces of dried spaghetti and chop them so that they are equal to the other two segments you measured.
- Take the first piece of spaghetti and attach one end of it to the left endpoint of the segment you created on your paper.
- Take the second piece of spaghetti and attach one end of it to the right endpoint of the segment you created on your paper.
- Move the pieces of chopped spaghetti until you get them to connect.
- Now, you have a triangle that matches the one you are copying.

1. Follow this set of directions to attempt to create a copy of the following triangle. (If you don't have dried spaghetti where you are doing your homework, you can use straws or even toothpicks.)

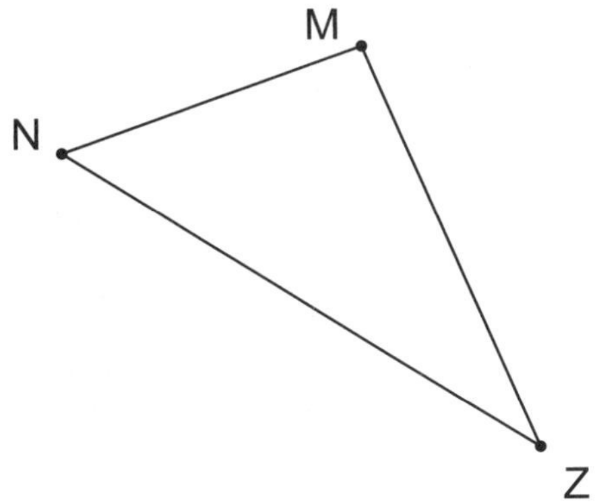

2. Attempt to create a triangle that is NOT a copy of the above triangle by using this student's directions. Were you successful?
3. Does this set of directions seem to work for the triangle above? Will this set of directions work for every triangle? Explain.

A different student in the same class created these directions.

- Measure ∠A and make a copy of it, making sure to extend both sides as far as you can. Call this ∠A'.
- Measure side AB.
- Start at point A' and create a segment on top of one of the sides of your angle that is equal in length to side AB. Label the second endpoint of the new segment as B'. (If you want, you can erase the rest of the ray, so that you only see $\overline{A'B'}$ and you don't get confused.)
- Measure side BC.
- Start at B' and create a segment that has the same length as $\overline{BC}$. Label the second endpoint as C', and you are done.

1. Follow this set of directions to attempt to create a copy of the following triangle.

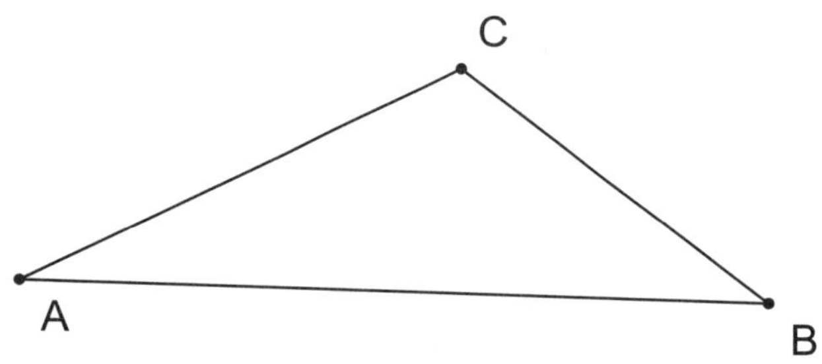

2. Attempt to create a triangle that is NOT a copy of the above triangle by using this student's directions. Were you successful?
3. Does this set of directions seem to work for the triangle above? Will this set of directions work for every triangle? Explain.

Assignment for Proving Triangles Congruent

Name: ______________________________

1. Use the naming convention from **Proving Triangles Congruent** to name the method of proving triangles congruent from the previous assignment. (Recall that only the first of the two methods actually worked.)

2. Use the three main methods for proving triangles congruent (SAS, ASA, and SSS) in order to show that the following sets of triangles are congruent. Make sure to mark the parts that are given as equal, and mark the parts that you prove equal as you progress through the explanation.

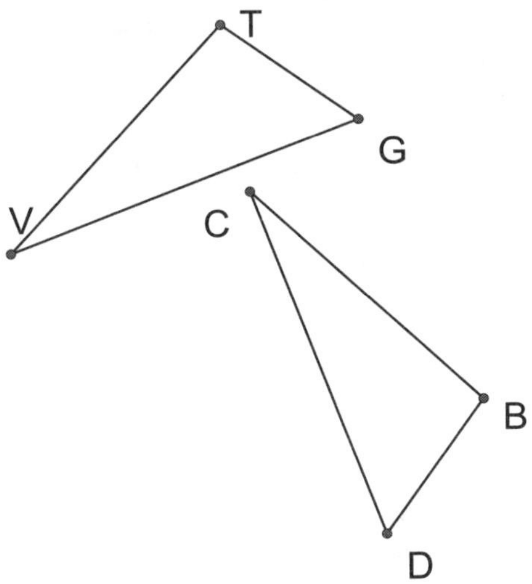

a. $m\angle C = m\angle V$; $m\angle G = m\angle D$; $CD = VG$;

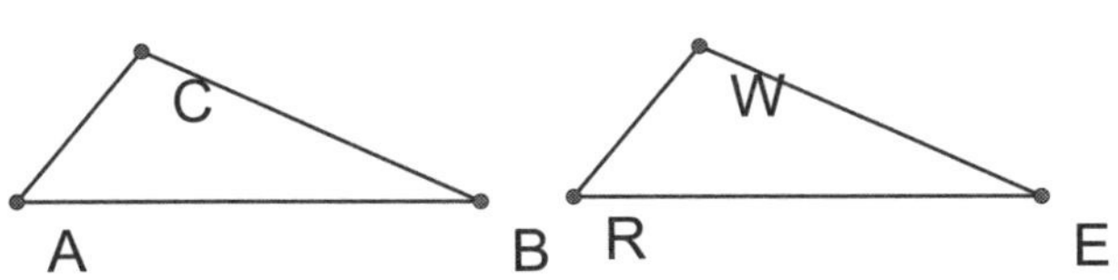

b. $AC = RW$; $BC = EW$; $BA = ER$

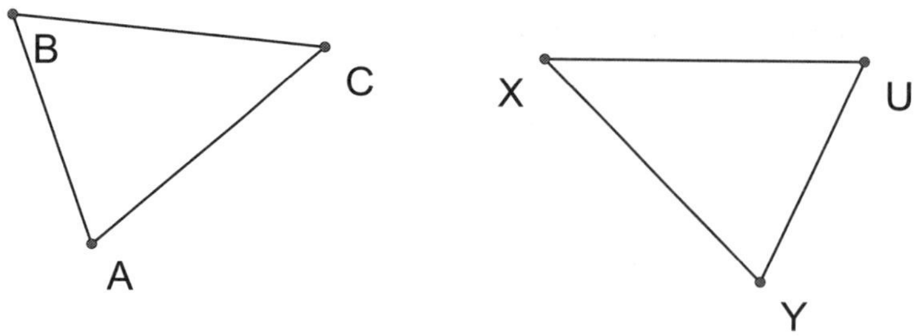

c. BC = UX; m∠B = m∠U; m∠Y = m∠A

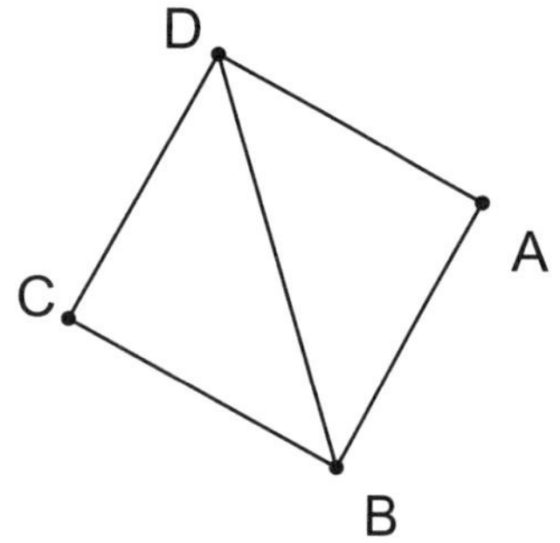

d. ABCD is a square.

3. Are these two triangles congruent? How do you know? Label the triangles. Write a congruence statement for these triangles.

4. If segment *GE* is parallel to segment *AM* and segment *GM* is parallel to segment *AR*, are the two triangles congruent? Why or why not?

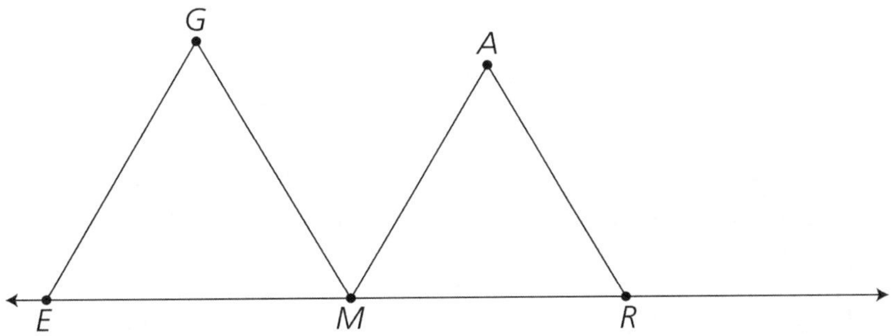

5. If segment *BG* bisects ∠*ABY*, is Δ*BAG* congruent to Δ*BYG*? Why or why not?

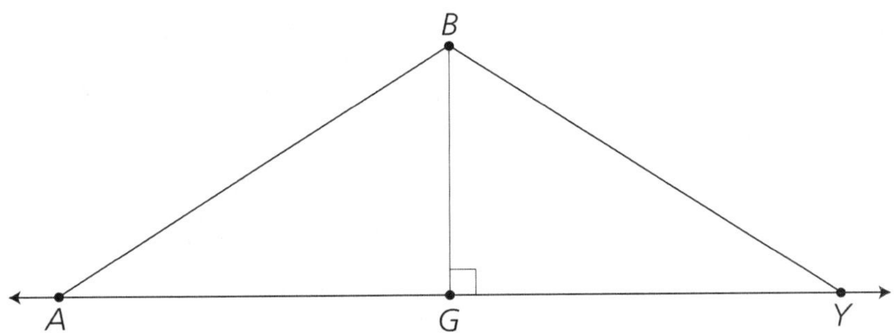

6. If segment *BG* is the perpendicular bisector of segment *AY*, is Δ*BAG* congruent to Δ*BYG*? Why or why not?

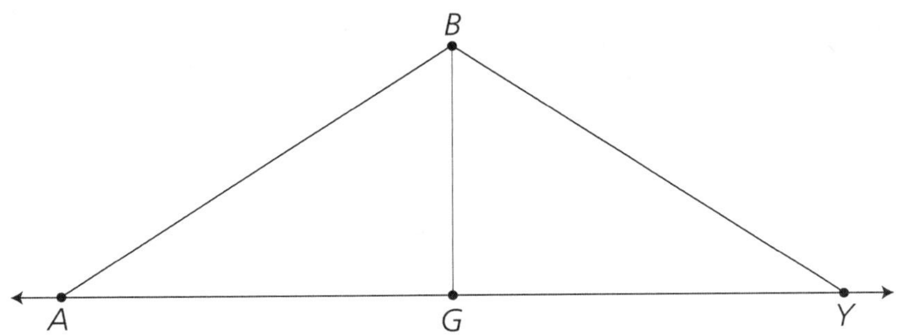

7. Are these two triangles congruent? Why or why not?

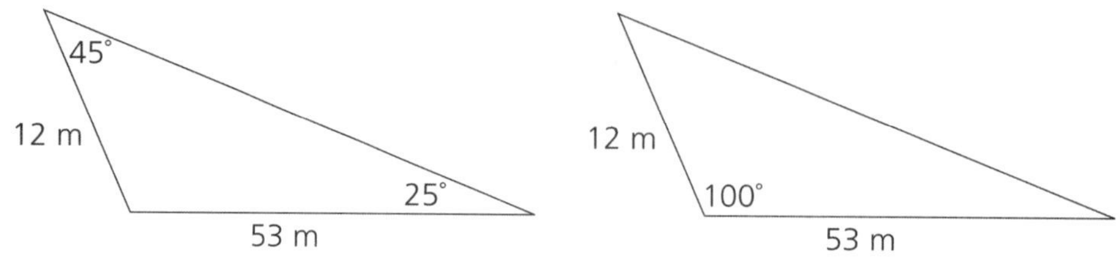

8. Prove or disprove the following statement. Drawing a sketch will help you. In an isosceles triangle, the bisector of the vertex angle divides the triangle into two congruent triangles.

9. If the perpendicular bisector of a side in any triangle intersects the opposite vertex, then it divides that triangle into two congruent triangles. Is this statement true or false? Justify your conclusion.

Assignment for Proving Triangles Similar

Name: ______________________________

1. Is ΔISA similar to ΔEBA? How do you know? Find $m\angle ISA$, $m\angle IAS$, $m\angle ABE$, and $m\angle BAE$. If the triangles are similar, write a similarity statement.

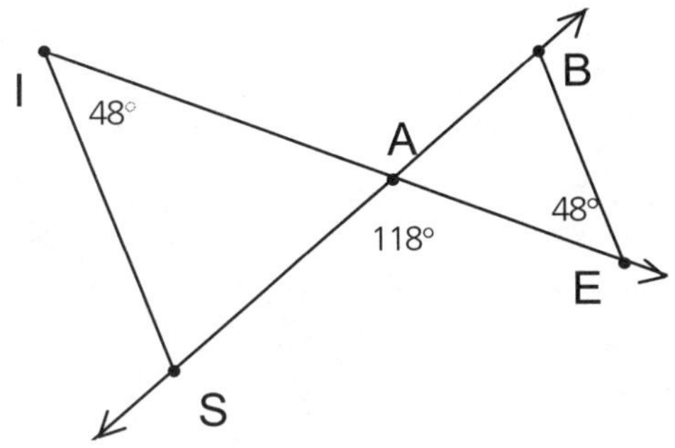

2. If segment GE is parallel to segment AM, are the two triangles similar? Why or why not?

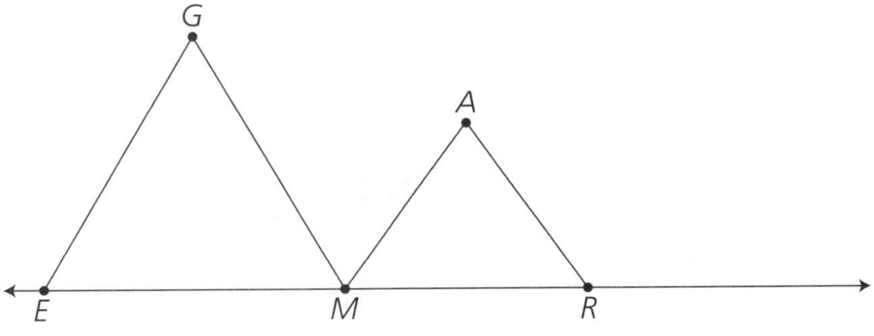

3. If $BA = 10$ cm, $AG = 8$ cm, and $GY = 4$ cm, is ΔBAG similar to ΔYBG? Why or why not?

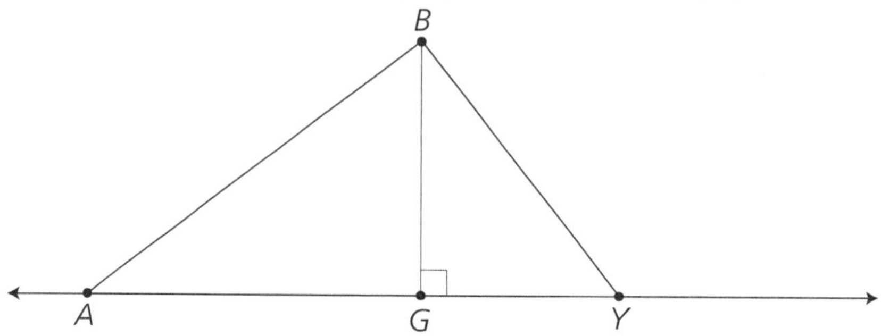

4. Are these triangles similar? Why or why not?

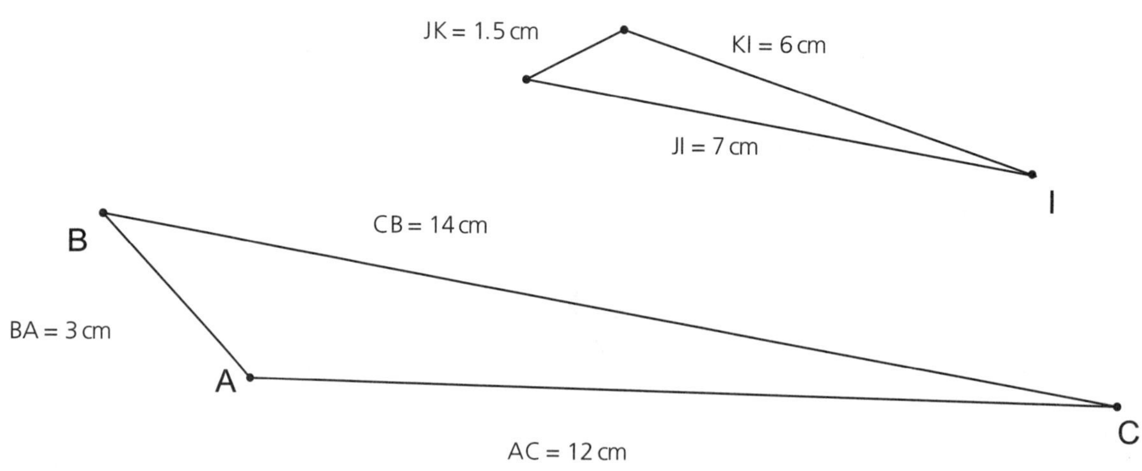

Assignment for Applications Involving Similar Triangles

Name: ____

1. Find the value of x. Note that this figure is not drawn to scale.

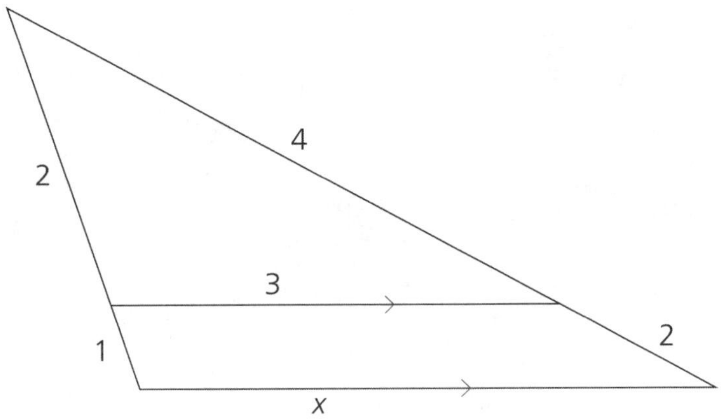

2. Find the values of x and y. Show your work.

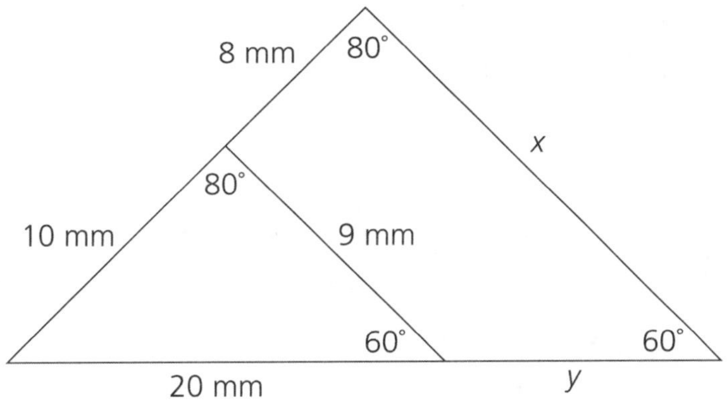

Assignment for Indirect Measurement

Name: ____

1. Use this picture to *write* and *solve* a problem about finding the length of a bridge across the lake. The solution should use similar triangles.

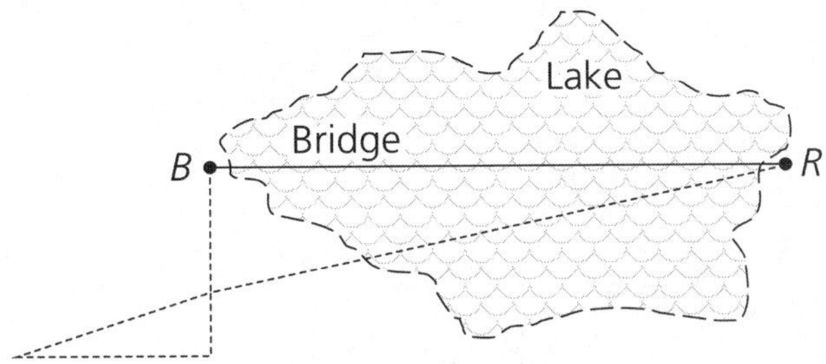

- Make up some given information using measurement units of your choice. It may help to add several landmarks to the picture.
- Describe the problem situation, including the given information and the information that is to be found.
- Solve your problem.

Description of Problem Situation:

Problem Solution:

Assignment for Midsegments of a Triangle

Name: ______________________________

Use this figure for problems 1–3.

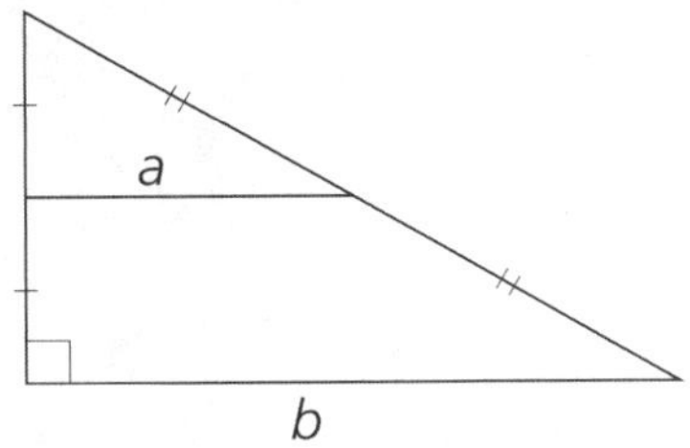

1. If $a = 35$ in, then $b =$
2. If $b = 19$ ft, then $a =$
3. If $a = x$ cm, then $b =$

4. Find the perimeter of the smaller triangle in this figure. Show your work.

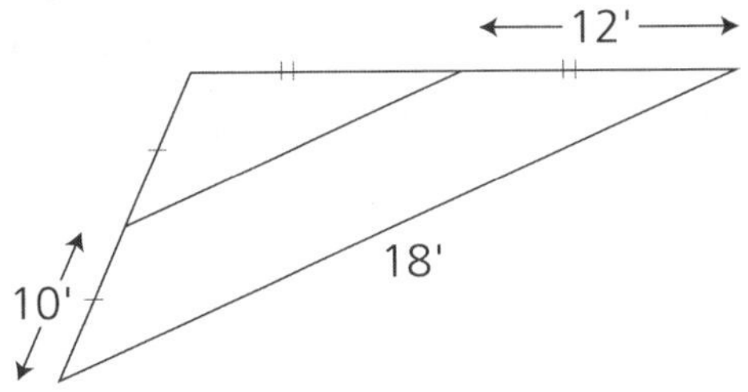

5. If $GT = 42$ feet and $TO = 54$ feet, find the perimeter of quadrilateral $MERT$.

6. If $GT = 36$ inches, $TM = 19$ inches, and $EO = 21$ inches, find the perimeter of ΔREM.

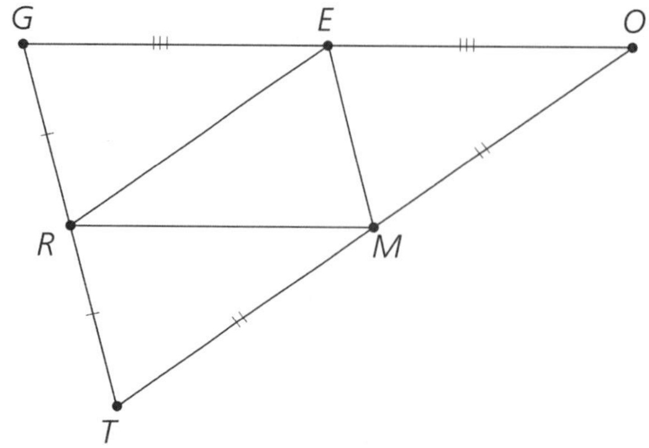

a. Is it true that $\Delta REM \sim \Delta OTG$? Explain.

b. Is it true that $\overline{RM} \parallel \overline{GO}$? Explain.

7. Summarize your findings about the midsegments in a triangle.

Assignment for Similarity with Two- and Three-Dimensional Figures

Name: ____

1. In the previous assignment, you found the perimeter of the smaller ΔREM, when given measurements for the lengths of the sides (or parts of the sides) of the larger ΔOTG.

   $GT = 36$ inches, $TM = 19$ inches, and $EO = 21$ inches.

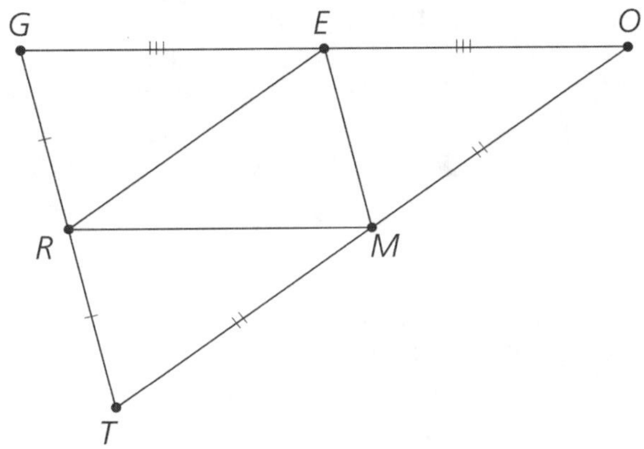

   a. What is the scale factor between these two similar triangles? How do you know?

   b. What is the ratio between the perimeters of these two triangles? How does this ratio compare to the scale factor? Explain.

   c. What is the ratio between the areas of these two triangles? (Note that these figures are not drawn to scale. Thus, you cannot use a measurement of the altitude to calculate area. Since it will be difficult to actually calculate these areas, attempt to answer this question based upon your learning from class and not the area measurements.)

d. How does the ratio of areas compare to the ratio of perimeters?

e. How does the ratio of areas compare to the scale factor?

f. Explain why the ratio of areas is not the same as the scale factor.

2. Pictured below are two square pyramids. One is a dilation of the other with respect to point F. The measurement of the front edge of each is provided. What is the scale factor between these two pyramids?

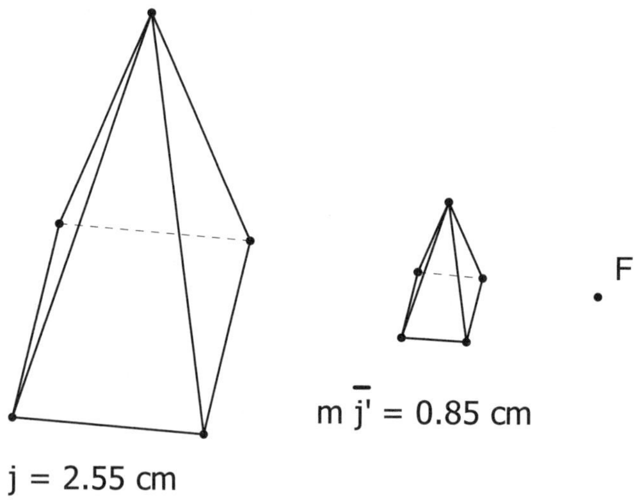

a. What is the ratio of the heights?

b. What is the ratio of the areas of the bases?

c. What is the ratio of the volumes?

Assignment for Koch Snowflake

Name: ______________________________

Pictured below is a cube that is one unit on each side. Consider this Stage 0.

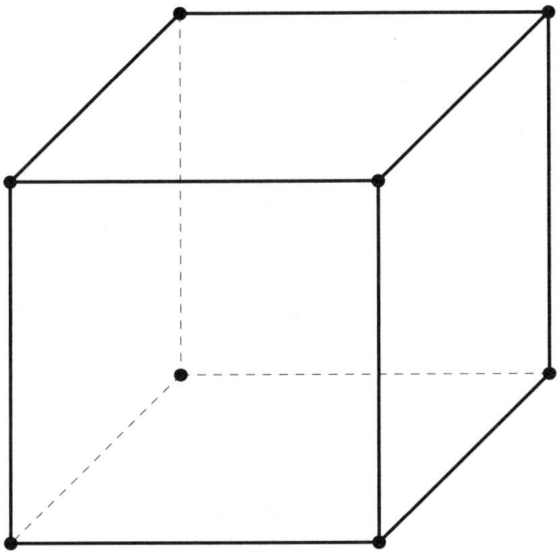

Divide each edge of the cube into thirds.

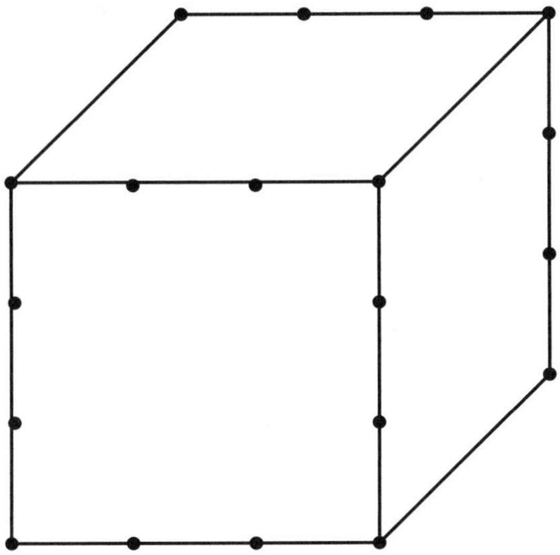

Connect the points that divided the edges into thirds, so that each face is divided into ninths. Your figure should now look like a Rubic's Cube®.

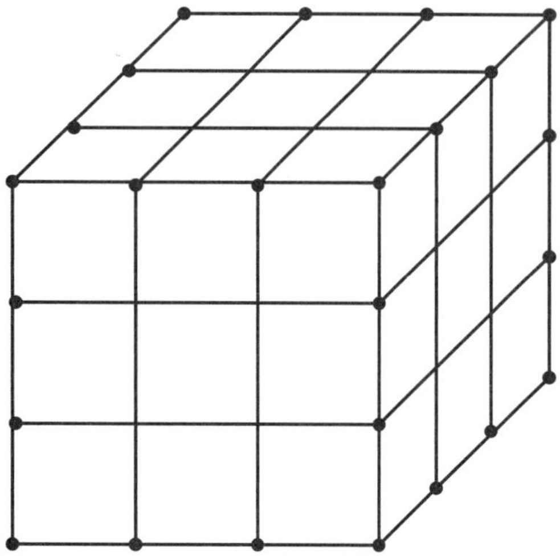

"Hollow out" the middle ninth of each face, by removing a small cube. This is Stage 1.

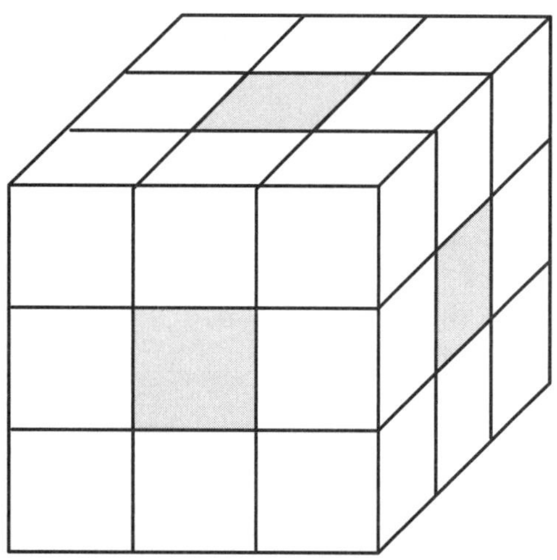

Repeat the above procedure:

- Divide the edges of each of the smaller squares on every face (even the ones 'inside' the cube) into thirds.
- Connect the opposite points to divide each of the smaller squares into ninths.
- "Hollow out" the middle ninth of each of these squares (even the ones 'inside' the cube).

What observations can you make about this fractal? Compare and contrast this fractal with the Koch Snowflake.

Assignment for Arithmetic Sequences and Series

Name: ______________________________

Your parents begin your weekly allowance at $3.00 the first week of the year. They agree to increase your allowance by $1.00 every week for the remainder of the year.

1. Write a recursive formula to compute your allowance for each week.

2. Write an explicit formula to compute your allowance for each week.

3. Compute your allowance during the:

   a. 6th week of the year.

   b. 27th week of the year.

   c. 33rd week of the year.

4. Compute the total allowance you received during the year up to and including the:

   a. 19th week of the year.

   b. 41st week of the year.

   c. 52nd week of the year.

Assignment for Geometric Sequences and Series

Name: ______________________________

You propose to your parents that they begin your weekly allowance at 1 penny for the first week of the year. They must triple your allowance every week for the remainder of the year.

1. Write a recursive formula to compute your weekly allowance each week.

2. Write an explicit formula to compute your weekly allowance each week.

3. Compute your allowance during the:
   a. 6th week of the year.

   b. 12th week of the year.

   c. 21st week of the year.

4. Do you think your parents would agree to your proposal? Explain why or why not.

Assignment for Computer Virus

Name: ________________________________

In Science class, you are studying bacterial growth. You begin by placing a single bacterium in a Petri dish. The number of bacteria triples every day.

1. Complete the table of values below.

Time	Number of bacteria
Days	**Bacteria**
0	
1	
2	
3	
4	
5	
6	
7	
8	
9	
10	

2. Write an equation to model the number of bacteria over time.

3. Graph the function below

4. What is the y-intercept of the graph? Describe what this means in terms of the problem situation. Is this realistic is this problem?

5. What is the x-intercept of the graph? Describe what this means in terms of the problem situation. Is this realistic in this problem situation?

Assignment for Double or Nothing

Name: ______________________________

1. Write each expression using positive exponents and then evaluate each expression.

   a. $2^{-3} \bullet 3^{2}$

   b. $\left(\frac{1}{2}\right)^{-3}\left(\frac{2}{3}\right)^{-2}$

   c. $2^{-3} \bullet 3^{-2} \bullet 4^{3}$

   d. $\left(\frac{1}{4}\right)^{-3}\left(\frac{1}{2}\right)^{-4} \bullet 8^{-3}$

   e. $\left(\frac{2}{5}\right)^{-3}\left(\frac{4}{5}\right)^{2}$

Assignment for Exponential Transformations

Name: ______________________________

1. Write the function that results when the following transformations are applied to the given parent function.

   a. The parent function $f(x) = 2^x$ is reflected about the x axis, stretched vertically by a factor of $\frac{1}{3}$, and shifted down 3 units.

   b. The parent function $f(x) = 6^x$ is stretched vertically by a factor of 4, and shifted up 4.75 units.

   c. The parent function $f(x) = 4^x$ is reflected about the x axis and stretched vertically by a factor of $\frac{2}{7}$.

2. Describe the transformations applied to the parent function that result in each function below.

   a. $f(x) = -3^x + 2$

   b. $f(x) = \frac{5}{6} \cdot 2^x - \frac{4}{3}$

   c. $f(x) = -10 \cdot 7^x + 12$

Assignment for Properties of Exponents

Name: ______________________________

Simplify the following expressions. Write the final answer using only positive exponents.

1. $\left(4a^3b^2\right)\left(4a^2b^5\right)$

2. $\left(-2y^3z\right)^3$

3. $\dfrac{3xy^2z^4}{9x^2yz^2}$

4. $\dfrac{\left(a^2b^5\right)^2}{a^3b^4}$

5. $\left(\dfrac{7t}{t^3}\right)^{-2}$

6. $\dfrac{18^{-1}u^{-4}w^{-2}}{6^{-1}u^{-6}w}$

7. $\left(\dfrac{8x^{-4}y^3}{4x^{-3}y^3}\right)^{-3}\left(xy^2\right)^2$

Assignment for College Fund

Name: ______________________________

You want to save money to buy a new car. You've managed to save $1000 so far. After looking at various bank ads in the newspaper, you find two that look promising. The Royal National Bank offers a 4 year Certificate of Deposit at 4.5% compounded annually. The Mount Haven Bank offers a 5 year Certificate of Deposit at 3.8% compounded annually.

1. Write a function to compute the balance of the CD through Royal National Bank.

2. Write a function to compute the balance of the CD through Mount Haven Bank.

3. Compute the balance of each CD after 1 year.

4. Compute the balance of each CD after 2 years.

5. Compute the balance of each matured CD.

6. Which bank would you choose? Explain.

Assignment for Compounding the Problem

Name: ______________________________

You and your best friend are both saving money to buy a new car. You've managed to save \$1000 and have deposited it in a savings account offering 4.5% interest compounded monthly. Your friend has managed to save \$850 and has deposited it in a savings account offering 8% interest compounded twice a year. You want to buy the new car in 5 years.

1. Write a function to compute your principal.

2. Write a function to compute your friend's principal.

3. Complete the table below.

Time	Your principal	Your friend's principal
Years	**dollars**	**dollars**
0		
1		
2		
3		
4		
5		

4. After 5 years, who will have more money, you or your friend?

Assignment for Properties of Logarithms

Name: ______________________________

1. Write the following logarithmic expressions in expanded form.

   a. $\log_3 3ab^2$

   b. $\log_2 \frac{x^3}{4y}$

   c. $\log\left(2t^3\right)^2$

2. Write the following logarithmic expressions using a single logarithm.

   a. $2\log_5 x + 2\log_5 y$

   b. $\log_4 3 + 2\log_4 x - \log_4 y$

   c. $2\ln x - 3\ln y + 5(\ln 2x - \ln y)$

Assignment for Graphs of Logarithms

Name: ______________________________

Answer the following questions for the function $f(x) = 3^x$.

1. Determine an equation for the inverse of $f(x)$.

2. Perform a change of base conversion to allow you to enter the inverse function into your calculator.

3. Create a graph of $f(x)$, $f^{-1}(x)$ and the line y=x.

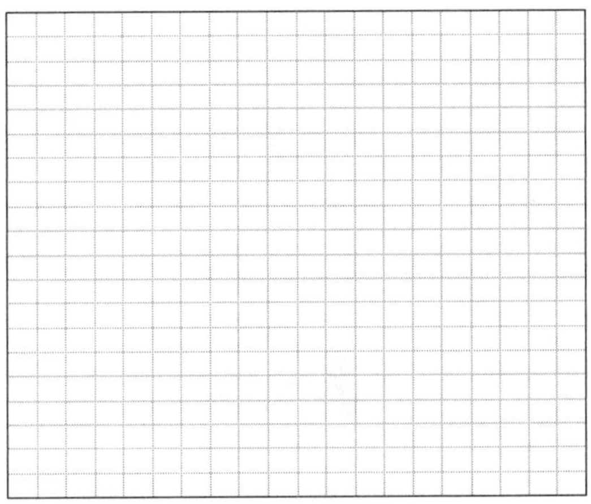

4. What is the y-intercept of $f(x)$? What is the x-intercept of $f^{-1}(x)$?

5. Is this consistent with what you know about inverses? Explain why or why not.

6. What is the equation for the asymptote for $f(x)$? What is the equation for the asymptote for $f^{-1}(x)$?

7. Is this consistent with what you know about inverses? Explain why or why not.

Assignment for Logarithmic Transformations

Name: ______________________________

1. Write the function that results when the following transformations are applied to the given parent function.

   a. The parent function $f(x) = \log x$ is reflected about the x axis, stretched vertically by a factor of $\frac{1}{4}$, and shifted left 2 units.

   b. The parent function $f(x) = \log_3 x$ is stretched vertically by a factor of 4, shifted right 2 units, and shifted up 1.5 units.

   c. The parent function $f(x) = \ln x$ is reflected about the x axis, stretched vertically by a factor of $\frac{4}{5}$, and shifted down 3 units.

2. Describe the transformations applied to the parent function that result in each function below.

   a. $f(x) = -\ln(x+1) - 6$

   b. $f(x) = 5\log_3(x-4) + 1$

   c. $f(x) = -\frac{2}{3}\log_6(x+1) + 4$

Assignment for Richter Scale

Name: ______________________________

The pH scale is a logarithmic scale used to measure acidity of chemicals. The pH scale ranges from 0-14 with 7 indicating neutrality (such as pure water), 0-7 indicating acidic (such as vinegar or lemon juice) and 7-14 indicating basic (such as laundry detergents or ammonia).

The formula to calculate pH is: $pH = -\log_{10}[H]$ where [H] is the concentration of hydrogen ions measured in moles per liter.

1. Complete the table of values to show the pH value for various concentrations of hydrogen ions.

Concentration of hydrogen ions	**Example**	**pH Measurement**
Moles per Liter		**pH Measurement**
1	Nitric acid	
0.01	Lemon juice	
0.0001	Soda	
0.000001	Milk	
0.00000001	Egg whites	
0.0000000001	Milk of magnesia	
0.000000000001	Oven cleaner	
0.00000000000001	Household Lye	

2. Graph the pH function below.

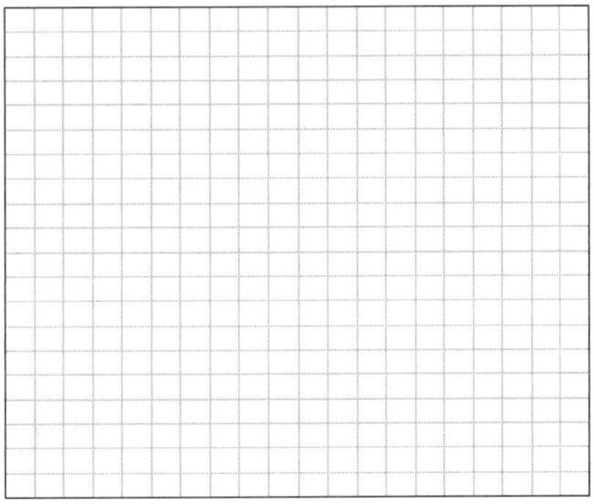

3. What is the pH measurement if the concentration of hydrogen ions is 0.045 moles per liter?

4. What is the pH measurement if the concentration of hydrogen ions is 0.00000007 moles per liter?

5. What is the concentration of hydrogen ions for a pH measurement of 3.5?

6. What is the concentration of hydrogen ions for a pH measurement of 11.2?

Assignment for Old Cars

Name: ______________________________

Your family recently won some money in the state lottery. It was not enough for everyone to quit their job, but it was enough to buy a new house.

The house that your family is considering sold for \$40,000 in 1968, \$64,000 in 1975, and \$83,250 in 1982.

1. Using your calculator, find an exponential equation of best fit to model the data. Write the equation.

2. According to the model, what will be the value of the home if your family buys it this year?

3. According to the model, what will be the value of the home 5 years after your family buys it?

4. According to the model, when will the value of the home reach \$250,000?

5. According to the model, when will the value of the home reach \$1,000,000?

Assignment for Animated gif Program: Adding a Banner

Name: ______________________________

You are a web page designer. A new client likes what you did with the dog owner's web page. She sells real estate and would like to have a web page that will attract people's attention. She wants the image of a house to appear, growing one inch for per second for five seconds. She also wants a fixed banner of her business card, measuring 5 inches by 4.5 inches, to appear when the page opens.

1. Complete the following table.

Labels	**Time**	**Banner Area**	**Image Area**	**Total Area**
Units	**Seconds**	**Square inches**	**Square inches**	**Square inches**
	0			
	1			
	2			
	3			
	4			
	5			
Expression				

2. Sketch the graph of the image area versus the time and the total area versus time.

3. How is the graph of the image area different from the graph of the total area?

4. Express the total area as a function of time.

5. As a final touch to the web page with the banner, you will add a sound bite of the agent welcoming the user. The sound bite is to begin when the total area is 34.75 square inches. Determine algebraically the time after the banner appears when the sound bite should start playing.

Assignment for Parent Functions and Vertical Transformations

Name: ______________________________

1. Write the resulting function if the following transformations are applied to the function $y = x^2$
   a. Shift up 3 units

   b. Shift down 1 unit

   c. Shift down 5 units

   d. Shift up 2 units

   e. Shift down 10 units

2. Describe the transformations applied to the function $y = x^2$ to result in each function
   a. $y = x^2 - 4$

   b. $y = x^2 + 7$

   c. $y = x^2 - 14$

   d. $y = x^2 + 1$

   e. $y = x^2 + 8$

Assignment for Animated gif Program: Time Changes

Name: ______________________________

Your client decides when the page opens she wants some music to be playing for three seconds. After the three second delay, she wants the image to begin to form. The image will still grow at 1 inch per second but this time for 9 seconds.

1. Complete the following table.

Labels	**Time**	**Image Length**	**Image Area**
Units	**Seconds**	**Square inches**	**Square inches**
	3		
	4		
	5		
	6		
	7		
	8		
	9		
Expression			

2. Graph the image length versus time and the image area versus time.

3. How is the graph of the image area different from the graph of $y = x^2$?

4. Express the total area as a function of time.

5. Determine algebraically how long it will take for the image to reach 30.25 square inches.

Assignment for Parent Functions and Horizontal Transformations

Name: ______________________________

1. Write the resulting function if the following transformations are applied to the function $y = x^2$
   a. Shift right 2 units

   b. Shift left 1 unit

   c. Shift right 8 units

   d. Shift left 2 units and shift up 1 unit

   e. Shift left 9 units and shift down 3 units

2. Describe the transformations applied to the function $y = x^2$ to result in each function
   a. $y = (x + 5)^2$

   b. $y = (x - 7)^2$

   c. $y = (x - 10)^2$

   d. $y = (x + 4)^2 + 1$

   e. $y = (x - 1)^2 - 6$

Assignment for Animated gif Program: Size Changes

Name: ______________________________

A year has passed and the dog owner wants a new animated gif. This time she is not sure if she wants the image to be larger or smaller. For the larger image she is thinking about a final side length of 8.25 inches. For the smaller image she is considering a final side length of 3.75 inches.

1. Make a table of values summarizing both options

Labels	**Time**	**Smaller Image Length**	**Smaller Image Area**	**Larger Image Length**	**Larger Image Area**
Units					
pression					
	1				
	2				
	3				
	4				
	5				

2. Answer the following based on the larger image

   a. The original image was 5 inches. By what factor will the new program change the length of the original image?

   b. Using this factor, express the length of the larger image as a function of time.

   c. Using this factor, express the area of the larger image as a function of time.

d. Sketch the graph of image length of the larger image versus time and the area as a function of time.

3. Answer the following based on the smaller image

a. The original image was 5 inches. By what factor will the new program change the length of the original image?

b. Using this factor, express the length of the smaller image as a function of time.

c. Using this factor, express the area of the smaller image as a function of time.

d. Sketch the graph of image length of the smaller image versus time and the area as a function of time.

4. Determine algebraically how long it will take for the smaller image to reach 1.1025 square inches.

5. Determine algebraically how long it will take for the larger image to reach 35 square inches.

Assignment for Square, and Squares, and More Squares

Name: ______________________________

Today in class you did an activity in which you formed squares with sides of 1 cm, 2 cm, and 3 cm. In this homework activity, you will theorize what would happen if we used cubes instead of squares.

1. Build or create a model of a cubic centimeter. (If you have blocks that are small cubes, you can use them to build. If not, you can draw the models instead.)

2. What is the length of each edge?

3. What is the perimeter of one of the faces?

4. What is the area of each face?

5. What is the total surface area?

6. What is the volume?

7. Build or create the next largest cube with full centimeter edges.

8. What is the length of each edge?

9. What is the scale factor between the first cube and this cube?

10. What is the perimeter of one of the faces?

11. What is the ratio of the perimeter of a face of cube 2 to the perimeter of a face of cube 1?

12. What is the area of each face in cube 2?

13. What is the ratio of the area of each face in cube 2 to the area of each face in cube 1?

14. What is the total surface area of cube 2?

15. What is the ratio of surfaces areas from cube 2 to cube 1?

16. What is the volume of cube 2?

17. What is the ratio of volumes between cube 2 and cube 1?

18. What ratio(s) should be the same as the scale factor? Why?

19. What ratio(s) are different than the scale factor?

20. How are these ratios related to the scale factor?

21. Can you make a generalization about how each of the ratios you calculated will relate to the scale factor? If so, verify your predictions using the next largest cube with full centimeter edges. If you are having a hard time creating a generalization, create the next largest cube. Calculate the same things you did above, create ratios between the lengths in this new cube and the lengths in previous cubes. Look for patterns. Create ratios between areas and look for patterns. Create ratios between volumes and look for patterns. State your generalization below.

Assignment for Parent Functions and Dilations and Reflections

Name: ______________________________

1. Write the resulting function if the following transformations are applied to the function $y = x^2$
   a. Dilate by a factor of 2

   b. Dilate by a factor of 1.35 and reflect about the x-axis

   c. Reflect about the x-axis and shift up 3 units

   d. Dilate by a factor of 3 and shift right 5 units

   e. Reflect about the x-axis, dilate by a factor of 4 and shift down 1 unit

2. Describe the transformations applied to the function $y = x^2$ to result in each function
   a. $y = -10x^2$

   b. $y = -(x + 2)^2$

   c. $y = 2(x - 7)^2 + 3$

   d. $y = -0.5x^2 - 1$

   e. $y = -6(x + 1)^2 - 3$

Assignment for Writing Quadratic Equations

Name:______________________________

1. Given $y = (x + 4)(x - 2)$, can you determine the following directly from the equation? If so, identify each. If not, explain how to calculate each.
   a. Transformations to $y = x^2$

   b. Vertex

   c. Axis of symmetry

   d. X-intercepts

   e. Y-intercepts

2. Write $y = (x + 4)(x - 2)$ in standard form.

3. Given $y = -(x + 1)^2 - 2$, can you determine the following directly from the equation? If so, identify each. If not, explain how to calculate each.

   a. Transformations to $y = x^2$

   b. Vertex

   c. Axis of symmetry

   d. X-intercepts

   e. Y-intercepts

4. Write $y = -(x + 1)^2 - 2$ in standard form.

5. Given $y = x^2 + 5x + 6$, can you determine the following directly from the equation? If so, identify each. If not, explain how to calculate each.

   a. Transformations to $y = x^2$

   b. Vertex

   c. Axis of symmetry

   d. X-intercepts

   e. Y-intercepts

6. Write $y = x^2 + 5x + 6$ in factored form

Assignment for Solving Quadratic Equations

Name:______________________________

Find the zeros, axis of symmetry, and vertex for the following quadratic equations

1. $x^2 = -5x + 6$

2. $2(x^2 + 2x) = 7x$

3. $5x^2 = -7x - 8$

4. $-12x^2 + 25x = 84$

5. $-x^2 + 9x = 4x - 12$

6. $4x^2 = 100$

7. $3(x^2 + 2x) - 90 = 2(x^2 + 21) - 5x$

Assignment for Solving Quadratic Inequalities

Name:________________________________

Solve each inequality

1. $x^2 - 4 < 0$

2. $x^2 + 2 \leq 0$

3. $(x + 3)^2 + 5 > 9$

4. $x^2 - 6x \geq 1$

5. $(x - 1)^2 \leq 0$

Assignment for Functions and Inverse Functions

Name:______________________________

1. Determine if the following relationships represent functions. If the relationship is a function, determine if it also one-to-one? Explain.

   a. The domain is the set of all passport holders and the range is the set of all identification numbers.

   b. The domain is the set of days of the year and the range is the set of people who have a birthday on that day.

   c. The domain is the set of all cars in the United States and the range is the current blue book value of the car.

2. Find the inverse of the following functions and determine if the inverse is a function.

   a. $y = 2x - 3$

   b. $y = x^2 + 3$

   c. $y = 2(x - 5)^2$

   d. $y = -(x + 1)^2 + 1$

Assignment for Federal Budget

Name: ______________________

The following table displays the average fuel efficiency for gasoline powered automobiles in the United States

Year	Fuel efficiency (miles/gallon)
1940	14.8
1950	13.9
1960	13.4
1970	13.5
1980	15.5
1990	20.0

1. Model the data using a scatter plot and determine the equation of best fit.

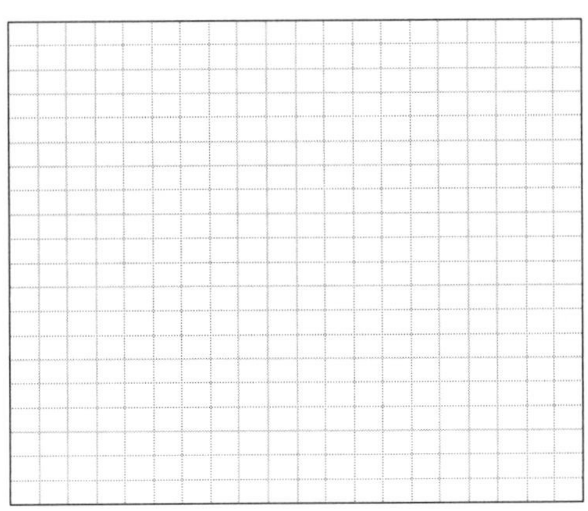

2. Based on the model, what will be the average fuel efficiency this year?

3. Based on the model, when will the average fuel efficiency be 30 miles/gallon? Does your answer make sense?

Assignment for Saving Up

Name: ______________________________

1. You have managed to save \$550 and plan on investing it in an account that will compound the interest annually. Your plan is to invest another \$550 per year every year for 3 years. Write the balance at the end of the three years in terms of the interest rate, r.

2. Graph the function below.

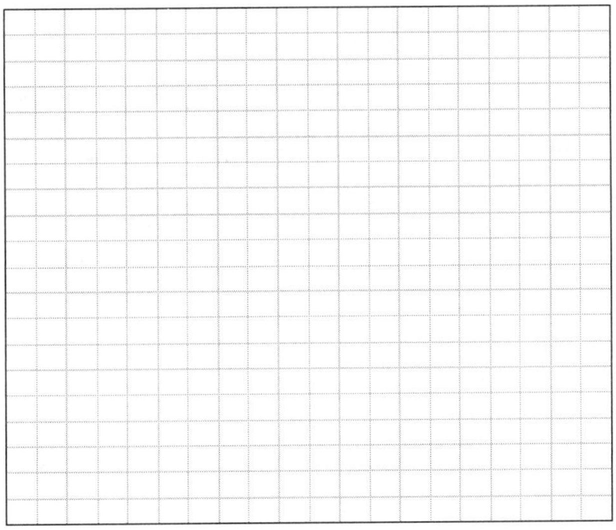

3. What will be your balance at the end of three years if you invest in an account that offers the following interest rates?

    a 4.8% interest

    b. 6.2% interest

4. What interest rate would be needed to have the following balances at the end of three years?

    a. $1900

    b. $2500

Assignment for Transforming Cubic Functions

Name: ____________________________

1. Write the resulting function if the following transformations are applied to the function $y = x^3$
   a. Shift right 3 units and shift up 2 units

   b. Reflect about the x-axis and shift left 5 units

   c. Dilate by a factor of 2, reflect about the x-axis and shift down 3 units

   d. Shift right 1 unit, shift up 3 units and dilate by a factor of 0.5

   e. Shift right 3 units, shift down 5 units, dilate by a factor of 2 and reflect about the x-axis

2. Describe the transformations applied to the function $y = x^3$ to result in each function.
   a. $y = -x^3 - 4$

   b. $y = \frac{1}{2}(x+1)^3$

   c. $y = -\frac{2}{3}x^3 - 5$

   d. $y = 3(x+11)^3 + 6$

   e. $y = -4(x-2)^3 - 3$

Assignment for Investigating Cubic Functions

Name: ____________________________

1. Suppose you cut an x-inch square out of each corner of an 8.5 inch by 11 inch sheet of paper. Fold the sides up to form a box without a top.

   a. Write an expression to calculate each dimension of the box

   b. Write an expression to calculate the volume of the box.

   c. Graph the volume using your graphing calculator.

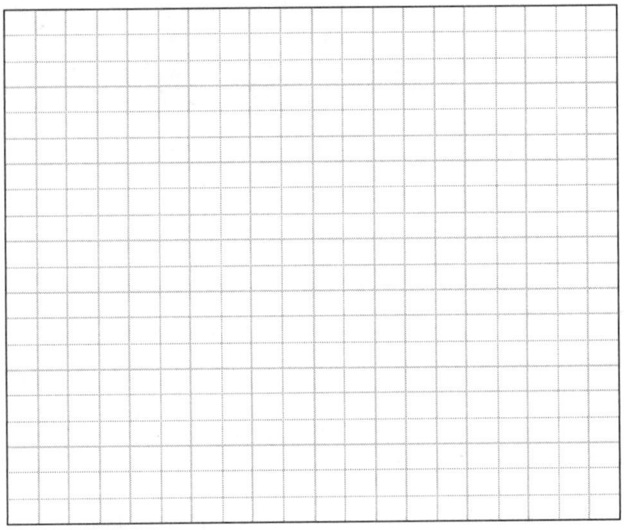

   d. Determine the dimension of the box that maximizes the volume and the maximum volume.

2. Write each polynomial function in standard form
   a. $y = (3x - 4)(x + 3)$

   b. $a = (12 - w)(7 - 2w)$

   c. $v = h(h + 3)(2h - 5)$

   d. $c = (2b + 5)(4b^2 - 10b + 25)$

3. Write each polynomial function in factored form
   a. $v = h^3 - 27$

   b. $y = 6x^2 - 27x - 15$

   c. $y = 2x^3 + x^2 - 3x$

   d. $p = 64q^3 + 343$

Assignment for Polynomial Functions

Name: ______________________________

1. For each of the following sketch the graph using a graphing calculator and determine the
   - x-intercept(s)
   - y-intercept(s)
   - Local Maximum
   - Local Minimum

   a. $y = x^3 - 4x$

   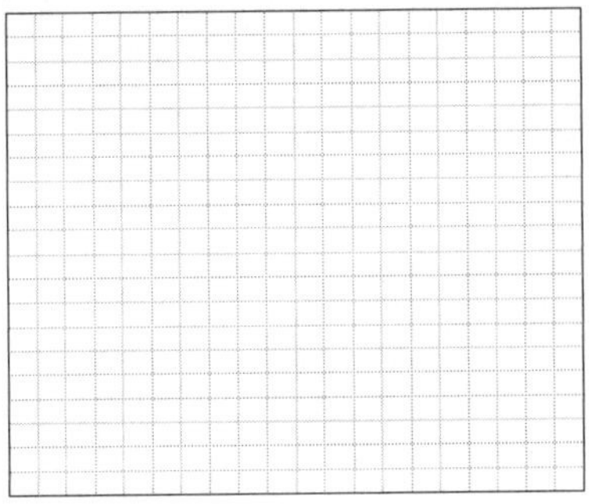

   b. $-x^3 - 2$

   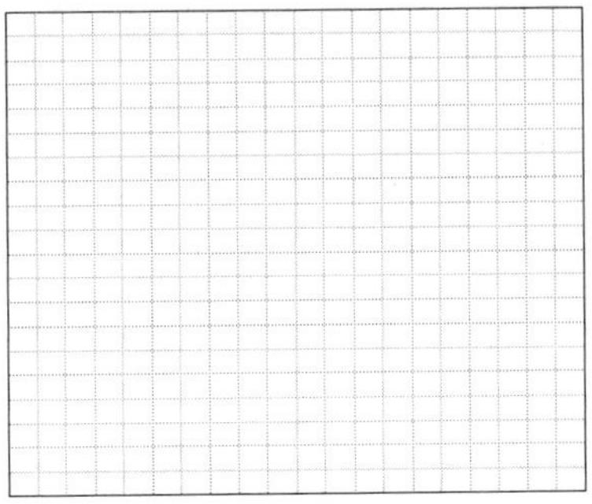

c. $y = x^3 - 6x^2$

d. $x^3 + 6x^2 + 9x$

Assignment for Modeling with Polynomials: Deer Population

Name: ______________________________

The number of female students in the US playing high school sports can be modeled using the polynomial function:

$$y = -38.88x^4 + 2{,}809.59x^3 - 67{,}574.84x^2 + 644{,}262.77x - 209{,}594.9$$

Where x is the time since 1970 in years and y is the number of females playing high school sports.

1. Sketch a graph of the function

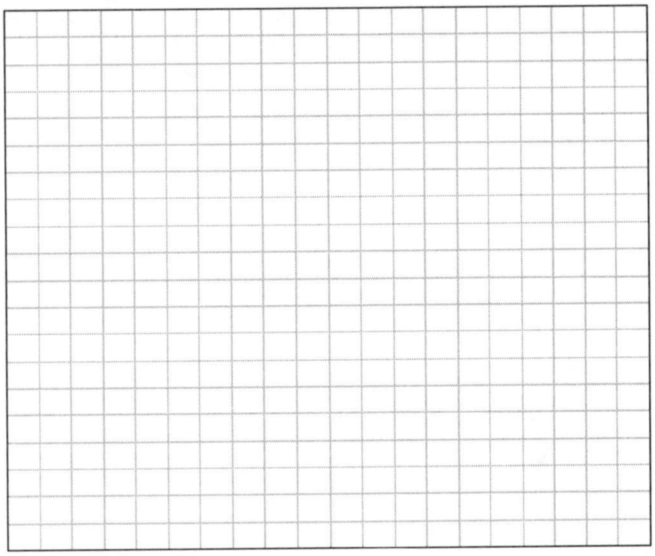

2. Write a paragraph to analyze female participation in high school sports since 1970. Be sure to include information related to local extrema, intervals of increase and decrease, intercepts and end behavior.

∞

∞

3. Based on your analysis, is the function a good model? Explain why or why not.

Assignment for Introduction to Rational Functions

Name: ______________________________

1. When will a rational function be discontinuous?

2. When will a rational function be continuous?

3. Determine any points of discontinuity for each rational function.

   a. $y = \frac{2}{x-3} + \frac{x}{x-4}$

   b. $f(x) = \frac{x-2}{x^2+7x+10}$

   c. $f(x) = \frac{(x-3)^2}{x^2-9}$

   d. $y = \frac{(x-3)(x-4)}{5}$

   e. $y = \frac{x}{x^2-3x-4} \cdot \frac{x-4}{x^2-x}$

Assignment for Analyzing Rational Functions

Name: ______________________________

1. Graph each rational function. Indicate all horizontal and vertical asymptotes and the domain and range.

   a. $y = \dfrac{1}{(x+3)(x-1)}$

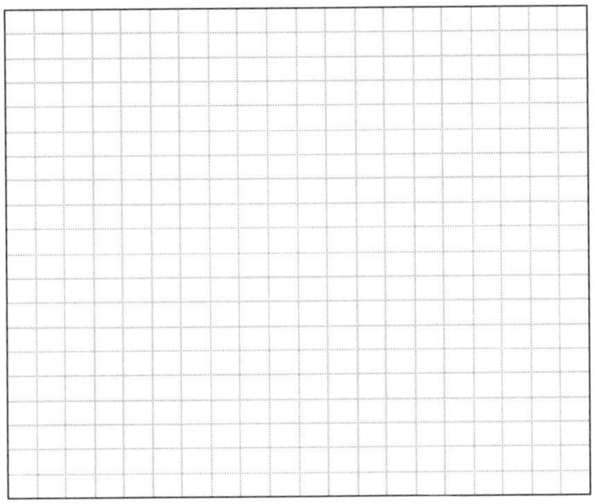

   b. $y = \dfrac{x^2}{x-1}$

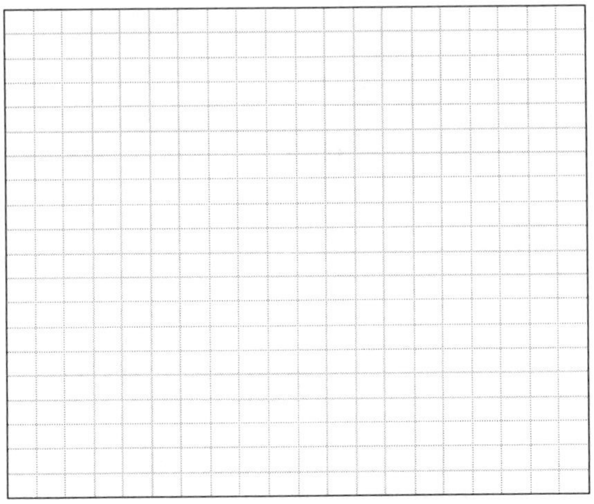

c. $y = \dfrac{10x}{x^2 + 1}$

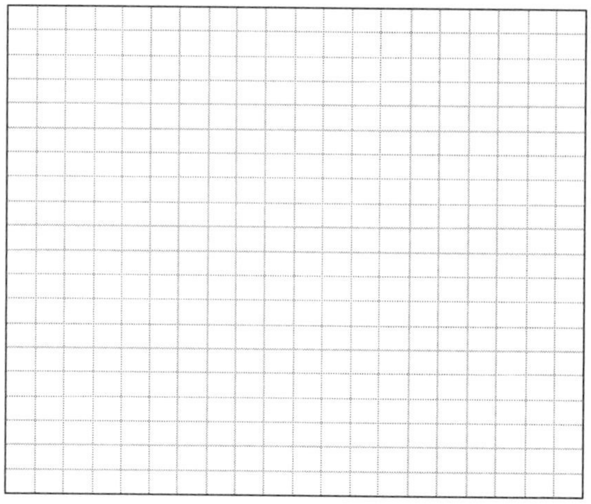

2. Find the inverse of each rational function.

a. $f(x) = \dfrac{2x - 2}{x - 2}$

b. $f(x) = \dfrac{x - 3}{4 - x}$

Assignment for Modeling with Rational Functions: Buffo Nutrition

Name: ____________________________

Buffo Nutrition stores have decided to produce a special can to commemorate the national weightlifting champion. The can will be cylindrical and hold 12 fluid ounces (approximately 21.6 cubic inches). The top and bottom will be made of aluminum costing 1.5 cents per square inch. The sides will be made of an imitation gold metal costing 4 cents per square inch.

1. Write an equation to determine the cost to produce the commemorative can based on the radius of the can.

2. Determine the minimum cost to produce the commemorative can.

3. Determine the radius and height that will result in the least cost.

4. Do the dimensions resulting in the minimum cost make sense? Explain why or why not.

Assignment for Solving Rational Functions

Name: ______________________________

Solve each rational equation

1. $\frac{2}{x-2} = 3$

2. $\frac{2}{t} = \frac{4}{t-2}$

3. $\frac{1}{x} + \frac{3}{2x} = 1$

4. $\frac{2}{r} + \frac{3}{r+1} = 0$